Feature Papers in Biobased and Biodegradable Metals

Feature Papers in Biobased and Biodegradable Metals

Guest Editor

Xiaobo Zhang

MDPI

Basel • Beijing • Wuhan • Barcelona • Belgrade • Novi Sad • Cluj • Manchester

Guest Editor
Xiaobo Zhang
School of Materials Science
and Engineering
Nanjing Institute
of Technology
Nanjing
China

Editorial Office
MDPI AG
Grosspeteranlage 5
4052 Basel, Switzerland

This is a reprint of the Special Issue, published open access by the journal *Metals* (ISSN 2075-4701), freely accessible at: https://www.mdpi.com/journal/metals/special_issues/3MG7209136.

For citation purposes, cite each article independently as indicated on the article page online and as indicated below:

Lastname, A.A.; Lastname, B.B. Article Title. *Journal Name* **Year**, *Volume Number*, Page Range.

ISBN 978-3-7258-5887-3 (Hbk)
ISBN 978-3-7258-5888-0 (PDF)
https://doi.org/10.3390/books978-3-7258-5888-0

Contents

About the Editor vii

Xiaobo Zhang
Feature Papers in Biobased and Biodegradable Metals
Reprinted from: *Metals* **2025**, *15*, 1198, https://doi.org/10.3390/met15111198 1

Kerolene Barboza da Silva, João Pedro Aquiles Carobolante, Roberto Zenhei Nakazato, Angelo Caporalli Filho and Ana Paula Rosifini Alves
Influence of a Novel Thermomechanical Processing Route on the Structural, Mechanical, and Corrosion Properties of a Biodegradable Fe-35Mn Alloy
Reprinted from: *Metals* **2025**, *15*, 462, https://doi.org/10.3390/met15040462 4

Lizhen Shi, Hui Liu, Houqing Liu, Cong Peng and Ling Ren
Effects of Mg Content and Pulsed Magnetic Field Treatment on Microstructure and Properties of As-Cast Biodegradable Zn-3Cu Alloy
Reprinted from: *Metals* **2025**, *15*, 175, https://doi.org/10.3390/met15020175 18

Andres Larraza, Shane Burke, Pedram Sotoudehbagha and Mehdi Razavi
Fabrication and Processing of Magnesium-Based Metal Matrix Nanocomposites for Bioabsorbable Implants
Reprinted from: *Metals* **2024**, *14*, 1318, https://doi.org/10.3390/met14121318 31

Alexandra-Tamara Șutic, Romeu Chelariu, Ramona Cimpoeșu, Ana-Maria Roman, Bogdan Istrate, Viorel Goanță, et al.
Corrosion Behavior and Mechanical Properties of Zn–Ti Alloys as Biodegradable Materials
Reprinted from: *Metals* **2024**, *14*, 764, https://doi.org/10.3390/met14070764 49

Michael Johanes, Vasuudhaa Sonawane and Manoj Gupta
A First-Time Investigation into Ecofriendly and Biocompatible Mg-Se Binary System for a Greener Earth
Reprinted from: *Metals* **2024**, *14*, 163, https://doi.org/10.3390/met14020163 69

Pulat Kadirov, Yury Pustov, Yulia Zhukova, Maria Karavaeva, Vadim Sheremetyev, Andrey Korotitskiy, et al.
Dependence of Electrochemical Characteristics of a Biodegradable Fe-30Mn-5Si wt.% Alloy on Compressive Deformation in a Wide Temperature Range
Reprinted from: *Metals* **2023**, *13*, 1830, https://doi.org/10.3390/met13111830 88

Bin Shi Jie Bryan, Kai Soon Fong, Chua Beng Wah, Sravya Tekumalla, Min Kyung Kwak, Eun Soo Park, et al.
Enhancing Mechanical and Biocorrosion Response of a MgZnCa Bulk Metallic Glass through Variation in Spark Plasma Sintering Time
Reprinted from: *Metals* **2023**, *13*, 1487, https://doi.org/10.3390/met13081487 102

Jiahao Jiang, Xue Geng and Xiaobo Zhang
Mechanical and Corrosion Properties of Mg–Gd–Cu–Zr Alloy for Degradable Fracturing Ball Applications
Reprinted from: *Metals* **2023**, *13*, 446, https://doi.org/10.3390/met13030446 118

António Fróis, Ana Cristina Santos and Cristina Santos Louro
Corrosion of Fixed Orthodontic Appliances: Causes, Concerns, and Mitigation Strategies
Reprinted from: *Metals* **2023**, *13*, 1955, https://doi.org/10.3390/met13121955 133

About the Editor

Xiaobo Zhang

Xiaobo Zhang, a Professor at Nanjing Institute of Technology, focuses his research on magnesium alloys for biomedical applications. He has successfully led multiple funded projects, including grants from the National Natural Science Foundation of China and the Natural Science Foundation of Jiangsu Province. With over 100 SCI/EI-indexed publications, he has been ranked among the world's top 2% scientists for six consecutive years.

metals

MDPI

Editorial

Feature Papers in Biobased and Biodegradable Metals

Xiaobo Zhang

School of Materials Science and Engineering, Nanjing Institute of Technology, Nanjing 211167, China; xbzhang@njit.edu.cn

1. Introduction and Scope

Biodegradable metals (Mg, Zn, Fe) represent a transformative approach for biomedical implants and special industrial applications. This Special Issue features nine cutting-edge studies advancing understanding of these systems, highlighting key material challenges: magnesium alloys exhibit excellent biocompatibility but suffer rapid corrosion and hydrogen evolution, zinc alloys offer moderate degradation rates yet have poor mechanical strength, and iron-based systems show superior mechanical properties but degrade too slowly for physiological needs. The research encompasses novel alloy design, thermomechanical processing, microstructure, mechanical properties, corrosion characterization, and in vitro degradation/cytocompatibility assessment. These interdisciplinary studies integrating metallurgy, materials science, electrochemistry, and biomedical engineering address clinical translation bottlenecks, providing critical insights to overcome current limitations and guide future development of biobased and biodegradable metals.

2. Contributions

This Special Issue comprises nine original contributions, featuring one review and eight original studies on biobased and biodegradable metals including Mg, Zn, and Fe.

Fróis et al. (contribution 1) addressed the intraoral degradation of fixed orthodontic appliances and pointed out current mitigation strategies using alloy design, process, and surface treatment. This paper sets a foundational framework for understanding degradation mechanisms in biomedical metallic systems, making it essential reading for researchers in both orthodontics and broader biomaterial fields.

Larraza et al. (contribution 2) presented a novel approach of ultrasonic melt treatment to enhance the strength and corrosion resistance of magnesium-based implants through the fabrication of bioactive glass–ceramic nanoparticle-reinforced metal matrix nanocomposites. Subsequent hot rolling enhanced strength but increased corrosion rates. The study demonstrated a critical trade-off between mechanical reinforcement and degradation control in Mg implants.

Bryan et al. (contribution 3) explored the fabrication of magnesium-based bulk metallic glasses using spark plasma sintering and investigated the influence of sintering duration (15–180 min) on the structural, mechanical, and corrosion behavior of $Mg_{65}Zn_{30}Ca_5$ ribbons. A key finding is that sintering at 150 °C for 90 min achieves a near-net amorphous structure with 98.2% densification and minimal crystallization, resulting in optimal mechanical properties. The study revealed that prolonged sintering beyond 90 min increased densification but compromised corrosion performance.

Johanes et al. (contribution 4) introduced selenium (Se) as a novel alloying element in magnesium, synthesizing a Mg-15Se binary alloy via powder metallurgy, including microwave sintering and hot extrusion. The alloy exhibited mechanical properties significantly superior to pure Mg: 57% higher hardness, 21–51% improvements in compressive

https://doi.org/10.3390/met15111198

strength and ductility, and a remarkable 76% increase in damping capacity, as well as a comparable corrosion rate to pure Mg. This work provides a basis for the use of selenium as an alloying element in biomedical materials.

Jiang et al. (contribution 5) reported a rapid biodegradable Mg-Gd-Cu-Zr alloy fracture balls for oil and gas extraction, where rapid degradation is a desired feature rather than a limitation. The alloy exhibited high ultimate tensile strength (>200 MPa) and elongation (>11.1%) at 93 °C. More strikingly, its corrosion rate in 3 wt.% KCl solution reached 1660.8–1955.1 mm/y. The findings underscore the versatility of Mg alloys and expand their utility into sustainable energy technologies, where temporary structural components that dissolve after use can reduce environmental impact and operational costs.

Sutic et al. (contribution 6) investigated the effects of titanium addition (0.10–1.00 wt.%) on the microstructure, mechanical properties, and corrosion behavior of zinc-based alloys. Microstructural refinement of α-Zn dendrites and eutectic grains led to improved mechanical performance. Electrochemical tests in simulated physiological conditions showed that increasing Ti content enhanced corrosion resistance, with lower degradation rates observed for Zn–1.00Ti.

Shi et al. (contribution 7) combined alloying and pulsed magnetic field treatment to enhance the performance of Zn-3Cu-xMg alloys. The addition of Mg (0.5–1.0 wt.%) promoted precipitation along grain boundaries, refining the microstructure and improving strength–ductility synergy. The Zn3Cu0.5Mg alloy achieved the best combination of strength and ductility, moderate corrosion rate, positive cytocompatibility, and antibacterial efficacy.

Silva et al. (contribution 8) presented a new processing route for a Fe-35Mn alloy involving arc melting, homogenization, hot swaging, and solution treatment. The alloy exhibited excellent mechanical properties: ultimate tensile strength of 533 MPa, elongation of 39%, and reduced Young's modulus of 171 GPa but a relatively low corrosion resistance. The study highlighted the role of Mn in reducing corrosion resistance due to its lower electrochemical potential, creating galvanic couples with the Fe matrix. The work demonstrated that tailored thermomechanical processing could optimize the mechanical–degradation trade-off in Fe-Mn alloys.

Kadirov et al. (contribution 9) investigated how compressive deformation at various temperatures (350–900 °C) affected the phase composition and corrosion behavior of Fe-30Mn-5Si alloy. It was found that deformation below 700 °C preserved a single-phase γ-austenite structure, while deformation at 900 °C induced a two-phase γ-austenite + ε-martensite structure, which enhanced biodegradation due to increased electrochemical activity. The corrosion rate ranged from 0.14 to 0.42 mm/year, depending on deformation conditions.

3. Conclusions and Outlook

The nine papers in this collection collectively illustrate the remarkable progress in biobased and degradable metallic alloys, showcasing innovative strategies to overcome the inherent limitations of Mg, Zn, and Fe systems from novel alloying concepts to advanced processing techniques. Future research should focus on the tunability of mechanical and degradation properties, in vivo validation, and long-term toxicity studies for clinical application. As these technologies mature, biodegradable metals are poised to revolutionize implantable medical devices, offering safer, more sustainable, and patient-centric solutions.

Acknowledgments: As Guest Editor, I would like to extend my sincere gratitude to all the reviewers and authors who contributed to this Special Issue, and to the Editorial staff for their dedicated support throughout the entire publication process.

Conflicts of Interest: The author declares no conflicts of interest.

List of Contributions

1. Fróis, A.; Santos, A.; Louro, C. Corrosion of Fixed Orthodontic Appliances: Causes, Concerns, and Mitigation Strategies. *Metals* **2023**, *13*, 1955. https://doi.org/10.3390/met13121955.
2. Larraza, A.; Burke, S.; Sotoudehbagha, P.; Razavi, M. Fabrication and Processing of Magnesium-Based Metal Matrix Nanocomposites for Bioabsorbable Implants. *Metals* **2024**, *14*, 1318. https://doi.org/10.3390/met14121318.
3. Bryan, B.; Fong, K.; Wah, C.; Tekumalla, S.; Kwak, M.; Park, E.; Gupta, M. Enhancing Mechanical and Biocorrosion Response of a MgZnCa Bulk Metallic Glass through Variation in Spark Plasma Sintering Time. *Metals* **2023**, *13*, 1487. https://doi.org/10.3390/met13081487.
4. Johanes, M.; Sonawane, V.; Gupta, M. A First-Time Investigation into Ecofriendly and Biocompatible Mg-Se Binary System for a Greener Earth. *Metals* **2024**, *14*, 163. https://doi.org/10.3390/met14020163.
5. Jiang, J.; Geng, X.; Zhang, X. Mechanical and Corrosion Properties of Mg–Gd–Cu–Zr Alloy for Degradable Fracturing Ball Applications. *Metals* **2023**, *13*, 446. https://doi.org/10.3390/met13030446.
6. Sutic, A.; Chelariu, R.; Cimpoesu, R.; Roman, A.; Istrate, B.; Goantă, V.; Benchea, M.; Moscu, M.; Alexandru, A.; Cimpoesu, N.; et al. Corrosion Behavior and Mechanical Properties of Zn–Ti Alloys as Biodegradable Materials. *Metals* **2024**, *14*, 764. https://doi.org/10.3390/met14070764.
7. Shi, L.; Liu, H.; Liu, H.; Peng, C.; Ren, L. Effects of Mg Content and Pulsed Magnetic Field Treatment on Microstructure and Properties of As-Cast Biodegradable Zn-3Cu Alloy. *Metals* **2025**, *15*, 175. https://doi.org/10.3390/met15020175.
8. Silva, K.; Carobolante, J.; Nakazato, R.; Filho, A.; Alves, A. Inffuence of a Novel Thermomechanical Processing Route on the Structural, Mechanical, and Corrosion Properties of a Biodegradable Fe-35Mn Alloy. *Metals* **2025**, *15*, 462. https://doi.org/10.3390/met15040462.
9. Kadirov, P.; Pustov, Y.; Zhukova, Y.; Karavaeva, M.; Sheremetyev, V.; Korotitskiy, A.; Baranova, A.; Prokoshkin, S. Dependence of Electrochemical Characteristics of a Biodegradable Fe-30Mn-5Si wt.% Alloy on Compressive Deformation in a Wide Temperature Range. *Metals* **2023**, *13*, 1830. https://doi.org/10.3390/met13111830.

metals

MDPI

Article

Influence of a Novel Thermomechanical Processing Route on the Structural, Mechanical, and Corrosion Properties of a Biodegradable Fe-35Mn Alloy

Kerolene Barboza da Silva, João Pedro Aquiles Carobolante, Roberto Zenhei Nakazato, Angelo Caporalli Filho and Ana Paula Rosifini Alves *

School of Engineering and Sciences, São Paulo State University (UNESP), Guaratinguetá 12516-410, Brazil; kerolene.barboza@unesp.br (K.B.d.S.); pedro.carobolante@unesp.br (J.P.A.C.); roberto.zenhei@unesp.br (R.Z.N.); angelo.caporalli@unesp.br (A.C.F.)

* Correspondence: paula.rosifini@unesp.br

Abstract: Recent studies have focused on developing temporary metallic implants made from biodegradable biomaterials, such as iron and its alloys, along with the associated manufacturing methods. These biomaterials allow the implant to gradually degrade after fulfilling its function, which reduces the risks of complications associated with permanent implants. Iron is particularly appealing from a structural standpoint, and adding manganese enhances its potential for use. The Fe-35Mn alloy demonstrates excellent mechanical properties and degradation characteristics, making it an ideal choice within the Fe-Mn system. As a result, new processing techniques can be applied to this alloy to further improve its performance. The objective of this research is to propose a new processing route and evaluate its impact on the properties of the Fe-35Mn alloy. The experimental alloy was produced using an arc melting furnace, followed by homogenization, hot swaging, and solution treatment. Alloy characterization was conducted using various techniques, including X-ray fluorescence (XRF), optical microscopy (OM), X-ray diffraction (XRD), differential scanning calorimetry (DSC), microhardness testing, tensile strength measurements, Young's modulus determination, and potentiodynamic polarization analysis. The microstructural evolution throughout the applied processing route was analyzed in relation to the alloy's mechanical performance and corrosion resistance. The typical microstructure of the Fe-35Mn alloy is primarily composed of austenitic grains stabilized at room temperature. Its mechanical properties—yield strength (297 MPa), ultimate tensile strength (533 MPa), and elongation to failure (39%)—are comparable to, or even surpass, those of conventional biomedical materials such as 316 L stainless steel and pure iron. The reduced Young's modulus (171 GPa), compared to other alloys, further underscores its potential for biomedical applications. Electrochemical testing revealed lower corrosion resistance than that of similar alloys reported in the literature, with a corrosion potential of −0.76 V and a current density of 3.88 $\mu A \cdot cm^{-2}$, suggesting an enhanced corrosion rate.

Keywords: biomaterials; biodegradable metals; iron-manganese alloy; processing

1. Introduction

In the group of metallic biomaterials, stainless steels, cobalt-chromium, and titanium alloys are currently the most used. Due to their superior mechanical properties, biocompatibility, and corrosion resistance, these materials are used commercially and occupy a dominant position in the implantable devices market. In general, they are intended to replace the original tissues and remain in the body permanently due to their high stability in the biological environment [1].

However, certain clinical applications, such as bone fractures or blood vessel blockages, specifically require only temporary support during the tissue healing process. In these instances, the presence of biomaterials that remain in the body as foreign objects can lead to various negative effects over time, including stress shielding, bone weakening, prolonged irritation, thrombosis, chronic inflammation, distortion of diagnostic images, and even the need for a second surgery to remove the implant, which carries significant risks and additional financial costs [2–4].

Biodegradable implants made from biomaterials have been developed as an alternative. These materials enable the implant to progressively degrade after serving its purpose. This biodegradation reduces the risk of complications associated with permanent implants, as previously mentioned [1,5–7]. Magnesium and iron-based alloys are among the biodegradable metallic systems recognized for their potential in biomedical applications, as these non-toxic elements exhibit suitable degradation behavior and have processing routes familiar to the human body, along with superior biocompatibility and appropriate mechanical properties [1,8]. However, despite its excellent properties, magnesium has low corrosion resistance, particularly in electrolytic and aqueous environments with high chloride content, where its degradation typically occurs rapidly, compromising the mechanical and functional integrity of the implant in a short timeframe [9–11]. Iron and its alloys are especially appealing from a structural perspective, as they demonstrate mechanical properties comparable to those of the 316 L stainless steel used as the reference standard for evaluating other alloys intended for biomedical applications. Recent studies indicate that these materials have shown significant promise for use in biodegradable implants [12], though their degradation rate in physiological media is considered very low [6,10,11].

In this context, the alloys of the iron-manganese system have been extensively examined in studies by Hermawan et al. [13]. The authors reported achieving adequate mechanical properties and a degradation rate that is superior to that of pure iron, adjustable based on manganese concentration [7]. Dargusch et al. [1] also highlight the positive impact that manganese has on the magnetic properties of iron-based alloys. The development of antiferromagnetic behavior enhances the material's compatibility with magnetic-based techniques, such as magnetic resonance imaging (MRI) exams.

Several authors have comprehensively evaluated a range of alloys within the Fe-Mn binary system (20–35 wt% Mn) and concur that the Fe-35Mn alloy demonstrates the best combination of mechanical, magnetic, and corrosion properties [14].

Increasing manganese content significantly expands the austenite phase field (γ) due to its stabilizing effect, which is less dense and less resistant than the martensite phase (ε) [13]. Consequently, the Fe-35Mn alloy exhibits the best performance against plastic deformation, showcasing the highest elongation among the alloys in the system, along with mechanical parameters comparable to 316 L steel [13,15]. According to Dehghan-Manshadi, Stjohn, and Dargusch [16], reasonable elongation values, along with adequate strength and modulus, suggest that the alloy containing 35% Mn could be a strong candidate for future biomedical applications. Increasing the manganese content decreases the alloy's magnetic susceptibility due to the lower saturation magnetization exhibited by the austenite phase (γ). As a less noble element with a lower electrochemical potential, manganese makes the iron matrix more susceptible to corrosion and increases its degradation rate in physiological environments [1,7,17].

Dargusch et al. [14] note that the Fe-35Mn composition shows the best combination of properties for use in biodegradable cardiovascular stents, however, its degradation rate remains lower than that of magnesium alloys and is considered slow for temporary implant applications. In addition to introducing new alloying elements, discussions about employing new processing techniques have uncovered an effective method for enhancing the mechanical and corrosion properties of biodegradable metallic alloys [11,18]. Most of the literature on biodegradable metallic alloys emphasizes their processing through powder metallurgy. Consequently, this research was developed to investigate the influence of a novel thermomechanical processing route—based on arc melting followed by homogenization, hot swaging, and solution treatment—on the structural, mechanical, and electrochemical properties of a Fe-35Mn biodegradable alloy. The goal is to assess the alloy's suitability for temporary biomedical implants by correlating its microstructural evolution with mechanical performance and corrosion behavior.

2. Materials and Methods

2.1. Fe-35Mn Alloy Preparation

In this study, the Fe-35Mn experimental alloy was prepared using an arc melting furnace under an argon atmosphere from Fe (99.8% purity) and Mn (99.9% purity) as raw materials. The elements were weighed according to the stoichiometry of Fe-35wt% Mn; however, a loss of Mn was observed during the process due to its high vapor pressure. To ensure the alloy's quantitative composition, this element was increased by approximately 5% by mass, as it was deposited on the crucible while being enveloped by Fe. The obtained ingots were remelted at least five times to enhance their chemical homogeneity. They were homogenized at 1000 °C for 86.4 ks, processed by hot swaging at 850 °C into rods with a diameter of 10 mm, and finally solubilized at 1000 °C for 10.8 ks, followed by quenching in water. The chemical compositions of the processed alloy were determined using an X-ray fluorescence (XRF) spectrometer (PANalytical Axios Max, Almelo, The Netherlands).

2.2. (Micro)Structural Characterization

Samples were cut into disks with a thickness of 3 mm from the ingots after each processing route step (melting, homogenization, swaging, and quenching) and prepared for characterization. Present phase identification in each sample was conducted by X-ray diffraction (XRD) using a diffractometer (Bruker D8 Advance, Baku, Azerbaijan) with Cu Kα radiation at an accelerating voltage of 40 kV, a current of 25 mA, and a step size of 0.02° in the scanning range of 20° to 100°. Diffraction patterns were analyzed using the Inorganic Crystal Structure Database (ICSD) and submitted to Rietveld structural refinement using the Total Pattern Analysis Solution (TOPAS) Academic version 5 software.

Differential scanning calorimetry (DSC) was used to assess the thermal behavior and transformations of apparent phases in the Fe-35Mn experimental alloy at the end of processing. Samples weighing approximately 80 mg were analyzed in a calorimeter (NETZSCH, STA 409 C), purged with argon at a controlled heating and cooling rate of 10 °C/min, from room temperature to 1000 °C, with an argon flow rate of 100 mL/min.

Before microstructural analysis, samples that were previously used for XRD were first ground with SiC sandpapers up to 1500 grit, polished with 0.25 μm diamond paste, and ultrasonically cleaned with ethanol to ensure the removal of abrasives from the surface. A 5% Nital solution was used as an etchant, and the microstructure of the alloy was examined using an Epiphot 200 optical microscope (OM) with reflected light (Nikon, Tokyo, Japan). The average grain size was measured from the obtained optical micrographs using open source software for processing and analyzing scientific images, ImageJ 1.53t version. Grain size measurements were based on the longest straight line segment inscribed within each grain.

2.3. Mechanical Properties

Microhardness values of the samples were measured using a Vickers microindentation tester (Wilson Instruments, 401 MVD), with a load of 1 kgf and a dwell time of 10 s, in accordance with ASTM E384 standard. At least twelve indentations were performed on each sample, and the mean values are reported. Statistical analysis was conducted using RStudio software (R version 4.4.1) with the rstatix package (version 0.7.2). Normality (Shapiro–Wilk) and homogeneity of variance (Levene) tests were applied. The non-parametric Kruskal–Wallis test was utilized to compare the groups, followed by Dunn's post hoc test with Bonferroni correction. In all cases, a significance level of 5% ($\alpha = 0.05$) was used for the tests performed.

Uniaxial tensile tests were conducted according to the ASTM E8M standard using a universal testing machine (INSTRON, 8801) equipped with a nominal 25 mm extensometer. The cylindrical specimens, measuring a total length of 90.0 mm, a gage length of 40.0 mm, a gage diameter of 5.0 mm, and a filet radius of 6.0 mm, were previously conditioned at room temperature. These were tested with a 50 kN load cell at a rate of 1 mm/min. The tests were performed in triplicate.

Young's modulus measurements were obtained using the impulse excitation technique (IET) with the Sonelastic® system (ATCP Physical Engineering), adhering to the ASTM E1876 standard. Average values were calculated from five measurements. Statistical analysis was conducted using RStudio (version 4.4.1) with the rstatix package (version 0.7.2). Normality was assessed using the Shapiro–Wilk test, and homogeneity of variance was evaluated with Levene's test when necessary, as a prerequisite for inferential analyses. To compare Young's modulus obtained from tensile tests and the IET, as well as to compare it with other properties of various metallic biomaterials reported in the literature, the non-parametric one-sample Wilcoxon signed-rank test, the Wilcoxon rank-sum test, and one-sample *t*-tests were employed. All tests were performed with Bonferroni correction, applying a significance level of 5% ($\alpha = 0.05$).

2.4. Corrosion Experiments

The corrosion performance of the Fe–35Mn alloy was evaluated using open circuit potential (OCP) and potentiodynamic polarization tests. Both tests were conducted in Modified Hank's solution (H1387, Sigma-Aldrich, Darmstadt, Germany), which has an ionic composition and concentration similar to that of human blood plasma. The testing solution was prepared according to the manufacturer's instructions, using deionized water at a temperature between 15 and 20 °C. Sodium bicarbonate was added to the solution, and NaOH was used to adjust the pH of the medium to 7.4.

Open circuit potential (OCP) and potentiodynamic polarization techniques were employed to evaluate the electrochemical corrosion behavior of the alloy. Before the tests, samples with a 1 cm^2 exposed surface area were cut, embedded in polyester resin, ground with SiC paper up to 1500 grit, and then ultrasonically cleaned with ethanol and dried. The tests were performed in Hank's solution at 37 $\pm$ 1 °C, using a potentiostat/galvanostat (EG&G, PAR 283) equipped with PowerSuite 2.40 software and a standard three-electrode cell. The samples were configured as the working electrode, while platinum and saturated calomel electrodes served as counter and reference electrodes, respectively. Open-circuit potential (OCP) measurements were recorded for 5400 s after the samples were immersed in the electrolyte. Following OCP stabilization, potentiodynamic polarization tests were conducted at a scan rate of 1 mV/s, varying from −300 mV (vs. OCP) to +700 mV. The tests were repeated at least three times. The data were analyzed using OriginPro 2016 software (OriginLab Corporation, Northampton, MA, USA), where the corrosion potential (E_{corr}) and corrosion current density (I_{corr}) were determined by means of the Tafel extrapolation method.

The statistical analysis was conducted using RStudio (version 4.4.1) with the rstatix package (version 0.7.2). Normality was assessed using the Shapiro–Wilk test, and homogeneity of variance was evaluated with Levene's test. The Wilcoxon rank-sum test was used for comparisons of corrosion properties. This test was conducted with Bonferroni correction, applying a significance level of 5% ($\alpha = 0.05$).

3. Results and Discussion

Table 1 shows the chemical composition of the Fe-35Mn alloy obtained from the arc melting furnace. The results confirm that the processed alloy achieved the desired chemical composition.

Table 1. Chemical composition of Fe-35Mn alloy by XRF expressed as a compound mass percentage, normalized to 100%.

Concentration (wt %)					
Alloy	Mn	Si	Al	S	Fe
Fe-35Mn	35.18	0.15	0.08	0.07	Balance

The presence of other elements, such as Si, Al, and S, in relatively low concentrations does not indicate metal contamination and is insufficient to cause a phase transformation, given the high concentration of Mn [13].

3.1. *(Micro)Structural Characterization*

XRD analysis was conducted to identify the phases present in each microstructure during alloy processing. Figure 1 illustrates the XRD patterns obtained after Rietveld refinement for as-cast, homogenized, swaged, and quenched samples. Typical peaks of CFC austenite (γ) and HCP martensite (ε) phases were observed in all analyzed samples. These phases were also described by Dargusch et al. [14] and Zhang and Cao [19].

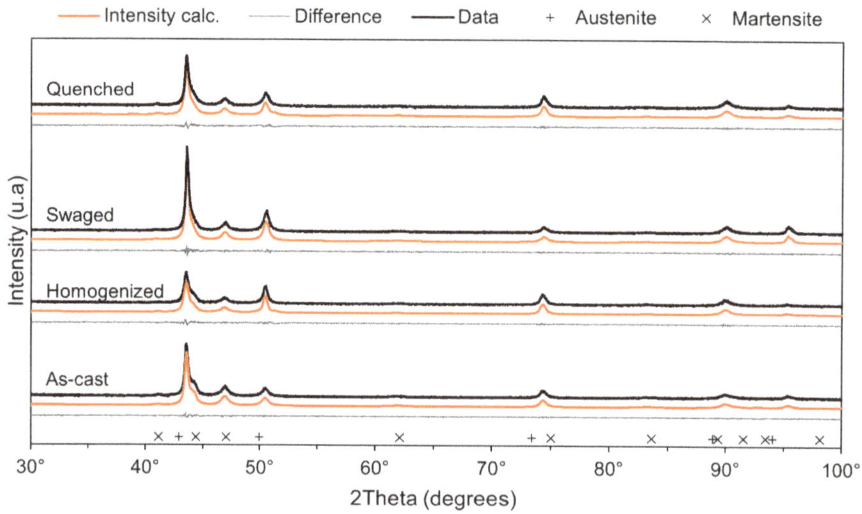

Figure 1. XRD refined patterns of the Fe–35Mn alloy for conditions as-cast, homogenized, swaged and quenched. The black line represents the experimental XRD pattern. The orange line represents the pattern calculated by Rietveld refinement. The gray line at the bottom of the curve presents differences between the experimental and calculated intensities. Markers + and × indicate XRD pattern of the austenite (ICSD 631730) and martensite (ICSD 103519) phases, respectively.

Previous studies [1,16,19] reported that the increase in Mn content significantly expanded the austenite (γ) phase field in Fe-Mn system alloys due to its stabilizing effect. Contents above 29% form a fully austenitic phase composition. However, in this work, the presence of the martensite (ε) phase in the as-cast alloy condition can be attributed to the lack of local homogeneity in the composition. According to the literature [20,21], Mn is easily volatilized at high temperatures because of its high vapor pressure, resulting in areas of Mn depletion with lower concentrations. For this reason, a greater amount of Mn was added during alloy preparation. After homogenization, a decrease in peak intensity related to the martensite (ε) phase and an increase in those associated with the austenite (γ) phase were observed, indicating a transformation and improved homogeneity of the composition. With swaging, peaks of the austenite (γ) phase were reduced but became more intense again after quenching. The formation of the martensite (ε) phase in Fe-Mn alloys can also be influenced by plastic deformation. The influence and peak intensity decrease as the Mn content increases [17]. However, in this study, due to successive swaging passes, the presence of the martensite phase was detected even with 35 wt% Mn in the composition. The Rietveld method was employed to determine the unit cell parameters and to quantify the composition of the different phases. The results and refinement factors for all alloy conditions are listed in Table 2.

Table 2. Unit cell parameters and austenite and martensite phase contents in the as-cast, homogenized, swaged, and quenched Fe-35Mn alloy, calculated from Rietveld refinement of the XRD patterns.

Sample	Phase	Content (wt %)	Lattice Parameters (Å)		Volume (A^3)	Refinement Factors		
			a	c		R_{wp}	GOF	R_{Bragg}
As-cast	γ	74.2	3.55	-	44.99	10.74	1.03	2.81
	ε	25.8	2.54	4.10	23.09			1.32
Homogenized	γ	98.1	3.56	-	45.29	11.90	1.07	4.32
	ε	1.9	2.54	4.10	22.96			1.97
Swaged	γ	20.4	3.58	-	46.20	9.94	1.03	0.953
	ε	79.6	2.44	4.26	22.03			1.12
Quenched	γ	96.4	3.57	-	45.84	11.11	1.11	3.80
	ε	3.6	2.55	4.10	23.18			1.97

As shown in Table 2, in the as-cast condition, 25.8% of the martensite phase in the composition was observed, attributed to the previously mentioned Mn depletion. After homogenization heat treatment, the homogeneity of the composition was corrected, restoring the predominance of the austenite phase and 1.9% of the martensite phase. The swaging process, as a form of plastic deformation, favored the martensitic transformation, increasing the martensite phase content to 79.6%. Finally, the solubilization heat treatment again promoted the austenitization of the alloy, which, after quenching, consisted mainly of the austenite phase with only 3.6% of martensite.

It can also be observed from Table 2 that the lattice parameters and unit cell volume of the found phases remained generally constant, with only minor changes due to deformation effects on the crystalline lattice during alloy processing.

The thermal behavior of the Fe-35Mn alloy was determined using DSC analysis. The sample in the quenched condition, which retained the austenite (γ) phase at room temperature, was investigated. DSC curves and transformation temperatures are shown in Figure 2.

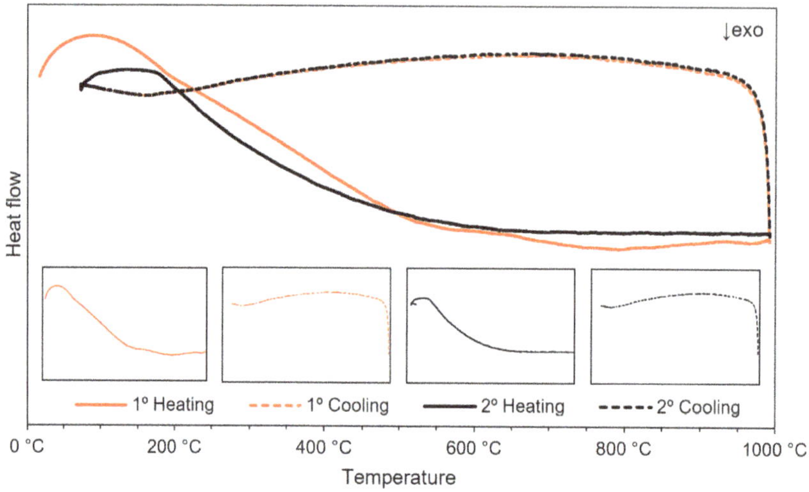

Figure 2. Thermogram of the Fe–35Mn alloy sample (quenched condition) on heating and cooling, measured with DSC from room temperature to 1000 °C.

No additional phase changes were detected during both the heating and cooling cycles examined. According to the literature data [22], adding high manganese content to iron tends to lower the transformation temperature and broaden the stability range of the austenite phase (γ). Witusiewicz, Sommer, and Mittemeijer [22] investigated the variation in the enthalpy of formation of γ-Fe-Mn alloys as a function of Mn content. The authors concluded that the transformations correspond to endothermic changes with elevated energy levels. Analyzing the heating curve for the first cycle (solid black line), more distinct and pronounced exothermic transitions are observed between temperatures of 500 and 1000 °C, possibly indicating the relief of internal stresses accumulated at the end of processing. During cooling (dashed black line), no changes were noted. In the second cycle, the heating (solid gray line) and cooling (dashed gray line) curves also showed no evidence of thermal events.

The micrographs in Figure 3 illustrate the evolution of the microstructure in the Fe-35Mn alloy through the applied processing route. In the as-cast Fe-35Mn alloy, the morphology obtained was qualitatively similar to the dendritic form resulting from the alloy's cooling inside a furnace equipped with a water-based refrigeration system (Figure 3a). With the homogenization heat treatment, the microstructure began to exhibit large, equiaxed grains (Figure 3b). The formation of twins and deformation bands can be observed after swaging and quenching (Figure 3c–d).

According to Hermawan, Dubé, and Mantovani [15], the Fe-Mn system can exhibit various microstructures based on the degree of plastic deformation and the Mn content present. The matrix featuring triangular structures suggests the presence of austenite (γ) and martensite (ε) phases, while their absence indicates only the austenite (γ) phase [15]. In Figure 3d, the typical microstructure of the Fe-35Mn alloy can be visualized, composed predominantly of austenitic grains with an average size of 102.38 μm (46.69 μm), uniformly distributed following the established processing route study.

The austenite (γ) phase, the main constituent of the Fe-35Mn alloy, exhibits relatively large grains, which could explain its superior ductility and low strength [15].

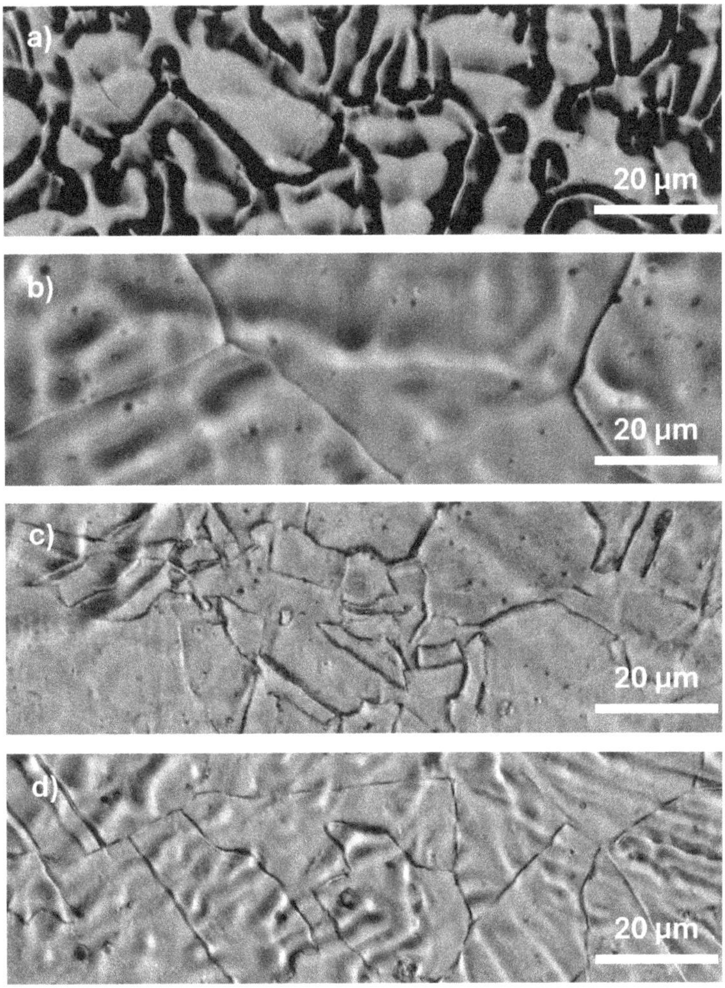

Figure 3. Typical optical micrographs of the Fe-35Mn alloy obtained during processing for conditions (**a**) as-cast, (**b**) homogeniszed, (**c**) swaged, and (**d**) quenched.

3.2. Mechanical Properties

The microhardness evaluation was performed after each processing step to verify the influence of the microstructural changes. The values found can be seen in Figure 4. The microhardness values showed variations during the alloy processing, however, they correspond with the changes introduced in the microstructure at each step. The martensite (ε) phase, which has an HCP structure, is harder and denser than the austenite (γ) phase with a CFC structure, contributing to the strengthening of the alloys in the Fe-Mn system [13,15]. Thus, after the melting and swaging steps, where there was an increase in the volume fraction of the martensite (ε) phase, the highest microhardness values are observed. Following the homogenization heat treatment and quenching, the microhardness values were relatively lower, with 113 HV being the microhardness obtained at the end of the alloy processing.

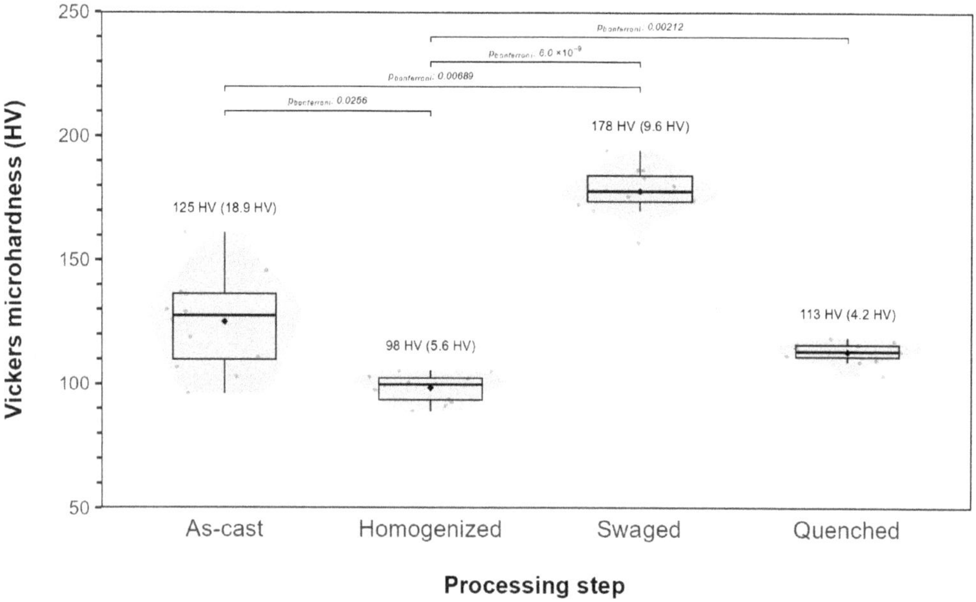

Figure 4. The Vickers microhardness values, Hv, of the Fe-35Mn alloy after each processing step—as-cast, homogenized, swaged, and quenched conditions. Boxplot with violin plot representation showing the distribution of the microhardness properties. The data points are represented by circles, while the diamond symbol indicates the mean value. Standard deviation in parentheses.

The Shapiro–Wilk test indicated that all groups followed a normal distribution (As-cast, p = 0.977; swaged, p = 0.870; homogenized, p = 0.444; quenched, p = 0.514). However, Levene's test showed heterogeneity of variance among the groups ($p = 5.52 \times 10^{-5}$), and two outliers were observed: one in the swaged group (157 HV) and another in the quenched group (104 HV). To compare the groups, the non-parametric Kruskal–Wallis test was applied, revealing significant differences in Vickers microhardness across the groups ($\chi^2_{(3)} = 37.9$; $p = 2.99 \times 10^{-8}$; $\eta^2_h = 0.793$). Dunn's test with a Bonferroni correction was performed as a post hoc analysis, demonstrating statistically significant differences between the groups: As-cast (median = 127 HV; IQR = 26.4) differed from swaged (median = 178 HV; IQR = 10.4 HV) ($p = 2.56 \times 10^{-2}$) and homogenized (median = 99.8 HV; IQR = 8.95 HV) ($p = 6.89 \times 10^{-3}$), and swaged differed from both homogenized ($p = 6.0 \times 10^{-9}$) and quenched (median = 113 HV; IQR = 4.85 HV) ($p = 6.89 \times 10^{-3}$ and $p = 2.12 \times 10^{-3}$).

Compared to the values obtained with annealed pure iron (119 HV), with 316 L stainless steel (195 HV)—considered reference standard in stent manufacturing—and with the same alloy composition obtained by powder metallurgy (143 HV), it is noted that there was a reduction in the microhardness value of the Fe-35Mn alloy processed in this research (113 HV) [23,24]. According to Nayak, Biswal, and Sahoo [23], this reduction indicates better performance for application in temporary implants.

The results obtained from the tensile test are presented in Table 3. This table shows the mechanical properties of the Fe-35Mn alloy after processing outlined in this study, including ultimate tensile strength, elongation to failure, yield strength, and Young's modulus.

Table 3. Mechanical properties of the Fe-35Mn alloy after processing. The data presented were obtained after performing statistical tests, indicating the significance of the variations in mechanical properties. The table shows the number of samples (n), mean, maximum value (Max), minimum value (Min), and standard deviation (SD) of each mechanical property evaluated.

Properties	n	Min	Max	Mean	SD
Young's modulus	3	154 GPa	189 GPa	171 GPa	17 GPa
Yield strength	3	144 MPa	457 MPa	297 MPa	157 MPa
Ultimate tensile strength	3	388 MPa	636 MPa	533 MPa	129 MPa
Elongation to failure	3	34%	46%	39%	6%

Although with an elongation to failure of 39% (6%) the experimental Fe-35Mn alloy processed in this study is similar to annealed pure iron (40%) and 316 L stainless steel (40%), the tensile test results indicate that the other analyzed mechanical properties are superior. In terms of ultimate tensile strength, the Fe-35Mn alloy achieved 533 MPa (129 MPa), while annealed pure iron supports 210 MPa, and 316 L stainless steel supports 490 MPa. The same is true for yield strength: the Fe-35Mn alloy has a yield strength of 297 MPa (157 MPa), compared to 150 MPa for annealed pure iron and 190 MPa for 316 L stainless steel. Regarding Young's modulus, the reduction observed in the value for the Fe-35Mn alloy, which stands at 171 GPa (17 GPa), compared to annealed pure iron (200 GPa) and 316 L stainless steel (193 GPa), reinforces its potential for biomedical applications, as it is closer to the Young's modulus of human bone (10–30 GPa) and helps avoid the stress shielding effect [10,13,25].

When compared to the alloy with the same nominal composition produced via powder metallurgy, an increase in mechanical resistance is also noted. Although this alloy exhibits properties comparable to 316 L stainless steel, which is currently used in most implanted stents, its values of ultimate tensile strength (428 MPa), yield strength (234 MPa), and elongation to failure (32%) are below those found for the Fe-35Mn alloy processed in this study [13,15]. The non-parametric one-sample Wilcoxon signed-rank test suggested that the mechanical properties of the Fe-35Mn alloy differed from those of pure iron, 316 L stainless steel, and the Fe-35Mn alloy processed by powder metallurgy.

The IET was used to confirm the value of Young's modulus determined from the tensile test. After a statistical study of the measurements, the data are presented in Table 4.

Table 4. Values of the young's modulus obtained by IET. The table presents the maximum and minimum values (Max and Min), mean, and standard deviation (SD). These parameters provide a comprehensive view of the distribution and variability of the data, allowing for a better interpretation of the results obtained in the statistical test.

Variable	n	Min	Max	Mean	SD
Young's modulus	5	177 GPa	186 GPa	182 GPa	3 GPa

An average value of 182 GPa (3 GPa) was obtained for the Young's modulus of the experimental alloy Fe-35Mn. When compared with the value derived from the previously presented tensile test, it is evident that the values were similar but exhibited a lower degree of dispersion among the measurements. The non-parametric Wilcoxon rank-sum test revealed no statistically significant difference between the methods used to measure the Young's modulus. The median values were 171 GPa (17 GPa) for the tensile test and 182 GPa (3 GPa) for the IET.

3.3. Corrosion Behavior

Open circuit potential measurements as a function of immersion time of the Fe-35Mn alloy samples were conducted in Hank's solution. The samples showed a rapid increase in potential during the initial exposure period, followed by stabilization with some fluctuations. The negative average value of −0.785 V (±0.016 V) for the OCP indicates low chemical stability of the alloy. This behavior can be attributed to the addition of Mn, which reduces the standard electrode potential of the system and renders the spontaneous passive film more thermodynamically unstable [26].

The potentiodynamic polarization curves of the Fe-35Mn alloy in Hank's solution are shown in Figure 5. The average corrosion-related parameters, such as OCP, corrosion potential, and current density determined by the Tafel extrapolation method, are displayed in Table 5.

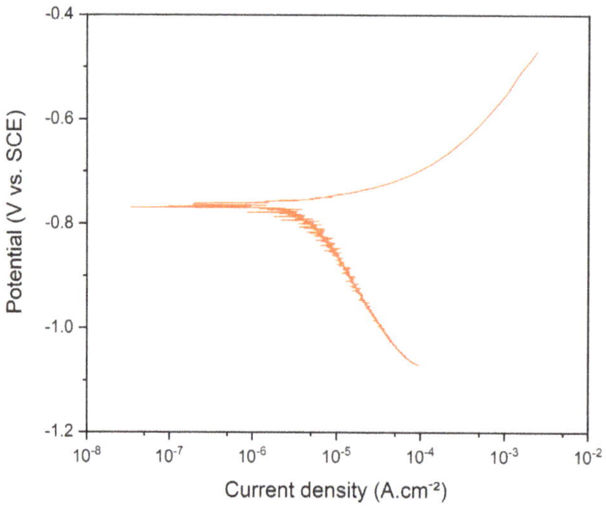

Figure 5. Average potentiodynamic polarization curve of Fe-35Mn alloy samples in Hank's solution at 37 °C.

Table 5. Corrosion average parameters of the Fe-35Mn alloy after route processing in arc melting furnace. The data presented were obtained after performing statistical tests, indicating the significance of the variations in the corrosion properties. The table shows the number of samples (n), mean, maximum value (Max), minimum value (Min), and standard deviation (SD) of each parameter evaluated.

Variable	n	Min	Max	Mean	SD
OCP	3	−0.804 V	−0.775 V	−0.785 V	0.016 V
E_{corr}	3	−0.770 V	−0.760 V	−0.763 V	0.006 V
I_{corr}	3	3.823 μA·cm^{-2}	3.956 μA·cm^{-2}	3.889 μA·cm^{-2}	0.066 μA·cm^{-2}

According to Wang et al. [27], corrosion occurs based on the potential (E_{corr}) in a thermodynamic system, where a higher potential corresponds to better corrosion resistance, while the corrosion rate is kinetically assessed by the current density (I_{corr}); the lower the current density, the lower the rate. Generally, a lower potential indicates a higher corrosion current density [26].

The curves demonstrated that all analyzed samples exhibited similar corrosion behavior, with no passivation occurring in the anodic region. The oxide films became unstable and discontinuous in the medium with increasing potential, rendering the material more susceptible to corrosion [28]. Dehestani et al. [17] reported that the addition of manganese to the structure of an iron-based material results in decreased corrosion resistance. As a less noble element with a lower electrochemical potential, manganese makes the iron matrix more prone to corrosion in physiological media. According to Liu et al. [29], galvanic corrosion mechanisms, in which one metal corrodes preferentially to another, can be introduced into the system due to the potential difference between the metallic elements, also increasing their corrosion rate.

The results obtained in this study, when compared to those indicated by Francis et al. [24] for 316 L steel and especially for as-cast pure iron, show that the addition of Mn led to reduced stability of the alloy in Hank's solution, resulting in decreased corrosion resistance. The 316 L stainless steel exhibits a corrosion potential of −0.26 V and a current density of 0.454 $\mu A.cm^{-2}$, while as-cast pure iron has a potential of −0.38 V and a current density of 0.652 $\mu A.cm^{-2}$. In contrast, the experimental Fe-35Mn alloy processed in this study showed a more negative potential of −0.76 V and a high current density of 3.88 $\mu A.cm^{-2}$. Previous studies by Shuai et al. [30] have already demonstrated that a lower corrosion potential indicates poor surface corrosion resistance, and an increased current density suggests an enhanced corrosion rate.

Compared to the corrosion parameters of the Fe–35Mn alloy produced via powder metallurgy, as reported by Dargusch et al. [1], a slight decrease in corrosion potential was observed, from −0.74 V to −0.76 V, indicating a slightly higher susceptibility to corrosion. However, the corrosion current density remained significantly lower at 3.88 $\mu A \cdot cm^{-2}$, in contrast to the 34 $\mu A \cdot cm^{-2}$ reported by the authors, suggesting an overall lower corrosion rate for the Fe–35Mn alloy processed in the present study.

The non-parametric one-sample Wilcoxon signed-rank test provided evidence that the corrosion properties of the Fe-35Mn alloy differ from those of as-cast pure iron, 316 L stainless steel, and the Fe-35Mn alloy processed by powder metallurgy.

4. Conclusions

In this study, a new processing route using an arc melting furnace was developed for the biodegradable Fe-35Mn alloy and its effects on the microstructure and properties, especially mechanical and corrosion, were evaluated. In addition, a statistical analysis was performed to check the correlation between variables and identify significant patterns in the data collected. Some conclusions are drawn as follows:

- The experimental Fe-35Mn alloy was successfully fabricated, and its microstructural characteristics after processing are consistent with the existing literature, showing the formation of a structure predominantly composed of austenite phase, with large, uniformly distributed grains.
- The determined mechanical properties, such as ultimate tensile strength, elongation to failure, yield strength, and Young's modulus, were similar to or superior to those of annealed pure iron and 316 L stainless steel, which were used as reference standards for evaluating other alloys intended for biomedical applications.
- The processed alloy proved to be more susceptible to corrosion. When compared to pure iron and 316 L stainless steel, the alloy exhibited a lower corrosion potential and an increased current density, suggesting, respectively, lower surface corrosion resistance and a higher corrosion rate. These results underscore the potential of the alloy for use in temporary biomedical devices, such as stents.

Although the results are promising, future studies could expand the characterization of the Fe–35Mn alloy processed via arc melting by conducting both in vitro and in vivo investigations, aiming to achieve a more comprehensive understanding of its behavior.

Author Contributions: Conceptualization, K.B.d.S., J.P.A.C. and A.P.R.A.; data curation, K.B.d.S., J.P.A.C., R.Z.N. and A.C.F.; funding acquisition, A.P.R.A.; methodology, K.B.d.S., R.Z.N., A.C.F. and A.P.R.A.; project administration, A.P.R.A.; resources, R.Z.N., A.C.F. and A.P.R.A.; supervision, A.P.R.A.; visualization, K.B.d.S., J.P.A.C. and R.Z.N.; writing—original draft, K.B.d.S. and J.P.A.C.; writing–review and editing, K.B.d.S., J.P.A.C., R.Z.N., A.C.F. and A.P.R.A. All authors have read and agreed to the published version of the manuscript.

Funding: This study was financed in part by the Coordenação de Aperfeiçoamento de Pessoal de Nível Superior—Brasil (CAPES)—Finance Code 001.

Data Availability Statement: The datasets presented in this article are not readily available because the data are part of an ongoing study. Requests to access the datasets should be directed to corresponding author.

Conflicts of Interest: The authors declare no conflict of interest.

References

1. Dargusch, M.S.; Choudhury, D.; Saha, T.; Pramanik, A.; Tran, P.A.; O'Mullane, A.P.; Brandt, M. Exploring the Role of Manganese on the Microstructure, Mechanical Properties, Biodegradability, and Biocompatibility of Porous Iron-Based Scaffolds. *ACS Biomater. Sci. Eng.* **2019**, *5*, 1686–1702. [CrossRef] [PubMed]
2. Li, H.; Zheng, Y.; Qin, L. Progress of Biodegradable Metals. *Prog. Nat. Sci. Mater. Int.* **2014**, *24*, 414–422. [CrossRef]
3. Sun, H.; Chen, F.; Chen, Z.-K. Current Status and Outlook on the Clinical Translation of Biodegradable Metals. *Mater. Today* **2019**, *23*, 57–71. [CrossRef]
4. Wang, Z.; Song, J.; Peng, Y. New Insights and Perspectives into Biodegradable Metals in Cardiovascular Stents: A Mini Review. *J. Alloys Compd.* **2024**, *958*, 175313. [CrossRef]
5. Rangel, R.d.C.R.; Rangel, A.L.R.; da Silva, K.B.; Escada, A.L.D.A.; Chaves, J.A.M.; Maia, F.R.; Pina, S.; Reis, R.L.; Oliveira, J.M.; Alves, A.P.R. Characterization of Iron Oxide Nanotubes Obtained by Anodic Oxidation for Biomedical Applications—In Vitro Studies. *Materials* **2024**, *17*, 3627. [CrossRef]
6. Rabeeh, V.P.M.; Hanas, T. Progress in Manufacturing and Processing of Degradable Fe-Based Implants: A Review. *Prog. Biomater.* **2022**, *11*, 163–191. [CrossRef]
7. AlMangour, B.; Sivasankaran, S.; Ammar, H.R.; Almufadi, A.A.; Alsaiari, N. Synthesis, Microstructures, Mechanical, and Corrosion Behavior of FeMn(35−x)–x (Cu, W, and Co)-Based Biodegradable Alloys Prepared by Mechanical Alloying and Selective Laser Melting. *Metall. Mater. Trans. A* **2023**, *54*, 3767–3780. [CrossRef]
8. Kiani, F.; Wen, C.; Li, Y. Prospects and Strategies for Magnesium Alloys as Biodegradable Implants from Crystalline to Bulk Metallic Glasses and Composites—A Review. *Acta Biomater.* **2020**, *103*, 1–23. [CrossRef]
9. Eliaz, N. Corrosion of Metallic Biomaterials: A Review. *Materials* **2019**, *12*, 407. [CrossRef]
10. Hu, T.; Yang, C.; Lin, S.; Yu, Q.; Wang, G. Biodegradable Stents for Coronary Artery Disease Treatment: Recent Advances and Future Perspectives. *Mater. Sci. Eng. C* **2018**, *91*, 163–178. [CrossRef]
11. Zhang, M.; Yang, N.; Dehghan-Manshadi, A.; Venezuela, J.; Bermingham, M.J.; Dargusch, M.S. Fabrication and Properties of Biodegradable Akermanite-Reinforced Fe35Mn Alloys for Temporary Orthopedic Implant Applications. *ACS Biomater. Sci. Eng.* **2023**, *9*, 1261–1273. [CrossRef] [PubMed]
12. Conti, M.C.; Aquilina, D.; Paternoster, C.; Vella, D.; Sinagra, E.; Mantovani, D.; Cassar, G.; Wismayer, P.S.; Buhagiar, J. Influence of Cold Rolling on In Vitro Cytotoxicity and Electrochemical Behaviour of an Fe-Mn-C Biodegradable Alloy in Physiological Solutions. *Heliyon* **2018**, *4*, e00926. [CrossRef] [PubMed]
13. Hermawan, H.; Dubé, D.; Mantovani, D. Degradable Metallic Biomaterials: Design and Development of Fe–Mn Alloys for Stents. *J. Biomed. Mater. Res. Part A* **2010**, *93*, 1–11. [CrossRef]
14. Dargusch, M.S.; Venezuela, J.; Dehghan-Manshadi, A.; Johnston, S.; Yang, N.; Mardon, K.; Lau, C.; Allavena, R. In Vivo Evaluation of Bioabsorbable Fe-35Mn-1Ag: First Reports on In Vivo Hydrogen Gas Evolution in Fe-Based Implants. *Adv. Healthcare Mater.* **2021**, *10*, 2000667. [CrossRef] [PubMed]
15. Hermawan, H.; Dubé, D.; Mantovani, D. Developments in Metallic Biodegradable Stents. *Acta Biomater.* **2010**, *6*, 1693–1697. [CrossRef]
16. Dehghan-Manshadi, A.; StJohn, D.H.; Dargusch, M.S. Tensile Properties and Fracture Behaviour of Biodegradable Iron–Manganese Scaffolds Produced by Powder Sintering. *Materials* **2019**, *12*, 1572. [CrossRef]
17. Dehestani, M.; Trumble, K.; Wang, H.; Wang, H.; Stanciu, L.A. Effects of Microstructure and Heat Treatment on Mechanical Properties and Corrosion Behavior of Powder Metallurgy Derived Fe–30Mn Alloy. *Mater. Sci. Eng. A* **2017**, *703*, 214–226. [CrossRef]

18. Mostaed, E.; Vedani, M.; Hashempour, M.; Bestetti, M. Influence of ECAP Process on Mechanical and Corrosion Properties of Pure Mg and ZK60 Magnesium Alloy for Biodegradable Stent Applications. *Biomatter* **2014**, *4*, e28283. [CrossRef]
19. Zhang, Q.; Cao, P. Degradable Porous Fe-35wt.% Mn Produced via Powder Sintering from NH4HCO3 Porogen. *Mater. Chem. Phys.* **2015**, *163*, 394–401. [CrossRef]
20. Liu, P.; Zhang, D.; Dai, Y.; Lin, J.; Li, Y.; Wen, C. Microstructure, Mechanical Properties, Degradation Behavior, and Biocompatibility of Porous Fe-Mn Alloys Fabricated by Sponge Impregnation and Sintering Techniques. *Acta Biomater.* **2020**, *114*, 485–496. [CrossRef]
21. Xu, Z.; Hodgson, M.A.; Cao, P. Effect of Immersion in Simulated Body Fluid on the Mechanical Properties and Biocompatibility of Sintered Fe–Mn-Based Alloys. *Metals* **2016**, *6*, 309. [CrossRef]
22. Witusiewicz, V.T.; Sommer, F.; Mittemeijer, E.J. Reevaluation of the Fe-Mn Phase Diagram. *J. Phase Equilib. Diffus.* **2004**, *25*, 346–354. [CrossRef]
23. Nayak, P.; Biswal, A.K.; Sahoo, S.K. Processing and Characterization of Fe-35Mn Biodegradable Metallic Materials. *Mater. Today Proc.* **2020**, *33*, 5284–5289. [CrossRef]
24. Francis, A.; Yang, Y.; Virtanen, S.; Boccaccini, A.R. Iron and Iron-Based Alloys for Temporary Cardiovascular Applications. *J. Mater. Sci. Mater. Med.* **2015**, *26*, 1–16. [CrossRef] [PubMed]
25. da Silva, K.B.; Carobolante, J.P.A.; Rajan, S.S.; Júnior, C.B.; Sabino, R.M.; Seixas, M.R.; Nakazato, R.Z.; Popat, K.C.; Claro, A.P.R.A. Mechanical Properties, Corrosion Behavior, and In Vitro Cell Studies of the New Ti-25Ta-25Nb-5Sn Alloy. *Materials* **2023**, *16*, 1970. [CrossRef]
26. Cheng, J.; Liu, B.; Wu, Y.H.; Zheng, Y.F. Comparative In Vitro Study on Pure Metals (Fe, Mn, Mg, Zn, and W) as Biodegradable Metals. *J. Mater. Sci. Technol.* **2013**, *29*, 619–627. [CrossRef]
27. Wang, H.; Zheng, Y.; Liu, J.; Jiang, C.; Li, Y. In Vitro Corrosion Properties and Cytocompatibility of Fe-Ga Alloys as Potential Biodegradable Metallic Materials. *Mater. Sci. Eng. C* **2017**, *71*, 60–66. [CrossRef] [PubMed]
28. Obayi, C.S.; Tolouei, R.; Mostavan, A.; Paternoster, C.; Turgeon, S.; Okorie, B.A.; Obikwelu, D.O.; Mantovani, D. Effect of Grain Sizes on Mechanical Properties and Biodegradation Behavior of Pure Iron for Cardiovascular Stent Application. *Biomatter* **2016**, *6*, e959874. [CrossRef]
29. Liu, R.-Y.; He, R.-G.; Xu, L.-Q.; Guo, S.-F. Design of Fe–Mn–Ag Alloys as Potential Candidates for Biodegradable Metals. *Acta Metall. Sin.* **2018**, *31*, 584–590. [CrossRef]
30. Shuai, C.; Yang, W.; Yang, Y.; Pan, H.; He, C.; Qi, F.; Xie, D.; Liang, H. Selective Laser Melted Fe-Mn Bone Scaffold: Microstructure, Corrosion Behavior, and Cell Response. *Mater. Res. Express* **2020**, *7*, 015404. [CrossRef]

metals

MDPI

Article

Effects of Mg Content and Pulsed Magnetic Field Treatment on Microstructure and Properties of As-Cast Biodegradable Zn-3Cu Alloy

Lizhen Shi [1,2,†], Hui Liu [1,2,†], Houqing Liu [3,4], Cong Peng [3,*] and Ling Ren [1,2,*]

[1] School of Materials Science and Engineering, University of Science and Technology of China, Shenyang 110016, China; lzshi22s@imr.ac.cn (L.S.); hliu17b@imr.ac.cn (H.L.)
[2] Shi-Changxu Innovation Center for Advanced Materials, Institute of Metal Research, Chinese Academy of Sciences, Shenyang 110016, China
[3] School of Aeronautical Manufacturing Engineering, Nanchang Hangkong University, Nanchang 330036, China; liuhq057@avic.com
[4] Avic Guizhou Anji Aviation Investment Casting Co., Ltd., Anshun 561000, China
* Correspondence: 70920@nchu.edu.cn (C.P.); lren@imr.ac.cn (L.R.)
† These authors contributed equally to this work.

Abstract: The microstructure, mechanical properties, corrosion behavior, cytocompatibility, and antibacterial properties of biodegradable Zn-3Cu-xMg (x = 0, 0.5, 1 wt.%) alloys with or without pulsed magnetic field treatment during casting were systematically investigated. Mg addition induced the formation of fine Mg_2Zn_{11} precipitated along the matrix grain boundaries. With the increase in Mg content, the precipitation of the Mg_2Zn_{11} phase increased, and the grain size became finer. Pulsed magnetic field treatment exacerbated the occurrence of this phenomenon. Under the combined action of the Mg_2Zn_{11} phase and refined grain size, Zn3Cu0.5Mg alloy with pulsed magnetic field treatment had the best strength–ductility match (σ_{UTS} = 181.46 ± 1.06 MPa, δ = 3.95 ± 0.07%), moderate corrosion rate (i_{corr} = 5.69 ± 3.96 μA/cm^2), positive cytocompatibility, and antibacterial properties. This study indicated that Zn3Cu0.5Mg alloy with pulsed magnetic field treatment had the greater potential to further improve its properties through subsequent conventional metal-forming processing and severe plastic deformation techniques to meet clinical requirements, compared to existing as-cast Zn alloys.

Keywords: Zn alloys; microstructure; mechanical properties; corrosion properties; biocompatibility; cytotoxicity

1. Introduction

Magnesium (Mg)-, iron (Fe)-, and zinc (Zn)-based alloys are considered biodegradable metallic materials for medical applications. However, the rapid degradation of Mg-based alloys in physiological environments can lead to a premature loss of mechanical integrity of the implant [1]. Fe-based alloys with degradation rates significantly below clinical requirements (complete bioabsorption within 12–24 months) may generate fragmentary embolism, as noted with other non-biodegradable implants [2]. Zn has a standard corrosion potential of 0.76 V, which is intermediate between that of Fe (0.44 V) and Mg (2.37 V) [3]. Therefore, in recent years, owing to moderate corrosion rates, the development of Zn-based biodegradable alloys has emerged as a research hotspot in biomedicine. Notably, Zn-Cu alloys have attracted more attention recently for their excellent ductility and strength enhanced by Cu solid solution, grain refinement, and dispersed $CuZn_5$ phase in the η-Zn matrix [4–6]. Tang et al. [7] reported that the elongation of as-extruded Zn-4Cu alloy reached 50.6 ± 2.8%, and the ultimate tensile strength reached 270 ± 0.5 MPa. Zn-Cu alloys exhibit moderate in vitro degradation rates (0.02–0.11 mm/year) [8,9]. Additionally, the alloying element Cu is an essential trace element for humans, and Zn-Cu alloys have been reported to stimulate endothelial cell proliferation, enhance angiogenesis, and possess antibacterial properties [7,10]. However, the mechanical properties of Zn-Cu alloys still fail to meet clinical strength standards (ultimate tensile strength (σ_{UTS}) > 300 MPa, yield strength (σ_{YS}) > 200 MPa [9]).

Mg has a superior strengthening ability for Zn alloys [11–14]. Pachla et al. [15] reported that the mechanical properties of Zn-xMg (x = 0, 0.5, 1, 1.5 wt.%) alloys continued to improve as the Mg content increased. Su et al. [16] incorporated various amounts (0, 0.5, 1, 2 wt.%) of Mg element into Zn-5Cu alloy. Compared to Zn-5Cu alloy, the tensile strength of Zn-5Cu-2Mg alloy increased by 35% to 216 MPa. The Mg_2Zn_{11} phase precipitates when the content of Mg is below 1.5 wt.% [13]. Dispersing the Mg_2Zn_{11} phase effectively impedes dislocation motion and, thus, improves the strength of Zn-Mg alloys [14]. Additionally, the effect of Mg addition on the corrosion behavior of Zn alloys has been studied by many scholars [17,18]. Jin et al. [17] reported that adding Mg improved the biocompatibility and degrading performance of Zn alloys. Ye et al. [11] demonstrated that the corrosion rate of Zn-Mg alloys varied due to the competition between the positive effect of grain refinement resulting from the addition of Mg and the negative effect of Mg_2Zn_{11} volume fraction increment. Furthermore, Mg is an essential element for humans, which can not only prevent acute thrombosis and regulate vascular dilation [19] but also stimulate bone growth and lower the risk of osteoporosis [20]. Therefore, adding Mg to Zn-Cu alloys is expected to optimize their mechanical properties, corrosion behavior, and biocompatibility.

In recent years, with the successful design of biodegradable Zn alloy compositions, researchers focused on further optimizing the microstructure and enhancing the mechanical properties of these alloys through subsequent conventional metal-forming processing, such as extrusion, drawing, rolling, forging, and severe plastic deformation techniques [4,6,7,10,15,16]. Additionally, bioactive coatings were prepared on the surface of biodegradable Zn alloys through a plasma electrolytic oxidation process, electrodeposition method, or microwave-assisted coating method [21,22] to regulate their corrosion rate and biocompatibility. However, the initial microstructure after solidification during casting, which also plays an important role in determining Zn alloys' final properties due to structural inheritance, has garnered relatively limited attention. Therefore, employing specialized techniques to control the solidification structure of biodegradable Zn alloys during casting to improve their properties is of great significance.

Pulsed magnetic field treatment is a commonly used and effective method in the field of alloy solidification processing. It can achieve microstructural homogenization, reduce casting defects, and refine the alloy grains. Zhang et al. [23] found that pulsed magnetic field treatment refined the solidification microstructure of Mg-7Zn alloy and reduced the volume fraction of the second phase. The σ_{UTS}, σ_{YS}, and elongation of Mg-7Zn treated with a 300 V pulsed magnetic field were increased by 40.4%, 38.2%, and 48.3%, respectively, compared to those of the untreated alloy. Li et al. [24] studied the effects of the high-voltage pulsed magnetic field on the microstructure of the Zn-Ag alloys and indicated that it refined the primary ε-$AgZn_3$ dendrites and distributed them homogeneously in the specimens. Li et al. [25] applied a 6 kV pulsed magnetic field to the solidification process of Zn-xCu (x = 1.5, 2.5, 4.0 wt.%) alloys and discussed the structure of the alloys refined due to the convection and Joule heat induced by the pulsed magnetic field in the melt. However, studies on the effect of pulsed magnetic field treatment on biodegradable Zn alloys have primarily focused on microstructures, and their effects on properties have not yet been systematically evaluated. Therefore, this study systematically investigated the effects of Mg addition and pulsed magnetic field treatment on the microstructure, mechanical properties, corrosion behavior, cytocompatibility, and antibacterial property of Zn-3Cu-xMg (x = 0, 0.5, 1 wt%) alloys, aiming to develop an as-cast Zn alloy with improved properties for subsequent conventional metal-forming processing (extrusion, drawing, rolling, or forging) and severe plastic deformation techniques.

2. Materials and Methods

Commercially pure Zn (99.99%), Cu (99.99%), and Mg (99.96%) were used to prepare the as-cast Zn-3Cu-xMg (x = 0, 0.5, 1 wt.%) alloys. According to the composition of the Zn-3Cu-xMg (x = 0, 0.5, 1 wt.%) alloys, the dried pure metals were placed into a refractory crucible and completely melted through electrical resistance furnace (SG2-5-10, Shanghai real research Electric Furnace Co., Ltd., Shanghai, China). Then, the molten alloy was held at 620 °C for 30 min. During this period, the molten alloy was continuously stirred, and impurities were promptly removed. Subsequently, the molten alloy was poured into a cylindrical graphite mold positioned in the pulsed magnetic field device, as shown in Figure 1. Immediately after pouring, the pulsed magnetic field device was activated until the metal solidified completely. To prevent oxidation of the alloy, the entire process of melting, pouring, and solidification was carried out under a protective gas atmosphere (CO_2: 99.5%, SF_6: 0.5%).

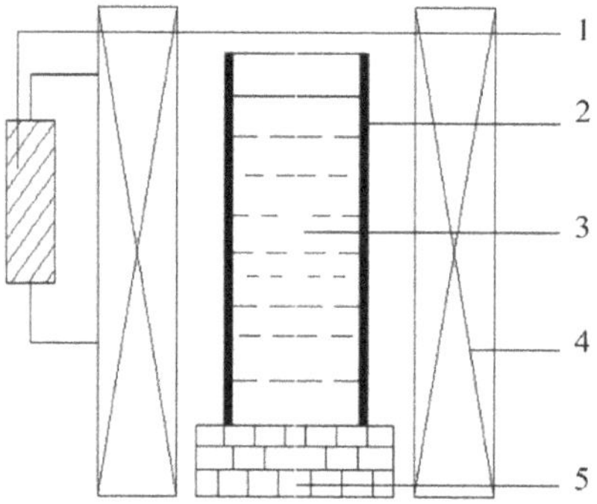

Figure 1. Schematic diagram of a pulsed magnetic field device. (1—Pulse Generator; 2—Graphite Mold; 3—Melt; 4—Coil; 5—Refractory Brick Base).

The pulsed magnetic field generates periodic pulsed forces within the melt, which combine with electromagnetic forces to form electromagnetic oscillations [26]. By controlling the pulsed voltage and frequency, the magnitude and period of the pulsed forces can be regulated. According to our previous study [27], the microstructure and mechanical properties of Zn-3Cu alloy were explored by setting different pulsed voltages and frequencies. It was demonstrated that when the pulsed voltage was 300 V and the frequency was 5 Hz, Zn-3Cu alloy had the finest and most homogeneous microstructure and the best mechanical properties. Based on this, in the solidification process of this study, an external pulsed magnetic field with a voltage of 300 V and a frequency of 5 Hz was directly applied. For the convenience of subsequent research description, the alloys without pulsed magnetic field treatment were named W-Zn3Cu, W-Zn3Cu0.5Mg, and W-Zn3Cu1Mg, respectively, and the alloys with pulsed magnetic field treatment were named P-Zn3Cu, P-Zn3Cu0.5Mg, and P-Zn3Cu1Mg, respectively.

For the microstructure observation, alloys were sandpapered to 5000 grit and then polished to a mirror finish. The polished section was etched by a solution of 1.0 g oxalic acid,

1.0 mL nitric acid, 1.0 mL acetic acid, and 150 mL distilled water. The elemental distribution on the surface of alloys was characterized using a scanning electron microscope (FEI Apreo, Thermo Fisher, Waltham, MA, USA) equipped with energy dispersive spectroscopy (EDS). Secondary electrons (SE) and backscattered electrons (BSE) were both used to characterize the surface morphology and phase distribution. In addition, the phase composition of alloys was detected by XRD techniques. The XRD scanning range was from 20 to 80°, and the scanning rate was 10°/min.

Tensile tests were conducted on a universal material testing machine (WDW-200D, Shandong Wanchen Testing Machine Co., Ltd, Jinan, China) at a strain rate of 1 mm/min. The dimensions of tensile samples were designed according to GB/T 228.1-2021 [28], with a gauge of Φ5 × 25 mm.

The electrochemical corrosion investigation was carried out in a standard three-electrode electrochemical flask cell with an electrochemical workstation (Reference 600 TM, Gamry Instruments, Inc., Warminster, PA, USA). A saturated calomel electrode (SCE), a platinum sheet, and a sample having dimensions of Φ10 × 2 mm were utilized as the reference, counter, and working electrodes, respectively. The exposed area of the sample was approximately 0.636 cm^2 after sealing the edges of sample with silicone. The simulated body fluid (SBF) solution was used in this study, which consisted of NaCl (8.035 g/L), $NaHCO_3$ (0.355 g/L), KCl (0.225 g/L), $K_2HPO_4 \cdot 3H_2O$ (0.231 g/L), $MgCl_2 \cdot 6H_2O$ (0.311 g/L), 1mol/L HCl (39 mL/L), $CaCl_2$ (0.292 g/L), Na_2SO_4 (0.072 g/L), and finally, the pH value was adjusted to 7.40 using Tris and 1mol/L HCl. After the open circuit potential (OCP) test for 3600 s, the electrochemical impedance spectroscopy (EIS) tests were performed with a sinusoidal voltage signal of 10 mV and in a frequency range of 100 kHz to 0.1 Hz at E_{OCP}. The potentiodynamic polarization (PD) test was performed from −0.2 $V_{vs\ OCP}$ to 0.3 $V_{vs\ OCP}$ at a scanning rate of 0.5 mV/s. All the experiments were conducted thrice to obtain reproducible results.

Human umbilical vein endothelial cells (HUVECs, KeyGEN Bio TECH, Nanjing, China) were used to assess the cytocompatibility of Zn-3Cu-xMg alloys. The experiment was designed with a negative control group, a positive control group, and experimental groups. Each group had five replicate wells in a 96-well plate. For the negative control group, 100 μL of medium was added to each well without culturing cells. For the positive control group, 100 μL of medium was added to each well to culture 5 × 10^3 cells. For the experimental group, 100 μL of medium containing 10% extract of the alloy [7] was added to each well to culture 5 × 10^3 cells. The medium used was RPMI 1640 medium (Hyclone, Logan, UT, USA) supplemented with 10% fetal bovine serum (Gibco, Langley, OK, USA), penicillin (80 U/mL), and streptomycin (0.08 mg/mL). The extract of the Zn-3Cu-xMg alloys was prepared in advance using a culture medium with an extraction ratio of alloy surface area to medium volume of 1.25 cm^2/mL in a humidified atmosphere with 5% CO_2 at 37 °C for 3 d. The HUVECs were cultured for 1, 3, and 7 d. At each time point, wells were cleaned with PBS solution, and 100 μL of medium containing 10% CCK-8 solution was added to each well. After that, the cells were cultured at a temperature of 37 °C for 2 h. The absorbance was measured by a plate reader (MuLtiskan GO, Thermo Scientific, Waltham, MA, USA) at 450 nm. The relative growth rate (RGR) is calculated by Formula (1):

$$RGR\ (\%) = (OD_{EG} - OD_{NC})/(OD_{PC} - OD_{NC}) \times 100\% = (OD_{Alloys}/OD_{Control}) \times 100\% \quad (1)$$

where OD_{Alloys} is the optical density of experimental group (OD_{EG}) minus negative control group (OD_{NC}), and $OD_{Control}$ is the optical density of positive control group (OD_{PC}) minus negative control group (OD_{NC}).

The antibacterial activity of Zn-3Cu-xMg alloys against *Staphylococcus aureus* (*S. aureus*, ATCC6538) was assessed by a plate count method according to ISO 10993-5:2009. The alloys were soaked in 1000 μL of bacterial suspension with a concentration of 10^6 CFU/mL and incubated at a temperature of 37 °C with 90% relative humidity. After 24 h, alloys with adherent bacteria were transferred to 500 μL of fresh phosphate-buffered saline (PBS) and shaken mechanically by vortex for 1 min. Then, 100 μL of the solution, after vortexing, was spread onto a nutrient agar plate and incubated at 37 °C for 24 h. Thereafter, the number of bacterial colonies on the nutrient agar plate was counted, and the antibacterial rate was calculated. The antibacterial rate is calculated by the Formula (2):

$$\text{Antibacterial rate (\%)} = (N_{CK} - N_{alloys})/N_{CK} \times 100\% \quad (2)$$

where N_{CK} and N_{alloys} denote the mean numbers of bacterial colonies on control check and Zn-3Cu-xMg alloys, respectively.

All experiments were performed three times, and the data were expressed as mean ± standard deviation. Student *t*-test was used to analyze the significant difference. * $p < 0.05$, ** $p < 0.01$, and *** $p < 0.001$ were considered statistically significant.

3. Results and Discussion

3.1. Microstructure

Figure 2(a_1–c_1) show the SEM images of Zn-3Cu-xMg (x = 0, 0.5, 1 wt.%) alloys without pulsed magnetic field treatment, and Figure 2(a_2–c_2) show the corresponding BSEM images. For W-Zn3Cu, the matrix and about 50 μm dendritic second phase were observed, and after adding Mg, about 5 μm lamellar phases precipitated along the grain boundaries of the matrix. Based on the EDS (Figure 2d) and XRD (Figure 2e) patterns, it could be concluded that the matrix was η-Zn, the dendritic second phase was $CuZn_5$, and the lamellar phase was Mg_2Zn_{11}. Based on the Zn-Cu binary phase diagram [29], a peritectic reaction Zn + L → $CuZn_5$ occurs at 425 °C. And a eutectic reaction L → Zn + Mg_2Zn_{11} occurs at 364 °C based on the Zn-Mg binary phase diagram [30]. Consequently, during the solidification process, the $CuZn_5$ phase initially precipitated from the melt as the temperature decreased to approximately 425 °C, followed by the precipitation of Mg_2Zn_{11} along the grain boundaries of the matrix at around 364 °C. The XRD (Figure 2e) patterns also show that higher intensity of the peaks of Mg_2Zn_{11} was detected with increasing Mg content. The volume fraction of the phase was statistically analyzed, and the results are shown in Figure 2f. It clearly showed that the volume fraction of the lamellar Mg_2Zn_{11} phase increased with increasing Mg content while that of the η-Zn phase decreased. Meanwhile, the volume fraction of the $CuZn_5$ phase first decreased and then increased.

Compared with Figure 2, the SEM, BSEM, EDS, and XRD results in Figure 3 all confirmed that applying a pulsed magnetic field during the solidification process, the phase composition of Zn-3Cu-xMg (x = 0, 0.5, 1 wt.%) alloys was the same as that of the alloys without pulsed magnetic field treatment. P-Zn3Cu consisted of the primary η-Zn matrix and the dendritic $CuZn_5$ phase. Similarly, P-Zn3Cu0.5Mg and P-Zn3Cu1Mg consisted of the η-Zn, $CuZn_5$ phase, and lamellar Mg_2Zn_{11} phase. Statistical analysis of the phase volume fraction for Zn-3Cu-xMg (x = 0, 0.5, 1 wt.%) alloys with pulsed magnetic field treatment was performed, with results compared to those of the alloys without pulsed magnetic field treatment (as indicated by the dashed line in Figure 3f). It showed that the pulsed magnetic field had a minimal effect on the volume fraction of η-Zn and $CuZn_5$ phase for Zn-3Cu alloy. However, for Zn-3Cu-0.5Mg and Zn-3Cu-1Mg alloys, pulsed magnetic field treatment resulted in the reduced volume fraction of the $CuZn_5$ phase, while that of the Mg_2Zn_{11} phase increased.

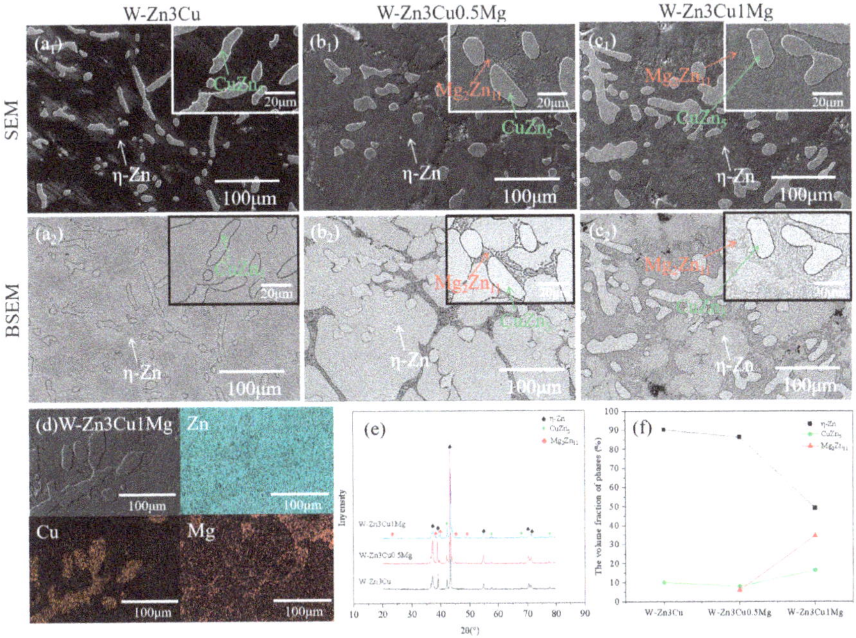

Figure 2. Microstructure of Zn-3Cu-xMg (x = 0, 0.5, 1 wt.%) alloys without pulsed magnetic field treatment: (**a**$_1$–**c**$_1$) SEM images; (**a**$_2$–**c**$_2$) the corresponding BSEM images; (**d**) EDS analysis; (**e**) XRD patterns; (**f**) phase volume fraction statistics.

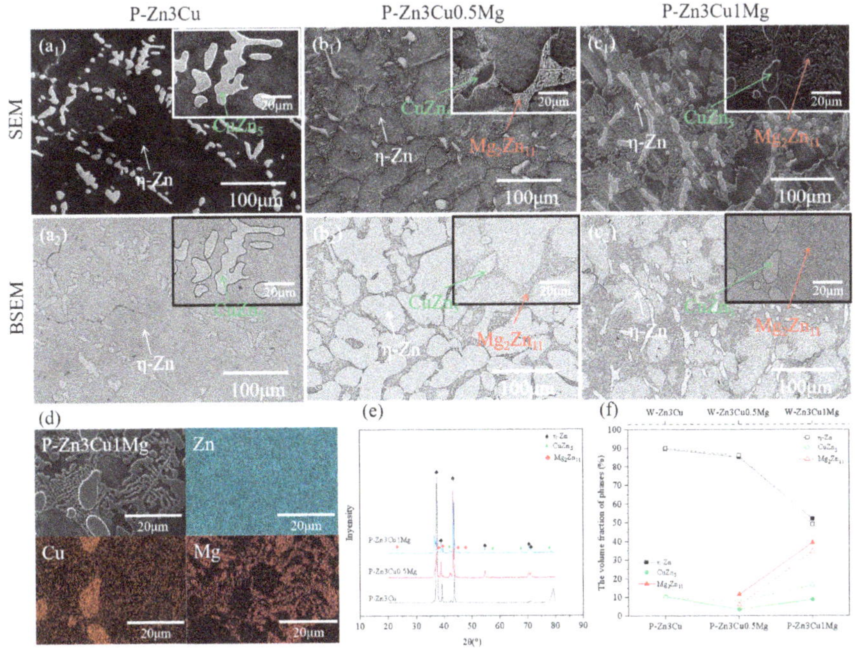

Figure 3. Microstructure of Zn-3Cu-xMg (x = 0, 0.5, 1 wt.%) alloys with pulsed magnetic field treatment: (**a**$_1$–**c**$_1$) SEM images; (**a**$_2$–**c**$_2$) the corresponding BSEM images; (**d**) EDS analysis; (**e**) XRD patterns; (**f**) phase volume fraction statistics.

In addition, the SEM and BSEM images in Figures 2 and 3 also indicated that the addition of Mg and pulsed magnetic field treatment affected the grain size of Zn-3Cu-xMg (x = 0, 0.5, 1 wt.%) alloys. Figure 4a–f show the optical microstructure (OM) images of Zn-3Cu-xMg (x = 0, 0.5, 1 wt.%) alloys with or without pulsed magnetic field treatment. The grain size of alloys was statistically analyzed and was shown in Figure 4g. As the Mg content increased, the grain size gradually decreased, and the pulse magnetic field treatment further refined the grain size. The refinement of grain size was mainly attributed to the addition of Mg, and pulsed magnetic field treatment increased the precipitation of the Mg_2Zn_{11} phase along the grain boundaries (Figures 2 and 3), which impeded the growth of the Zn matrix. And pulsed magnetic fields influenced the nucleation and growth processes of grains during solidification by increasing the number of nucleation sites [31], further refining grain size. Thus, P-Zn3Cu1Mg alloys had the smallest grain size of 59.90 ± 4.22 μm among all the Zn-3Cu-xMg (x = 0, 0.5, 1 wt.%) alloys.

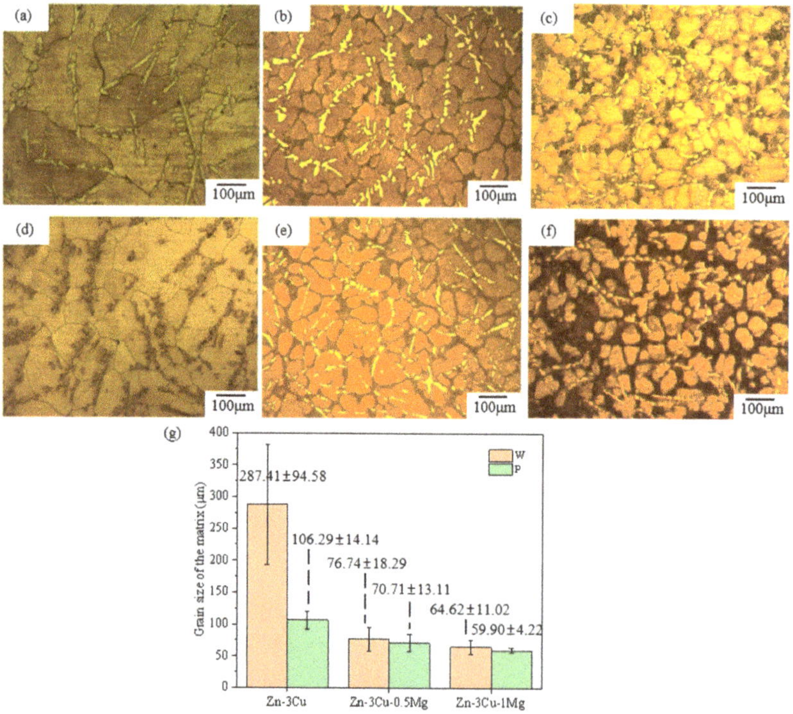

Figure 4. The optical microstructure (OM) images of Zn-3Cu-xMg (x = 0, 0.5, 1 wt.%) alloys: (**a**) W-Zn3Cu; (**b**) W-Zn3Cu0.5Mg; (**c**) W-Zn3Cu1Mg; (**d**) P-Zn3Cu; (**e**) P-Zn3Cu0.5Mg; (**f**) P-Zn3Cu1Mg; (**g**) grain size statistics.

3.2. Mechanical Properties

Figure 5a shows the room temperature stress–strain curves of the Zn-3Cu-xMg (x = 0, 0.5, 1 wt.%) alloys with or without pulsed magnetic field treatment. All alloys exhibited a rapid decline in the stress–strain curve after reaching the maximum stress, almost with no plastic deformation before fracture. The ultimate tensile strength (σ_{UTS}) and elongation (δ) are shown in Figure 5b,c. As the Mg content increased, σ_{UTS} gradually increased while δ decreased. The improvement of σ_{UTS} could be mainly attributed to the massive precipitation of the Mg_2Zn_{11} phase (Figures 2f and 3f), blocking the dislocation sliding [14] and strengthening the alloys. Massive precipitation of brittle Mg_2Zn_{11} phase also led to the stress concentration, resulting in cracking, thus decreasing the ductility. It also showed that the pulsed magnetic field treatment increased both σ_{UTS} and δ of Zn-3Cu-xMg alloys. The improvements in strength and ductility were attributed to the grain refinement [4] after pulsed magnetic field treatment (Figure 4g). Therefore, under the combined action of the Mg_2Zn_{11} phase and refined grain size, P-Zn3Cu0.5Mg alloy had the best strength–ductility match (σ_{UTS} = 181.46 ± 1.06 MPa, δ = 3.95 ± 0.07%). Compared with other types of as-cast Zn alloys [32] (Figure 5d), the P-Zn3Cu0.5Mg alloy had the greater potential to further improve its mechanical properties through subsequent conventional metal-forming processing such as extrusion, drawing, rolling, forging, and severe plastic deformation techniques to meet clinical requirements.

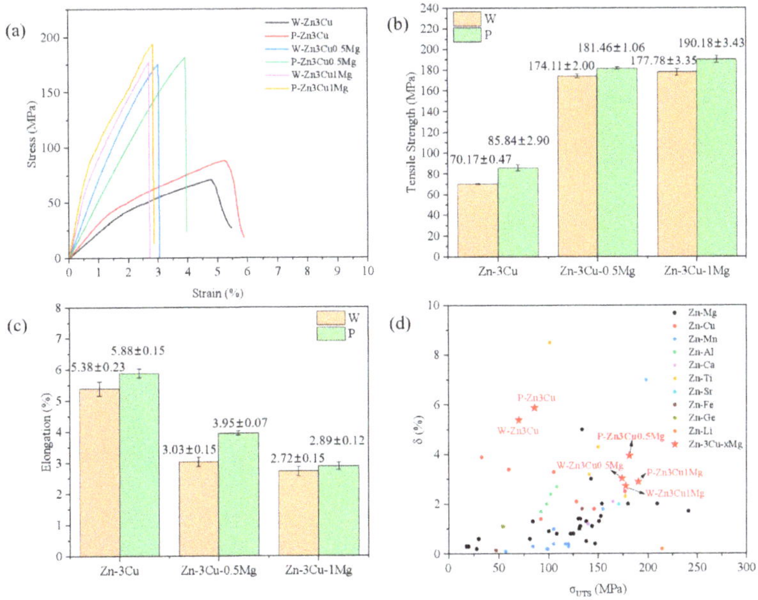

Figure 5. (**a**) Stress–strain curves, (**b**) tensile strength, (**c**) elongation of Zn-3Cu-xMg (x = 0, 0.5, 1 wt.%) alloys, and (**d**) comparison of tensile properties among different as-cast Zn alloys (the data of as-cast Zn alloys is adapted from Ref. [32]).

3.3. Corrosion Properties

The corrosion behavior of Zn-3Cu-xMg (x = 0, 0.5, 1 wt.%) alloys with or without pulsed magnetic field treatment in SBF solution was also evaluated systematically. Potentiodynamic polarization (PD) curves proved that all the alloys showed spontaneously passive behaviors in the SBF solution (Figure 6a). The corrosion current density (i_{corr}) is depicted in Figure 6b. It can be concluded that the value of icorr increased as the Mg content increased, and pulsed magnetic field treatment further increased icorr. The corrosion rate (CR) can be evaluated using ASTM G102-23 [33]:

$$CR = 3.27 \times 10^{-3} \times i_{corr} \times EW/\rho \quad (3)$$

where EW is the equivalent weight (EW = 32.43 g/eq), and ρ is the density (ρ = 6.5 g/cm^3) of the Zn-3Cu-xMg(x = 0, 0.5, 1 wt.%) alloys.

The CR value was calculated and shown in Figure 6c. It indicated that the addition of Mg and pulsed magnetic field treatment both accelerate the corrosion rate of Zn-3Cu-xMg (x = 0, 0.5, 1 wt.%) alloys. Precipitate phases influence the corrosion behavior of the material. The

Mg_2Zn_{11} phase and η-Zn matrix phase could form a micro-battery cell due to the difference in their corrosion potentials [11,16], resulting in galvanic corrosion. In addition, the Mg_2Zn_{11} phase precipitated along the grain boundary of the matrix had a low potential, causing the grain boundaries to be preferentially corroded during corrosion [34]. The addition of Mg and pulsed magnetic field treatment both promoted the precipitation of the Mg_2Zn_{11} phase along grain boundaries (Figures 2f and 3f) and increased the number of grain boundaries (Figure 4g) by refining grain size, thus escalating galvanic corrosion and grain boundary corrosion, accelerating the corrosion rate of Zn-3Cu-xMg (x = 0, 0.5, 1 wt.%) alloys.

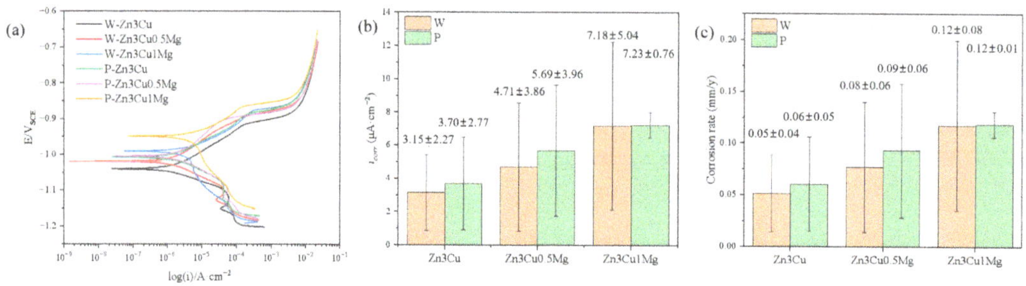

Figure 6. Zn-3Cu-xMg (x = 0, 0.5, 1 wt.%) alloys: (**a**) potentiodynamic polarization curves; (**b**) corrosion current density i_{corr}; (**c**) corrosion rate.

Although the addition of Mg and pulsed magnetic field treatment accelerated the corrosion rate of Zn-3Cu alloy, these rates remained between 0.02 and 0.11 mm/year [8,9], and were lower than that of Mg alloys (0.42–4.83 mm/year) [35], retaining the characteristic of the moderate degradation rate of Zn alloys.

Figure 7 shows that as the Mg content increased, both high-frequency capacitive arc radius (related to the charge transfer) and low-frequency capacitive arc radius (related to the diffusion process) [36] decreased and were even smaller after pulsed magnetic field treatment. Smaller capacitive arc radii indicate poorer corrosion resistance. The $R_s(CPE_1R_1)(CPE_2R_2)$ [36] model was used to analyze the EIS spectra, and the results are shown in Table 1, where R_s represents the solution resistance, R_1 and CPE_1 represent the resistance and capacitance of the charge transfer layer, and R_2 and CPE_2 represent the resistance and capacitance of the corrosion product layer. It could be seen that both R_1 and R_2 values decreased in the order of Zn-3Cu > Zn-3Cu-0.5Mg > Zn-3Cu-1Mg, and the pulsed magnetic field treatment also decreased R_1 and R_2 values, indicating that adding Mg and pulsed magnetic field treatment made the charge transfer easier and the protective performance of corrosion product weaker, accelerating the corrosion rate of Zn-3Cu-xMg alloys.

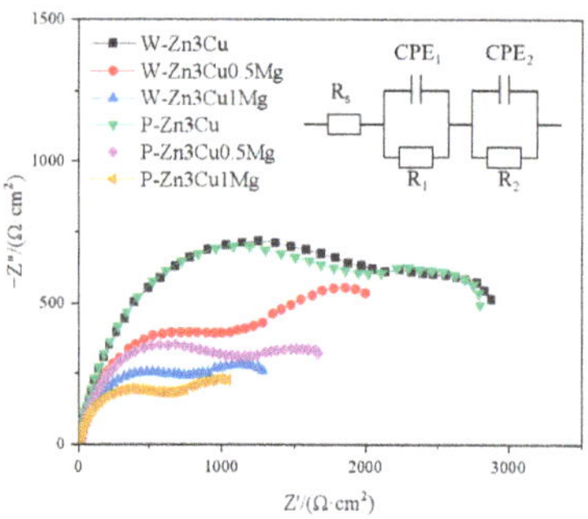

Figure 7. Nyquist diagrams at OCP of Zn-3Cu-xMg (x = 0, 0.5, 1 wt.%) alloys.

Table 1. Simulated EIS data for Zn-3Cu-xMg (x = 0, 0.5, 1 wt.%) alloys.

Sample	R_s /Ω·cm²	CPE_1 /μF·cm⁻²	n_1	R_1 /Ω·cm²	CPE_2 /μF·cm⁻²	n_2	R_2 /Ω·cm²	X^2
W-Zn3Cu	11.54 ± 1.25	5.02 ± 0.36	0.87 ± 0.03	1219.00 ± 154.15	102.63 ± 12.02	0.73 ± 0.03	1571.50 ± 405.17	3.09×10^{-3}
W-Zn3Cu0.5Mg	11.75 ± 1.21	5.25 ± 1.70	0.86 ± 0.03	1070.47 ± 495.62	228.54 ± 102.80	0.72 ± 0.05	1177.77 ± 539.01	1.37×10^{-3}
W-Zn3Cu1Mg	11.64 ± 1.21	5.04 ± 1.50	0.83 ± 0.02	765.05 ± 332.27	210.71 ± 60.39	0.68 ± 0.02	976.00 ± 91.92	2.56×10^{-3}
P-Zn3Cu	12.91 ± 1.09	122.82 ± 38.31	0.75 ± 0.03	1524.70 ± 496.79	4.73 ± 0.79	0.86 ± 0.01	1320.70 ± 447.63	1.97×10^{-3}
P-Zn3Cu0.5Mg	11.81 ± 0.42	4.93 ± 0.75	0.79 ± 0.04	1326.63 ± 939.22	220.10 ± 45.61	0.72 ± 0.10	1132.57 ± 264.25	1.97×10^{-3}
P-Zn3Cu1Mg	10.40 ± 2.64	179.28 ± 58.01	0.69 ± 0.13	1029.63 ± 610.66	6.13 ± 0.24	0.79 ± 0.14	883.73 ± 127.43	7.68×10^{-3}

3.4. Cytocompatibility and Antibacterial Activity

The deficiency in mechanical properties is one of the primary issues faced by Zn alloys at present. Considering the enhancement of strength and the acceleration of corrosion rate in Zn-3Cu-xMg alloys after pulsed magnetic field treatment, their responses to HUVECs and *S. aureus* were further investigated. Figure 8a,b show that Zn-3Cu-xMg alloys promoted cell proliferation and had no cytotoxicity compared with the control group. As the Mg content increased, the level of cell proliferation increased. Differences in cell viability can be attributed to the ion concentration of the extract, which is closely related to the alloy degradation process. The corrosion rate of Zn-3Cu-xMg alloy accelerated as the Mg content increased, releasing more Zn^{2+} and Mg^{2+} ions; thus, biocompatibility was improved. Li et al. [37] reported the in vivo corrosion rates of the Zn-1X (X = Mg, Ca, Sr) alloys as 0.17 mm/y, 0.19 mm/y, and 0.22 mm/y for Zn-1Mg, Zn-1Ca, and Zn-1Sr, respectively. And the radiographs of mice with Zn-1X alloy implants at 0, 1, 2, 3, 4, 6, and 8 weeks showed no inflammation and gas shadows around the implantation site, and all alloys exhibited no cytotoxicity in vivo. The corrosion rates of Zn-3Cu-xMg (x = 0, 0.5, 1 wt.%) alloys were 0.06–0.12 mm/y (Figure 6c), lower than those of Zn-1X alloys. It could be concluded that the Zn-3Cu-xMg (x = 0, 0.5, 1 wt.%) alloys should not cause cytotoxicity in vivo over extended periods. Figure 8c shows that the antibacterial rates of Zn-3Cu-xMg alloys were all up to 99.9%, owing to the combination of positively charged Zn^{2+}, Cu^{2+}, and Mg^{2+} ions with the negatively charged thiol, carboxyl, and amino groups on the bacterial membrane, leading to complete enzyme inactivation, protein destruction, and subsequent bacterial death [38].

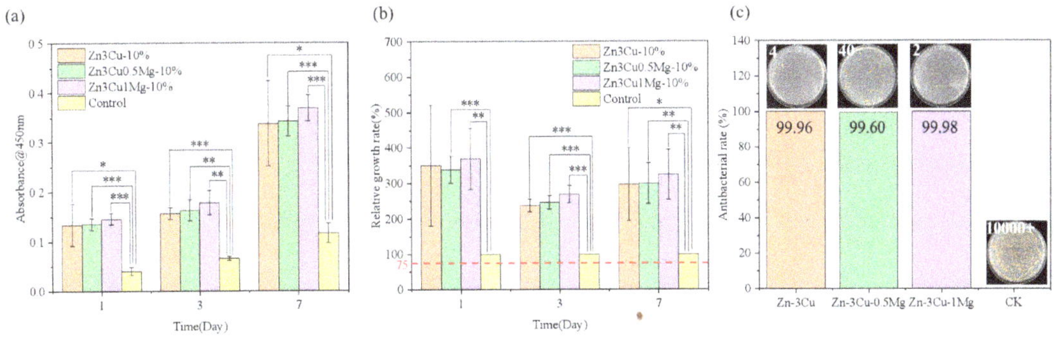

Figure 8. Zn-3Cu-xMg alloys: (**a**) cell proliferation; (**b**) cell relative growth rate; (**c**) antibacterial rate. * $p < 0.05$, ** $p < 0.01$, *** $p < 0.001$.

4. Conclusions

This study investigated the effects of Mg addition and pulsed magnetic field treatment on the microstructure and properties of as-cast biodegradable Zn-3Cu-xMg (x = 0, 0.5, 1 wt.%) alloys. Mg addition induced the precipitation of fine Mg_2Zn_{11} along matrix grain boundaries. With the increase in Mg content, the precipitation of the Mg_2Zn_{11} phase increased, and the grain size became finer. Pulsed magnetic field treatment exacerbated the occurrence of this phenomenon. Under the combined action of Mg_2Zn_{11} phase and refined grain size, Zn3Cu0.5Mg alloy with pulsed magnetic field treatment had the best strength–ductility match (σ_{UTS} = 181.46 ± 1.06 MPa, δ = 3.95 ± 0.07%), moderate corrosion rate (i_{corr} = 5.69 ± 3.96 μA/cm^2), cytocompatibility and antibacterial property, which makes it a promising as-cast Zn alloy with improved properties for subsequent conventional metal-forming processing (extrusion, drawing, rolling, or forging) and severe plastic deformation techniques.

Author Contributions: Conceptualization, C.P., L.R. and H.L. (Hui Liu); methodology, H.L. (Houqing Liu); validation, L.S. and H.L. (Houqing Liu); formal analysis, L.S.; investigation, L.S. and H.L. (Houqing Liu); data curation, H.L. (Hui Liu); writing—original draft preparation, L.S.; writing—review and editing, L.S., H.L. (Hui Liu) and H.L. (Houqing Liu); visualization, L.S., H.L. (Hui Liu) and L.R.; supervision, C.P. and L.R.; project administration, H.L. (Hui Liu); funding acquisition, C.P. and L.R. All authors have read and agreed to the published version of the manuscript.

Funding: This research was funded by the National Key Research and Development Program of China (No. 2022YFC2406000), Liaoning Provincial Science and Technology Program—Excellent Youth Fund Program (No. 2023JH3/10200002), National Natural Science Foundation of China (No. 52301308), IMR Innovation fund (No. 2023-PY15), Natural Science Foundation of Jiangxi Provincial Education Department (No. DA202103161).

Data Availability Statement: The original contributions presented in this study are included in the article. Further inquiries can be directed to the corresponding authors.

Acknowledgments: We would like to express our gratitude to Ke Yang (Shi-changxu Innovation Center for Advanced Materials, Institute of Metal Research, Chinese Academy of Sciences, Shenyang 110016, China; kyang@imr.ac.cn), Hai Wang (Shi-changxu Innovation Center for Advanced Materials, Institute of Metal Research, Chinese Academy of Sciences, Shenyang 110016, China; hwang13s@imr.ac.cn) and Yang Hong (Department of Neurosurgery, Shengjing Hospital, China Medical University, Shenyang 110122, China; hongyangcmu@hotmail.com) for their valuable guidance and assistance in overcoming the challenges encountered during the research work of this project.

Conflicts of Interest: Author Houqing Liu was employed by the company Avic Guizhou Anji Aviation Investment Casting Co., Ltd. The remaining authors declare that the research was conducted in the absence of any commercial or financial relationships that could be construed as a potential conflict of interest.

References

1. Banerjee, P.C.; Al-Saadi, S.; Choudhary, L.; Harandi, S.E.; Singh, R. Magnesium implants: Prospects and challenges. *Materials* **2019**, *12*, 136. [CrossRef]
2. Peuster, M.; Hesse, C.; Schloo, T.; Fink, C.; Beerbaum, P.; von Schnakenburg, C. Long-term biocompatibility of a corrodible peripheral iron stent in the porcine descending aorta. *Biomaterials* **2006**, *27*, 4955–4962. [CrossRef]
3. Mostaed, E.; Sikora-Jasinska, M.; Drelich, J.W.; Vedani, M. Zinc-based alloys for degradable vascular stent applications. *Acta Biomater.* **2018**, *71*, 1–23. [CrossRef]
4. Lin, J.; Tong, X.; Wang, K.; Shi, Z.; Li, Y.; Dargusch, M.; Wen, C. Biodegradable Zn-3Cu and Zn-3Cu-0.2Ti alloys with ultrahigh ductility and antibacterial ability for orthopedic applications. *J. Mater. Sci. Technol.* **2021**, *68*, 76–90. [CrossRef]
5. Lan, C.; Hu, Y.; Wang, C.; Li, W.; Gao, X.; Lin, X. Phase formation and strengthening mechanism of Zn-2 wt%Cu alloy fabricated by laser powder bed fusion. *Addit. Manuf.* **2024**, *85*, 104153. [CrossRef]
6. Tong, X.; Shen, T.; Zhou, X.; Zeng, J.; Tao, J.; Munir, K.; Li, Y.; Huang, S.; Wu, X.; Ma, J.; et al. Biodegradable Zn-Cu-Li alloys with ultrahigh strength, ductility, antibacterial ability, cytocompatibility, and suitable degradation rate for potential bone-implant applications. *Smart Mater. Manuf.* **2023**, *1*, 100012. [CrossRef]
7. Tang, Z.; Niu, J.; Huang, H.; Zhang, H.; Pei, J.; Ou, J.; Yuan, G. Potential biodegradable Zn-Cu binary alloys developed for cardiovascular implant applications. *J. Mech. Behav. Biomed. Mater.* **2017**, *72*, 182–191. [CrossRef] [PubMed]
8. Yang, H.; Jia, B.; Zhang, Z.; Qu, X.; Li, G.; Lin, W.; Zhu, D.; Dai, K.; Zheng, Y. Alloying design of biodegradable zinc as promising bone implants for load-bearing applications. *Nat. Commun.* **2020**, *11*, 401. [CrossRef] [PubMed]

9. Liu, J.; Wang, D.; Liu, B.; Li, N.; Liang, L.; Chen, C.; Zhou, K.; Baker, I.; Wu, H. Microstructural evolution, mechanical properties and corrosion mechanisms of additively manufactured biodegradable Zn-Cu alloys. *J. Mater. Sci. Technol.* **2024**, *186*, 142–157. [CrossRef]
10. Duan, J.; Li, L.; Cao, F.; Suo, Y.; Yang, Q.; Qin, J.; Wang, X.; Yang, Y. An in vitro and in vivo study of biodegradable Zn-Cu-Li alloy with high strength and ductility fabricated by hot extrusion combined with room-temperature ECAP. *J. Mater. Res. Technol.* **2024**, *33*, 4226–4242. [CrossRef]
11. Ye, L.; Huang, H.; Sun, C.; Zhuo, X.; Dong, Q.; Liu, H.; Ju, J.; Xue, F.; Bai, J.; Jiang, J. Effect of grain size and volume fraction of eutectic structure on mechanical properties and corrosion behavior of as-cast Zn-Mg binary alloys. *J. Mater. Res. Technol.* **2022**, *16*, 1673–1685. [CrossRef]
12. García-Mintegui, C.; Córdoba, L.C.; Buxadera-Palomero, J.; Marquina, A.; Jiménez-Piqué, E.; Ginebra, M.-P.; Cortina, J.L.; Pegueroles, M. Zn-Mg and Zn-Cu alloys for stenting applications: From nanoscale mechanical characterization to in vitro degradation and biocompatibility. *Bioact. Mater.* **2021**, *6*, 4430–4446. [CrossRef] [PubMed]
13. Venezuela, J.; Dargusch, M.S. The influence of alloying and fabrication techniques on the mechanical properties, biodegradability and biocompatibility of zinc: A comprehensive review. *Acta Biomater.* **2019**, *87*, 1–40. [CrossRef]
14. Liu, S.; Yao, L.; Wang, Y.; Li, Y.; Jia, Y.; Yang, Y.; Li, N.; Hu, Y.; Kong, D.; Dong, X.; et al. Immunomodulatory hybrid micro-nanofiber scaffolds enhance vascular regeneration. *Bioact. Mater.* **2019**, *21*, 464–482. [CrossRef] [PubMed]
15. Pachla, W.; Przybysz, S.; Jarzębska, A.; Bieda, M.; Sztwiertnia, K.; Kulczyk, M.; Skiba, J. Structural and mechanical aspects of hypoeutectic Zn-Mg binary alloys for biodegradable vascular stent applications. *Bioact. Mater.* **2021**, *6*, 26–44. [CrossRef]
16. Su, L.; Liu, W.; Wang, Y.; Jiang, Y.; Li, Z.; Wang, M.; Liu, G. Corrosion behavior, antibacterial properties and in vitro and in vivo biocompatibility of biodegradable Zn-5Cu-xMg alloy for bone-implant applications. *Biomater. Adv.* **2024**, *165*, 214000. [CrossRef] [PubMed]
17. Jin, H.; Zhao, S.; Guillory, R.; Bowen, P.K.; Yin, Z.; Griebel, A.; Schaffer, J.; Earley, E.J.; Goldman, J.; Drelich, J.W. Novel high-strength, low-alloys Zn-Mg (<0.1 wt% Mg) and their arterial biodegradation. *Mater. Sci. Eng. C* **2018**, *84*, 67–79. [CrossRef]
18. Vojtěch, D.; Kubásek, J.; Šerák, J.; Novák, P. Mechanical and corrosion properties of newly developed biodegradable Zn-based alloys for bone fixation. *Acta Biomater.* **2011**, *7*, 3515–3522. [CrossRef] [PubMed]
19. Qin, Y.; Liu, A.; Guo, H.; Shen, Y.; Wen, P.; Lin, H.; Xia, D.; Voshage, M.; Tian, Y.; Zheng, Y. Additive manufacturing of Zn-Mg alloy porous scaffolds with enhanced osseointegration: In vitro and in vivo studies. *Acta Biomater.* **2022**, *145*, 403–415. [CrossRef] [PubMed]
20. Castiglioni, S.; Cazzaniga, A.; Albisetti, W.; Maier, J.A.M. Magnesium and osteoporosis: Current state of knowledge and future research directions. *Nutrients* **2013**, *5*, 3022–3033. [CrossRef]
21. Bordbar-Khiabani, A.; Ebrahimi, S.; Yarmand, B. In-vitro corrosion and bioactivity behavior of tailored calcium phosphate-containing zinc oxide coating prepared by plasma electrolytic oxidation. *Corros. Sci.* **2020**, *173*, 108781. [CrossRef]
22. Shoeib, M.A.; Abdel-Gawad, S.A. High performance nano hydroxyapatite coating on zinc for biomedical applications. *J. Mater. Sci.* **2023**, *58*, 740–756. [CrossRef]
23. Zhang, L.; Hu, P.H.; Zhou, Q.; Zhan, W.; Jin, F. Effects of pulsed magnetic field on microstructure, mechanical properties and bio-corrosion behavior of Mg-7Zn alloy. *Mater. Lett.* **2017**, *193*, 224–227. [CrossRef]
24. Li, L.; Liang, W.; Ban, C.; Suo, Y.; Lv, G.; Liu, T.; Wang, X.; Zhang, H.; Cui, J. Effects of a high-voltage pulsed magnetic field on the solidification structures of biodegradable Zn-Ag alloys. *Mater. Charact.* **2020**, *163*, 110274. [CrossRef]
25. Li, L.; Liang, W.; Yang, L.; Cao, F.; Sun, K.; Ban, C.; Cui, J. Structure Refinement and Homogenization of Zn-Cu Alloys Induced by a High-Voltage Pulsed Magnetic Field During the Solidification Process. *Int. J. Met.* **2023**, *17*, 399–413. [CrossRef]
26. Wang, B.; Yang, Y.-S.; Sun, M.-L. Microstructure refinement of AZ31 alloy solidified with pulsed magnetic field. *Trans. Nonferrous Met. Soc. China* **2010**, *20*, 1685–1690. [CrossRef]
27. Liu, H.; Peng, C.; Yuan, Y.; Zhou, Q.; Chen, L.; Xu, Y. Effect of pulsed magnetic field on solidification microstructure and properties of Zn-3Cu alloy. *Trans. Mater. Heat Treat.* **2024**, *45*, 93–101. Available online: https://link.cnki.net/doi/10.13289/j.issn.1009-6264.2023-0275 (accessed on 5 February 2025).
28. *GB/T 228.1-2021*; Metallic Materials—Tensile Testing—Part 1: Method of Test at Room Temperature. Standards Press of China: Beijing, China, 2021.
29. Okamoto, H. Comment on Mg-Zn (magnesium-zinc). *J. Phase Equilibr.* **1994**, *15*, 129–130. [CrossRef]
30. Su, Y.; Lin, X.; Wang, M.; Huang, W. Phase and microstructure pattern selection of Zn-rich Zn-Cu peritectic alloys during laser surface remelting. *J. Mater. Sci.* **2021**, *56*, 14314–14332. [CrossRef]
31. Zhang, Y.; Cheng, X.; Zhong, H.; Xu, Z.; Li, L.; Gong, Y.; Miao, X.; Song, C.; Zhai, Q. Comparative Study on the Grain Refinement of Al-Si Alloy Solidified under the Impact of Pulsed Electric Current and Travelling Magnetic Field. *Metals* **2016**, *6*, 170. [CrossRef]
32. Kabir, H.; Munir, K.; Wen, C.; Li, Y. Recent research and progress of biodegradable zinc alloys and composites for biomedical applications: Biomechanical and biocorrosion perspectives. *Bioact. Mater.* **2021**, *6*, 836–879. [CrossRef]

33. *ASTM G102-23*; Standard Practice for Calculation of Corrosion Rates and Related Informationfrom Electrochemical Measurements. ASTM International: West Conshohocken, PA, USA, 2023.
34. Afshari, V.; Dehghanian, C. Effects of grain size on the electrochemical corrosion behaviour of electrodeposited nanocrystalline Fe coatings in alkaline solution. *Corros. Sci.* **2009**, *51*, 1844–1849. [CrossRef]
35. Johnston, S.; Shi, Z.; Atrens, A. The influence of pH on the corrosion rate of high-purity Mg, AZ91 and ZE41 in bicarbonate buffered Hanks' solution. *Corros. Sci.* **2015**, *101*, 182–192. [CrossRef]
36. Jamesh, M.I.; Wu, G.; Zhao, Y.; McKenzie, D.R.; Bilek, M.M.M.; Chu, P.K. Electrochemical corrosion behavior of biodegradable Mg-Y-RE and Mg-Zn-Zr alloys in Ringer's solution and simulated body fluid. *Corros. Sci.* **2015**, *91*, 160–184. [CrossRef]
37. Li, H.F.; Xie, X.H.; Zheng, Y.F.; Cong, Y.; Zhou, F.Y.; Qiu, K.J.; Wang, X.; Chen, S.H.; Huang, L.; Tian, L.; et al. Development of biodegradable Zn-1X binary alloys with nutrient alloying elements Mg, Ca and Sr. *Sci. Rep.* **2015**, *5*, srep10719. [CrossRef]
38. Mwaanga, P.; Carraway, E.R.; van den Hurk, P. The induction of biochemical changes in Daphnia magna by CuO and ZnO nanoparticles. *Aquat. Toxicol.* **2014**, *150*, 201–209. [CrossRef]

metals

MDPI

Article

Fabrication and Processing of Magnesium-Based Metal Matrix Nanocomposites for Bioabsorbable Implants

Andres Larraza [1,2], Shane Burke [1], Pedram Sotoudehbagha [1] and Mehdi Razavi [1,2,3,*]

1 Biionix (Bionic Materials, Implants & Interfaces) Cluster, Department of Medicine, College of Medicine, University of Central Florida, Health Sciences Campus at Lake Nona 6900 Lake Nona Blvd, Orlando, FL 32816, USA; andres.larraza@ucf.edu (A.L.); shane.burke@ucf.edu (S.B.); pedram.sotoudehbagha@ucf.edu (P.S.)

2 Department of Material Sciences and Engineering, University of Central Florida, Orlando, FL 32816, USA

3 Biomedical Engineering Program, Department of Mechanical and Aerospace Engineering, University of Central Florida, Orlando, FL 32816, USA

* Correspondence: mehdi.razavi@ucf.edu; Tel.: +1-407-266-1551

Abstract: A novel magnesium (Mg)-based metal matrix nanocomposite (MMNC) was fabricated using ultrasonic melt treatment to promote the de-agglomeration of the bioactive glass–ceramic nanoparticles and the homogenization of the melt. The cast samples were then heat treated, machined, and hot rolled to reduce grain size and remove structural defects. Standard mechanical and electrochemical tests were conducted to determine the effect of fabrication and processing on the mechanical and corrosion properties of MMNCs. Compression tests, potentiodynamic polarization tests, electrochemical impedance spectroscopy, and static immersion testing were conducted to determine the characteristics of the MMNCs. The results showed that the combination of ultrasonic melt processing and thermomechanical processing caused the corrosion rate to increase from 8.7 mmpy after 10 days of immersion to 22.25 mmpy when compared with the ultrasonicated MMNCs but remained stable throughout the immersion time, showing no statistically significant change during the incubation periods. These samples also experienced increased yield stress (135.5 MPa) and decreased elongation at break (21.92%) due to the significant amount of grain refinement compared to the ultrasonicated MMNC (σ_Y = 59.6 MPa, elongation = 40.44%). The MMNCs that underwent ultrasonic melt treatment also exhibited significant differences in the corrosion rate calculated from immersion tests.

Keywords: magnesium; metal matrix composites; ultrasonic melt processing; biodegradable implants

1. Introduction

Metal implants are predominantly used for bone-fixing devices due to their high mechanical properties [1,2]. Common alloys consist mainly of bioinert permanent materials like titanium alloys (Ti-6Al-4V) and stainless steels. These metals provide mechanical rigidity and chemical stability, which the bone needs to heal properly [3–6]. However, since these metals do not degrade over time, revisionary surgeries to remove the implants when the wound is fully healed are often necessary. Similarly, the release of metallic particles and ions from the surfaces poses a risk of adverse effects [7,8]. To combat these risks, developing a biocompatible, biodegradable metal can eliminate the need for revisionary surgery and overcome limitations associated with permanent implants [9].

Magnesium has shown significant potential in biodegradable implant applications due to its high biocompatibility; however, its potential is hindered by rapid corrosion rates and the generation of hydrogen gas as a corrosion byproduct [2,7]. Efforts to improve corrosion resistance through alloying and surface coating have yielded significant advances, but these enhancements often compromise biocompatibility [10–12]. WE43 is a commercially available biocompatible magnesium alloy that utilizes rare earth metals to reduce corrosion, rendering it nearly bioinert. Recent research is focused on fine-tuning corrosion resistance while maintaining biodegradability and enhancing osteogenic and angiogenic properties [13]. The addition of scandium (Sc) and strontium (Sr) has been shown to provide corrosion resistance in vitro while also providing biocompatible environments for osteoblast differentiation and offering antibacterial properties [14,15].

Current methods for fabricating magnesium alloys or composites typically involve either stir casting or powder metallurgy. In stir casting, an impeller is submerged into the molten material and rotated at high speeds to promote thorough mixing. Alternatively, powder metallurgy involves milling raw materials into fine powders, compacting them, and then sintering to achieve the desired density and porosity. A major drawback of stir casting is the introduction of gas bubbles into the melt, which leads to porosity during solidification [16–18]. This results in reduced mechanical strength and corrosion resistance, issues that are challenging to address within the casting process itself. In powder metallurgy, porosity is reduced by sintering the compacted sample, typically under high temperatures and pressures. This process increases sample density; however, achieving zero porosity remains challenging [19].

The application of melt processing techniques, such as ultrasound treatment, enhances the de-agglomeration and dispersion of additive particles, influences the grain morphology of as-cast materials, and eliminates structural defects like porosity and non-uniformities [19–21]. The stimulation of the melt with acoustic waves induces acoustic streaming, a phenomenon where the fluid solution moves in a regular oscillatory motion, which can result in cavitation of gas bubbles and de-agglomeration of reinforcing nanoparticles in the melt [22,23]. Post-melt processing via hot rolling also significantly affects the grain structure, which in turn affects the mechanical properties and corrosion behavior of the metals [24,25].

Although the individual alloying of magnesium with strontium (Sr) and scandium (Sc) has been studied for improvements in corrosion resistance, mechanical properties, and biocompatibility [26–29], the synergistic potential of these elements in a ternary Mg-Sr-Sc alloy system remains largely unexplored. Moreover, to our knowledge, no substantial studies have investigated the integration of this alloy with bioactive glass–ceramic nanoparticles. By developing this novel composite, we aim to harness the complementary benefits of Sr and Sc alloying together with bioactive glass–ceramic reinforcement, creating a multifunctional implant material that meets both the structural and biological demands of bone regeneration.

Incorporating diopside ($CaMgSi_2O_6$) bioactive glass–ceramic nanoparticles (BG) into a magnesium-based metal matrix nanocomposite (MMNC) offers significant potential benefits for implant biocompatibility and bioactivity. The sustained release of calcium, magnesium, and silicon ions from these nanoparticles can enhance osteogenesis and angiogenesis and improve biocompatibility [30–32].

In this study, we fabricated a novel MMNC by incorporating Sc and Sr as alloying elements and BG as nano-reinforcements using ultrasonic melt treatment, followed by a hot rolling process. We aim to understand the effects of ultrasonic melt processing and rolling on the microstructure, as well as the mechanical properties and corrosion resistance. The results were compared with those of magnesium, unprocessed MMNCs, and WE43, a commercial biocompatible magnesium-based alloy, as a control.

2. Materials and Methods

2.1. *Synthesis of Diopside BG*

Diopside ($CaMgSi_2O_6$) powder was synthesized using a sol–gel method: 2.1 g of calcium nitrate tetrahydrate (Acros Organics, 99+%, Geel, Belgium) and 2.0 g of magnesium chloride hexahydrate (Acros Organics, 99%) were dissolved in 200-proof ethanol (Decon Labs, King of Prussia, PA, USA). The solution was magnetically stirred at 250 rpm for 30 min at room temperature. Then, 20 mL of tetraethyl orthosilicate (Acros Organics, 98%) was added to the solution and stirred at 450 rpm for 24 h at 80 °C until a gel was formed. The gel was then dried at 100 °C for 5 days to obtain a xerogel, which was then manually milled with a mortar and pestle until an evenly fine powder was reached. The powder was calcinated in a furnace at 850 °C for 2 h.

2.2. Casting of MMNC

A negative sand mold was made using an oil-based sand (PMC Supplies LLC, Lake Katrine, NY, USA) with a riser and sprue to reach approximate laminar flow into the cavity. For the melting, magnesium (ThermoScientific, Waltham, MA, USA, 99.8% pure), scandium (Luciteria Science, Olympia, WA, USA 99.95% pure, 3 wt%), and the BG (0.5 wt%) were added to a graphite crucible and heated to 930 °C (1700 °F) under a constant argon atmosphere. The strontium (Luciteria Science, 99.9% pure, 0.3 wt%) was added last to reduce its oxidation. A graphite rod was inserted into the melt and stirred for 30 s to begin mixing. As a control, magnesium was cast using 99.8% pure Mg. WE43 (Goodfellow, Huntingdon, UK, Mg93-Y4-Nd3) was sourced as a commercialized control.

2.3. Ultrasonic Melt Processing

Ultrasound processing of the melt was conducted using a Hielscher UP200st (Hielscher Ultrasonics, Teltow, Germany), equipped with a ceramic sonotrode and an air circulation system to keep the transducer cool during the process. Prior to insertion, the temperature of the furnace was increased to 980 °C (1800 °F), and the sonotrode was pre-heated above the open furnace to prevent significant melt solidification on the probe. The probe was inserted into the melt approximately one centimeter from the bottom of the crucible. The melt underwent a 1 min treatment consisting of 6 cycles of 6 s on and 4 s off at a power output of 1 watt at 50% amplitude (maximum amplitude 35 μm). Shortly after the treatment was completed, slag was removed, and the entire melt was cast into a sand mold and allowed to cool for 2 h before extraction. The riser and sprue were cut and excluded from further analysis due to their non-uniform microstructure and inclusions.

2.4. Heat Treatment

Heat treatment was performed at 400 °C for 6 h to homogenize the microstructure. Samples were placed in quartz tubes topped with aluminum foil to prevent surface oxidation. After heating, samples were cooled to room temperature under ambient conditions while remaining in the quartz tubes.

2.5. Hot Rolling

Samples were machined to 6.3 mm diameter and pre-heated for one hour at 400 °C in aluminum foil-topped quartz tubes. Samples were removed from the tubes and rolled to 6.0 mm diameter. A second pre-heating and rolling to 5.0 mm diameter was performed to further reduce grain sizes.

2.6. Microstructural Characterization

Samples were cut using a wafering saw (Dimost, Suzhou, China, CT-2300), cleaned with 70% ethanol, and polished using an EXAKT 400 CS Microgrinder (Exakt Technologies, Inc., Oklahoma City, OK, USA) with ascending grits, finishing with 1200 grit. Polishing was performed with a clean felt pad and 0.5 μm alumina particles. After polishing, samples were cleaned with 70% ethanol and rinsed, then sonicated for 3 min in acetone to remove any surface contamination or organic materials. Samples were then etched in a nitric acid, acetic acid, and polyethylene glycol solution for approximately thirty seconds before rinsing with ethanol and deionized water to stop the etching process. A Keyence VHX microscope was used for microstructural imaging at 200× and 700× magnification.

After etching, images were analyzed for grain size measurements using ASTM E112 standards [33]. Mean lineal intercepts were calculated using ImageJ (version 153k, ImageJ, Bethesda, MD, USA) for measurements. In total, 11 replicates were randomly selected from various metallography images, and calculated grain sizes were recorded.

Further microstructural analysis was performed using scanning electron microscopy (SEM; Zeiss, Oberkochen, Germany, ULTRA-55 FEG) equipped with an energy-dispersive spectrometer (EDS; Noran System 7). The accelerating voltage and aperture size were 20 kV and 30 μm. Samples were coated with gold (Au) using a sputter coater (Quorum,

Lewes, UK, EMS150T ES) to avoid surface charge on the samples. Furthermore, the morphology and size of the BG nanoparticles were observed using transmission electron microscopy (TEM, JEM-1011; JEOL, Tokyo, Japan) to examine the morphology and size of the synthesized BG nanoparticles after calcination.

2.7. Compression Testing

Compression testing was conducted in accordance with ASTM E9 standards [34]. Samples were cut to a length-to-diameter ratio of 2:1, assuring that the two faces were as parallel as possible. Tests were conducted on a TestResources model 313 with a crosshead speed of 0.5 mm/mm/min. Yield stress (σ_Y) was determined by identifying the linear elastic region of the compression test curve and applying a 0.2% offset. Ultimate compressive strength (UCS) was calculated as the maximum stress recorded during the compression test. The maximum compressive strain at the point of fracture was measured and reported as the elongation at break (% elongation).

2.8. Electrochemical Corrosion Testing

Samples were subject to electrochemical analysis using a potentiostat (Biologic, Sp-50e, Biologic, Seyssinet-Pariset, France). Samples were cut to 3 mm segments from the rods and sequentially polished with 320-, 600-, and 1200-grit silicon carbide paper. Samples for corrosion testing were cut from different regions to evaluate the homogeneity of the microstructure and composition across the rods using corrosion data. WE43, Mg, ultrasonically treated (UST) MMNC, and unprocessed MMNC were loaded into a conductive sample holder with 0.5 cm^2 of surface area exposed for corrosion testing. However, the 5.3 mm diameter of the UST Rolled MMNC sample could not be fit into the commercial sample holder without causing leakage. As a result, these samples were wrapped with conductive wire and mounted in resin before polishing. The exposed surface area for the UST Rolled MMNC composition was 0.19 cm^2.

Samples were submerged in 200 mL of Hanks Balanced Salt Solution (HBSS) with Ca and Mg (Cytiva) adjusted to pH 7.4 ± 0.05. The solution temperature was maintained at 37 ± 1 °C via water bath. Corrosion properties were investigated using a traditional three-electrode electrochemical cell. Each experimental condition involved the working electrode connected to a graphite counter electrode and an Ag/AgCl reference electrode attached via a bridge tube.

Electrochemical impedance spectroscopy (EIS) and potentiodynamic polarization tests (PDP) were captured and analyzed using the EC-Lab®v11.43 (Biologic, France) software. The test began with a 55 min open circuit voltage (OCV) test. Single sine EIS analysis immediately followed OCV at frequency ranging from 100,000 Hz to 0.01 Hz and amplitude of 10 mV. EIS data were reported as a Nyquist plot and analyzed using EC-Labs software v11.43 (Biologic, France) with the equivalent circuit. PDP was conducted at a scanning rate of 10 mV/min and from −0.2 V to 0.2 V of OCV immediately after the EIS test. Tafel fit lines derived from slopes at ±20 mV and ±50 mV were used to extrapolate the corrosion current density (I_{corr}), and corrosion potential (E_{corr}) in the EC-Labs software v11.43. The corrosion rate was calculated with Equation (1) established in ASTM G59 [35]:

$$\text{Corrosion Rate(mmyr)} = 3.27 \times 10^{-3} \frac{I_{corr} EW}{\rho} \tag{1}$$

where I_{corr} is the corrosion current density ($\mu A/cm^2$) of the sample, EW is the sample equivalent weight of 12.15 g/eq for Mg based on oxidation from Mg to Mg^{2+}, and ρ is the density of Mg considered to be 1.74 g/cm^3. Six samples were analyzed in each condition.

2.9. Immersion Testing

Immersion testing was conducted in HBSS to measure corrosion rate. Samples were machined and cut to approximate diameters of 5 mm and lengths of 10 mm. After machining and cutting, samples were polished with 1200-grit sandpaper and washed in a sonicated acetone bath for 3 min to remove any oils or organic residues on the surface. Following ASTM G31 guidelines, the samples were weighed to an accuracy of ±0.001 g and immersed in pH-corrected HBSS (7.4 ± 0.05) at a volume-to-surface area ratio of 0.2 mL/mm^2 [36]. Samples were immersed for 1, 3, 7, or 10 days in HBSS in glass beakers covered with Parafilm. Beakers were immersed in a water bath at 37 °C to mimic body temperatures and HBSS was refreshed every 72 h.

Corroded samples were removed from HBSS at the end of incubation and dried in an oven at 100 °C to stop further corrosion. Corroded weights were measured using a LEADZM electronic balance and recorded to an accuracy of ±0.001 g. Corrosion products were removed using 200 g/L chromic acid immersion for 10 min followed by a 5 min rinse in 200-proof ethanol. Samples were weighed after washing, and the change in mass was used to calculate the corrosion rate via mass loss percent following Equation (2), where Δm is the change in mass calculated after corrosion products were removed. A is the surface area of the samples, T is the time immersed in hours, and D is the density of the samples.

$$\text{Corrosion Rate}(\text{mmyr}) = (87{,}600 \times \Delta\text{m})/(\text{A} \times \text{T} \times \text{D}) \quad (2)$$

2.10. Statistical Analysis

Metallographic imaging was performed on 1–2 samples per group, selected from different sections of the cast rods, and representative images were chosen. Mechanical testing samples were cut from randomized sections of the cast rods, with 5 replicates per group. Electrochemical testing was conducted with 6 replicates per group. For immersion testing, 3 replicates per group were used for each time point. The results are presented as mean ± standard deviation. Statistical analyses were carried out using one-way ANOVA with Tukey's post hoc analysis in GraphPad Prism 10.2.3 software (San Diego, CA, USA). Statistical significance was considered at $p \leq 0.05$, with results reported as follows: * $p \leq 0.05$, ** $p \leq 0.01$, *** $p \leq 0.001$, **** $p \leq 0.0001$.

3. Results

Figure 1 shows the polished and etched surfaces of each material. In UST and UST Rolled MMNCs, the increased concentration of dark regions along the grain boundaries and within the grains likely corresponds to diopside bioactive glass–ceramic (BG) nanoparticles. Rolling led to a significant reduction in grain size and induced deformation twinning within the MMNC matrix. The rolled samples exhibit increased segregation of BG along the grain boundaries, as the grain deformation during hot rolling drives the nanoparticles toward these regions. Compared to Mg, UST MMNCs showed a more refined grain microstructure, with BG nano-reinforcements dispersed throughout the Mg matrix. However, when rolling is performed on UST MMNCs, the grain size is reduced. Moreover, the simultaneous application of high temperature and pressure enhanced the nucleation of intermetallics, primarily along the grain boundaries and at the sites of nano-reinforcements. Overall, both UST MMNCs and UST Rolled MMNCs displayed a significantly more refined grain structure, with smaller and more uniform grain sizes compared to Mg. UST MMNCs and unprocessed MMNCs showed similar grain sizes, with UST MMNC having smaller grains (112.2 ± 40.2, 148.4 ± 60.75, respectively, $p > 0.05$). UST Rolled MMNCs showed a grain size of 51.78 ± 26.56, significantly lower than unprocessed MMNCs ($p < 0.0001$) and UST Rolled MMNCs ($p < 0.01$).

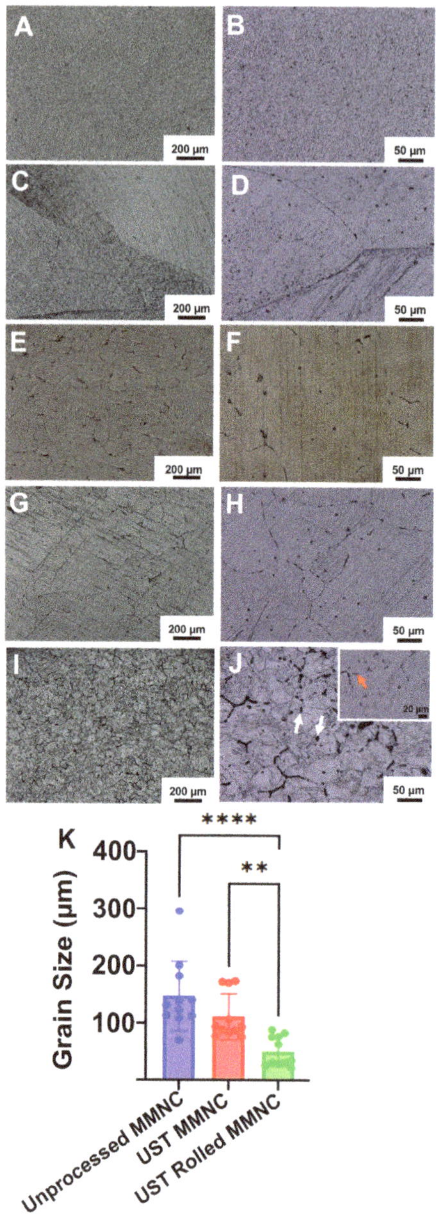

Figure 1. Microstructures of WE43 (**A**,**B**), Mg (**C**,**D**), unprocessed MMNC (**E**,**F**), and UST MMNC (**G**,**H**), UST Rolled MMNC (**I**,**J**) at 200× (**A**,**C**,**E**,**G**,**I**) and 700× (**B**,**D**,**F**,**H**,**J**) magnification. The images show grain sizes, intermetallics, and nano-reinforcements, with white arrows indicating nano-reinforcements and red arrows intermetallic formation at grain boundaries. The inset in panel J shows a higher magnification image (1500×). (**K**) Grain size measurements conducted according to ASTM E112 standards, calculated as mean lineal intercepts from 11 randomly selected replicates obtained from the metallographic images. ** $p \leq 0.01$, **** $p \leq 0.0001$.

Figure 2 presents the SEM images of unprocessed MMNC, UST MMNC, and UST Rolled MMNC and the results of EDS analysis. In Figure 2A, precipitates are observed accumulating along the grain boundaries, which exhibit a discontinuous interface, indicating poor matrix integrity. The inset in Figure 2A is enlarged in Figure 2B, which highlights the EDS analysis area predominantly composed of Mg, O, and Sr (Figure 2C). For UST MMNC, the dispersion of particles within the matrix appears more homogeneous, demonstrating improved integrity compared to the unprocessed sample (Figure 2D,E). The selected inset in Figure 2E reveals a significant presence of Si (0.3 ± 0.2 wt.%) and Ca (2.2 ± 0.2 wt.%), indicating the dispersion of BG nanoparticle aggregates. However, some BG nanoparticles remain aggregated and exhibit irregular shapes after the UST process. Conversely, UST Rolled MMNC displays smaller, round particles with a more uniform distribution (Figure 2G,H). The EDS results for UST Rolled MMNC (Figure 2I) indicate the presence of Si (0.3 ± 0.1 wt.%), indicating the incorporation of BG. In addition, the Sc concentration was highest in the following order: UST MMNC (1.3 ± 0.3 wt.%), UST Rolled MMNC (0.5 ± 0.1 wt.%), and unprocessed MMNC (0.1 ± 0.1 wt.%). Figure 2J demonstrates the TEM image of BG nanoparticles after calcination at 850 °C for 2 h. It is composed of 3–5 nanoparticles with varying morphologies, ranging from elongated to semi-spherical shapes.

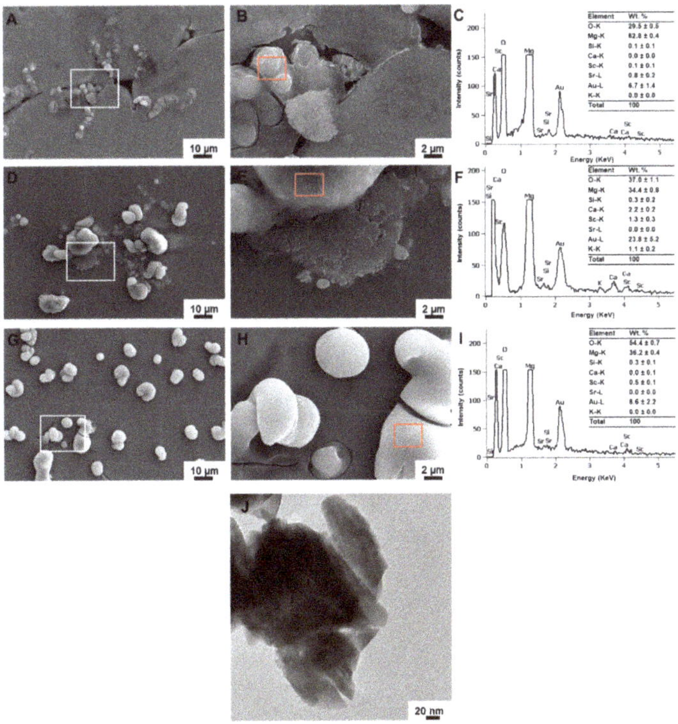

Figure 2. SEM images and chemical point analysis of unprocessed MMNC (**A**–**C**), UST MMNC (**D**–**F**), and UST Rolled MMNC (**G**–**I**). The white insets in (**A**,**D**,**G**) are enlarged in (**B**,**E**,**H**), respectively. The red insets in (**B**,**E**,**H**) indicate the analyzed points using EDS represented in (**C**), (**F**), and (**I**), respectively. The maximum intensity of the EDS spectra was decreased to 200 counts for a better illustration of trace elements. TEM image of BG nanoparticles after synthesis and calcination at 850 °C for 2 h (**J**).

Compression testing was performed on the samples to evaluate their potential as load-bearing bone implants. WE43 was used as a control due to its well-documented properties and its established commercial use in bone implants. WE43 has been reported to have a yield stress, σ_y, of about 240 MPa, an ultimate compressive stress, UCS, of approximately 400 MPa, and a percent elongation at break of about 17% [37]. These reported values are comparable to our WE43 measurements of σ_y (209 $\pm$ 8.71), UCS (393 $\pm$ 2.8), and percent elongation (17.02 $\pm$ 0.74) (Figure 3 and Table 1). Compression testing showed that the Mg had a significantly lower ($p < 0.05$) yield stress than the UST MMNCs and UST Rolled MMNCs, but not compared to unprocessed MMNC. The hot rolling process led to a significantly higher UCS compared to the UST MMNCs, while having no significant effect on the yield strength. Similarly, the percent elongation at break for rolled MMNCs was significantly lower ($p < 0.05$), showing a reduction of 45% compared to the unrolled UST samples.

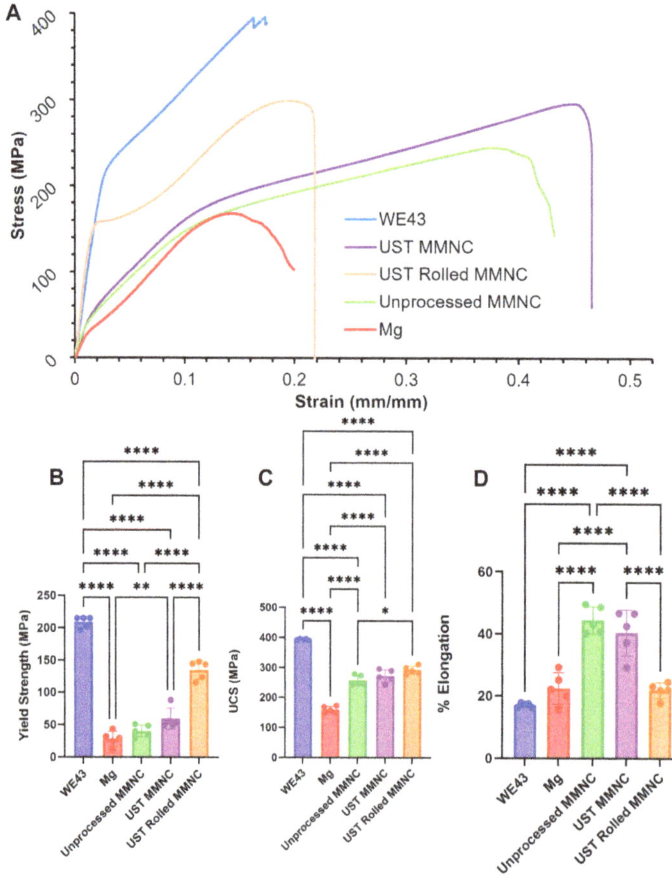

Figure 3. (**A**) Stress–strain curves for WE43, Mg, unprocessed MMNC, UST MMNC, and UST Rolled MMNC, along with their corresponding calculated values for yield stress (**B**), ultimate compressive strength (UCS) (**C**), and % elongation at break (**D**). Nanocomposite fabrication significantly enhanced UCS, with minimal impact on yield stress. Ultrasonication of the MMNC melt improved mechanical properties compared to Mg but showed no significant difference compared to unprocessed MMNCs. The introduction of hot rolling increased yield stress and reduced % elongation, while applying minimal influence on UCS. One-way ANOVA: * $p \leq 0.05$, ** $p \leq 0.01$, **** $p \leq 0.0001$; $n = 5$ per group.

Table 1. Yield stress, ultimate compressive stress, and % elongation at break of WE43, Mg, unprocessed MMNC, UST MMNC, and UST Rolled MMNC (n = 5 per group).

Samples	UCS (MPa)	Yield (MPa)	Elong (%)
WE43	393 ± 2.8	209 ± 8.71	17.02 ± 0.74
Mg	159 ± 11.42	27.99 ± 12.05	22.49 ± 5.168
UST MMNC	272.2 ± 21.58	59.6 ± 16.3	40.44 ± 7.367
UST Rolled MMNC	292.2 ± 13.24	135.5 ± 14.49	21.92 ± 2.508
Unprocessed MMNC	257.5 ± 17.79	40.6 ± 9.198	44.44 ± 4.346

Figure 4A displays representative PDP curves selected based on proximity to the average corrosion rate of six replicates. Figure 4B–D display the average corrosion rates, corrosion current densities (I_{corr}), and corrosion potentials (E_{corr}) of the experimental groups. Commercially developed WE43 and Mg were analyzed as control groups to compare the findings of the experimental MMNCs subjected to UST and rolling processes.

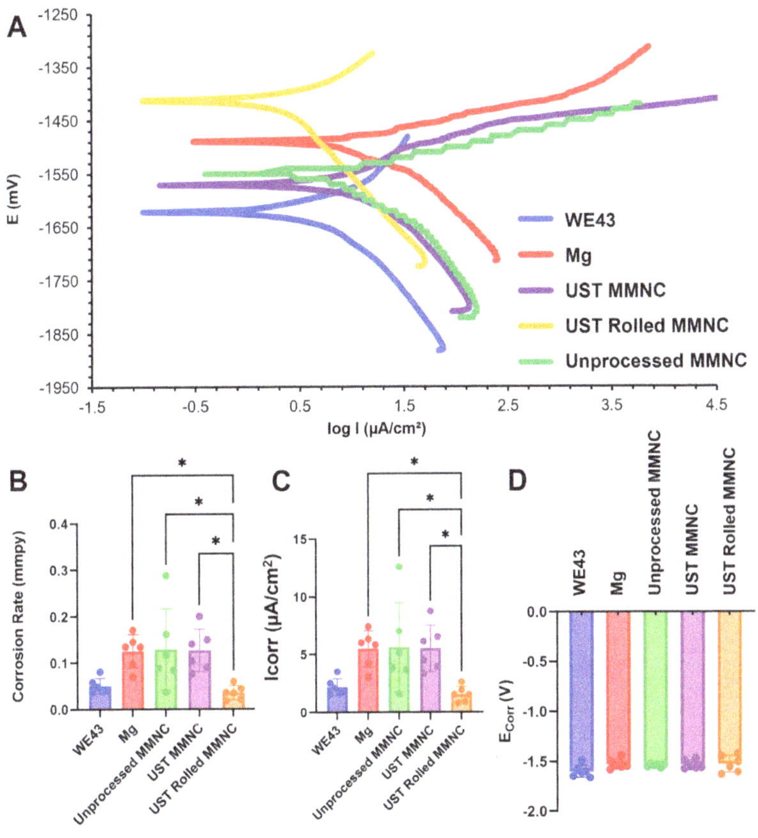

Figure 4. (**A**) PDP curves for WE43, Mg, unprocessed MMNC, UST MMNC, and UST Rolled MMNC, illustrating the average corrosion behavior of samples during electrochemical testing. (**B**) Corrosion rate (in mm per year, mmpy), showing a significant reduction for UST Rolled MMNCs compared to other MMNCs. (**C**) Corrosion current density (Icorr) follows similar trends, with a significant decrease for UST Rolled MMNCs. (**D**) Corrosion potential (Ecorr) is elevated in UST Rolled MMNCs relative to the other samples. One-way ANOVA: * $p \leq 0.05$; n = 6 per group.

Corrosion current density indicates the rate of corrosion, while corrosion potential represents the equilibrium between oxidation and reduction reactions at the metal's surface [38]. Samples with higher, or more positive, corrosion potentials are more likely to gain electrons rather than lose them, reducing their tendency to corrode [38]. Furthermore, a lower corrosion current density suggests a lower corrosion rate [38]. Both the unprocessed MMNC and UST MMNC were characterized by corrosion potentials near that of Mg and higher than WE43. The UST Rolled MMNCs displayed the highest E_{corr} of -1.44 ± 0.08 V, compared to other groups, including WE43 which displayed an E_{corr} of -1.6 ± 0.06 V.

Tafel-derived corrosion current densities show that there are significant differences ($p < 0.05$) between I_{corr} of the UST Rolled MMNC and that of the UST MMNC. The I_{corr} of the UST Rolled MMNC was 1.55 ± 0.67 $\mu A/cm^2$ compared to the I_{corr} of 5.53 ± 2.01 $\mu A/cm^2$ in the UST MMNC. Significant differences also exist between the UST Rolled MMNC and Mg (I_{corr} = 5.48 ± 1.58 $\mu A/cm^2$) and the unprocessed MMNC, where I_{corr} was 5.63 ± 3.85 $\mu A/cm^2$. The near 110 mV difference in E_{corr}, along with these findings, suggests that the changes induced by the rolling process are significant, leading to increased corrosion resistance in the UST Rolled MMNC. These modifications result in the UST Rolled MMNC exhibiting a current density similar to that of commercialized WE43, although the difference between the two groups is not statistically significant ($p > 0.05$).

The UST Rolled MMNC displayed the lowest corrosion rate of all groups. The corrosion rate of the UST Rolled MMNC was 0.03 ± 0.02—a 4-fold reduction in corrosion rate compared to rates of 0.13 mm per year (mmpy) in the UST MMNC, Mg, and unprocessed MMNC. No significant difference ($p > 0.05$) was determined between corrosion rates of UST Rolled MMNC and WE43. The trend in corrosion rates corresponds directly to the trend in current density and corrosion potential. Furthermore, the trends are justified by the interrelation of corrosion rate and current density through the formula established in ASTM G59 (Equation (1)) [35]. It is also notable that the unprocessed MMNC exhibited greater variation in corrosion, likely due to non-uniform microstructures across samples. This underscores the role of UST in producing more uniform microstructures and consistent corrosion rates.

Figure 5A,B depict the results of electrochemical impedance spectroscopy (EIS) for the representative curves closest to the average for each condition. The simulated equivalent circuit used to evaluate each Nyquist plot can be found inlayed in Figure 5A. This equivalent circuit was drafted based on the presence of both high- and low-frequency capacitive loops in the Nyquist plots of all samples [39]. Mg alloys characterized by this type of impedance behavior exhibit two porous layers, the outer layer of which can indicate the formation of a protective corrosion layer [39,40].

The layers are modeled using two elements: electrical resistance (R) and a capacitance/constant phase element (Q), with an impedance defined as $Z_Q = Q_0(i.\omega)^n$ (Q_0 = constant; i = imaginary number; ω = angular frequency; and n = coefficient between 0.5 and 1). When n approaches 1, it indicates near-ideal capacitance, suggesting an effective barrier to solution penetration [41]. The accuracy of the fitting models was evaluated by comparing the reduced chi-square ($\chi^2/|Z|$), which represents the ratio of observed experimental variance to theoretical variance. Although $\chi^2/|Z|$ is less than 1, indicating minimal noise in the data, the model effectively fits the raw Nyquist data.

Among the values represented by the equivalent circuit, R_1 indicates the solution resistance (R_s) characteristic of the electrolytes in HBSS. R_2 and Q_1 indicate the resistance and capacitance, respectively, between the protective outer porous layer and solution. The charge transfer resistance and capacitance at the inner layer of the alloy are represented by R_3 and Q_2, respectively. EIS data extrapolated from equivalent circuits using EC-Lab software v11.43 are summarized in Table 2, with polarization resistance (R_p) being of particular significance. All six replicates of a single group were analyzed sequentially before proceeding to the next sample.

Q_1 for Mg (6.21 ± 2.07 μF·cm^{-2}·s^n), UST MMNC (6.88 ± 4.02 μF·cm^{-2}·s^n), and UST Rolled MMNC (6.25 ± 6.34 μF·cm^{-2}·s^n) shows smaller values compared to WE43 (8.38 ± 3.55 μF·cm^{-2}·s^n) and unprocessed MMNC (9.89 ± 7.72 μF·cm^{-2}·s^n), while the n_1 values are quite similar in all groups (0.68–0.71). On the other hand, Q_2 is significantly higher for UST Rolled MMNC (155.97 ± 375.88 μF·cm^{-2}·s^n) compared to the other groups and is a less ideal capacitance (n_2 = 0.75 ± 0.33). The outer- and inner-layer resistance for UST Rolled MMNCs (R_2 = 1365.69 ± 2545.17 Ohm·cm^2 and R_3 = 20,213.85 ± 13,413.07 Ohm·cm^2) show relatively higher values compared to the other groups (Table 2). Consequently, the UST Rolled MMNC displayed R_p of 21,579.54 ± 11,839.93 Ohm·cm^2, almost two-fold higher than the next highest value of 13,006.51 ± 1072.27 Ohm·cm^2 in the WE43 condition. These R_p values in Table 2 also correspond to the size of the capacitive loops in the Nyquist plot of each sample as larger capacitive loops correspond to higher resistance values and better protection against corrosion. The data in Table 2 show that UST rolling produced the most pronounced increase in the polarization resistance of the sample. Resistance at the outer porous layer of the samples was also highest in the UST Rolled MMNC group, at 1365.69 ± 2545.17 Ohm·cm^2 compared to all other conditions within a range of 95.11–169.02 Ohm·cm^2.

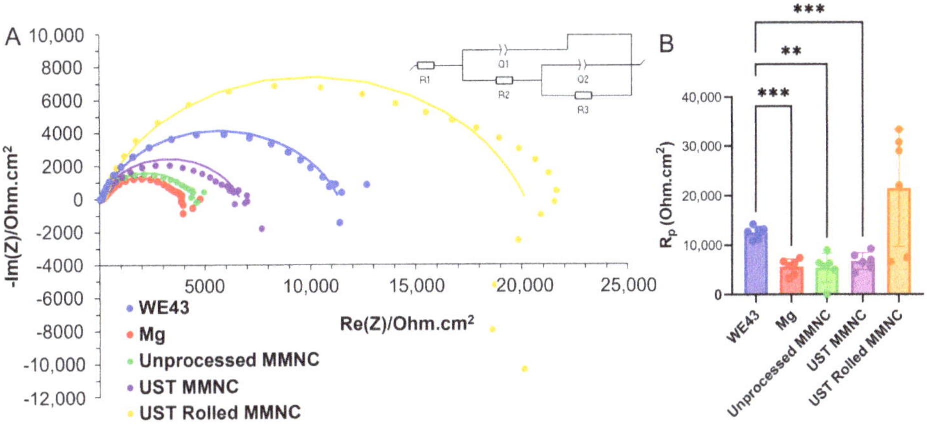

Figure 5. (**A**) Electrochemical impedance spectroscopy (EIS) curves for WE43, Mg, unprocessed MMNC, UST MMNC, and UST Rolled MMNC. The inset shows the simulated equivalent circuit used to analyze the Nyquist plots. (**B**) Corrosion resistance (R_p) of the samples, with UST Rolled MMNCs exhibiting a significant increase compared to the other samples. One-way ANOVA: ** $p \leq 0.01$, *** $p \leq 0.001$; n = 6 per group.

Although UST Rolled MMNC exhibited a significantly higher corrosion rate than all other samples at the initial immersion, its corrosion rate remained statistically constant over the incubation period. In contrast, other groups showed increasing corrosion rates with longer immersion times (Figure 6, Table 3). After 7 days of immersion, unprocessed MMNC exhibited a significant increase in corrosion rate, rising ten-fold. Similarly, between days 3 and 7, UST MMNC showed a 3.8-fold increase in corrosion rate and began showing statistically significant differences in corrosion rates compared to Mg.

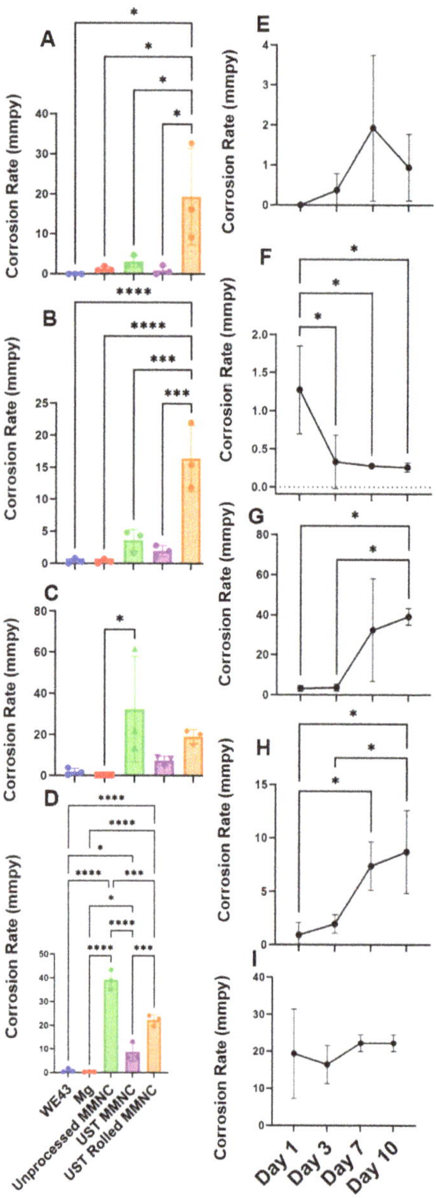

Figure 6. (**A–D**) Corrosion rate (in mm per year) calculated from mass loss percentage according to ASTM G31 for 1-day (**A**), 3-day (**B**), 7-day (**C**), and 10-day (**D**) immersion tests. (**E–I**) Corrosion rates within each sample group (n = 3), illustrating changes in corrosion behavior over time. UST Rolled MMNCs (**I**) demonstrated consistently stable corrosion rates, which were higher than the control WE43 (**E**) and Mg (**F**), and generally higher than UST MMNCs (**H**). Unprocessed MMNCs (**G**) exhibited a rapid increase in corrosion rate with immersion time, while UST MMNCs showed a similar trend but at one-fifth of the magnitude. In contrast, UST Rolled MMNCs maintained a more stable corrosion rate throughout the tests. One-way ANOVA: * $p \leq 0.05$, *** $p \leq 0.001$, **** $p \leq 0.0001$; n = 3 per group.

Table 2. Electrochemical corrosion parameters extrapolated from EIS analysis (n = 6 per group). Definitions of the values are the following: R_s: solution resistance; R_p: polarization resistance; Q_1: capacitance of the outer oxide layer; n_1: coefficient of Q_1; R_2: resistance at the outer oxide layer/solution interface; Q_2: capacitance of the inner layer of the alloy; R_3: charge transfer resistance at the inner layer of the alloy; n_2: coefficient of Q_2.

Composition	R_s (Ohm·cm^2)	R_p (Ohm·cm^2)	Q_1 (μF·cm^{-2}·s^n)	R_2 (Ohm·cm^2)	n_1	Q_2 (μF·cm^{-2}·s^n)	R_3 (Ohm·cm^2)	n_2	χ^2/|Z|
WE43	31.88 ± 11.94	13,006.51 ± 1072.27	8.38 ± 3.55	133.31 ± 62.45	0.69 ± 0.04	11.61 ± 2.37	12,873.2 ± 1074.57	0.85 ± 0.03	0.13 ± 0.12
Mg	36.63 ± 14.77	5613.18 ± 1651.33	6.21 ± 2.07	169.02 ± 57.86	0.68 ± 0.06	15.05 ± 4.21	5444.17 ± 1612.98	0.81 ± 0.04	0.50 ± 0.58
Unprocessed MMNC	17.16 ± 9.39	5474.16 ± 2969.57	9.89 ± 7.72	95.11 ± 37.3	0.7 ± 0.08	7.58 ± 2.75	5379.05 ± 2950.74	0.87 ± 0.05	0.27 ± 0.14
UST MMNC	24.74 ± 13.08	6808.95 ± 1757.51	6.88 ± 4.02	133.28 ± 72.78	0.71 ± 0.12	10.3 ± 3.15	6675.67 ± 1721.07	0.83 ± 0.04	0.28 ± 0.29
UST Rolled MMNC	91.08 ± 84.70	21,579.54 ± 11,839.93	6.25 ± 6.34	1365.69 ± 2545.17	0.71 ± 0.08	155.97 ± 375.88	20,213.85 ± 13,413.07	0.75 ± 0.33	0.19 ± 0.08

Table 3. Corrosion rates (in mm per year) calculated from mass loss percentage (n = 3 per group).

Samples	Day 1	Day 3	Day 7	Day 10
WE43	0 ± 0	0.37 ± 0.42	1.93 ± 1.82	0.94 ± 0.83
Mg	1.27 ± 0.58	0.33 ± 0.35	0.28 ± 0.01	0.26 ± 0.06
UST MMNC	0.92 ± 1.15	1.94 ± 0.86	7.38 ± 2.25	8.71 ± 3.89
UST Rolled MMNC	19.35 ± 12.06	16.43 ± 5.14	19.18 ± 3.39	22.25 ± 2.28
Unprocessed MMNC	3.10 ± 1.44	3.61 ± 1.68	32.37 ± 25.67	39.2 ± 4.09

4. Discussion

This study demonstrates the significant impact of ultrasonic melt treatment (UST) on the microstructure, mechanical properties, and corrosion resistance of a novel magnesium-based metal matrix nanocomposite (MMNC). We also investigated the effects of rolling on UST MMNC, finding improvements in the consistency of corrosion rates, an increase in yield stress, and a reduction in percent elongation at break. Additionally, regardless of the processing techniques, adding alloying elements and nano-reinforcements led to increased ultimate compressive strength and yield stress compared to Mg.

Ultrasonication of the melt was selected as the treatment method due to its ability to induce acoustic streaming and bubble cavitation in the melt [22,42,43]. Preliminary testing revealed inconsistent corrosion results with adding bioactive glass–ceramic nanoparticles (BG) at significantly higher weight percentages than the current 0.5 wt.%. Reducing the BG concentration minimized agglomeration, as the lower overall amount of BG reduced the tendency for clustering. Subsequently, ultrasonic melt treatment was used to de-agglomerate and disperse the BG uniformly within the melt while also degassing the melt through acoustic streaming and bubble cavitation effects. This uniformity in microstructure is evident in the metallography images shown in Figure 1. However, the UST MMNC still exhibits large grains, potentially due to the low solidification rate of the melt in the sand mold. Previous research has shown that grain size significantly affects the corrosion rate of Mg [44,45]. Future work may involve optimizing the mold dimensions and mold material to enhance solidification rates. In addition, further refinement of the ultrasonic melt treatment process could help reduce grain sizes and achieve more controlled corrosion properties. Hot rolling was selected to reduce grain size and enhance the integration between the Mg matrix and BG reinforcements due to its simplicity and accessibility. However, this method is limited by the number of cycles achievable, as it necessitates a significant reduction in cross-sectional area during the process.

Ultrasonication of the melt did not lead to a significant reduction in grain size. The primary goal of melt sonication is to de-agglomerate and disperse BG nanoparticles while inducing cavitation of trapped gas, thereby reducing structural defects and casting porosity. The enhanced dispersion of BG was expected to promote grain nucleation, potentially resulting in smaller grain sizes. However, due to the solidification rate of the melt in the sand mold and the subsequent heat treatment, the impact of ultrasonication on grain size was negligible. Emadi et al. demonstrated that ultrasonication of an AZ91E Mg alloy melt significantly influenced grain size after a 24-h heat treatment [46]. In contrast, metallographic imaging of AZ91E in their study revealed a considerable presence of secondary beta phase throughout the SEM images. At higher amplitudes, ultrasonication led to a reduction in the segregation of large beta-phase areas. In our case, however, the minimal presence of secondary phases in MMNCs due to the low levels of alloying elements resulted in a more negligible effect of ultrasonication on grain size.

The SEM/EDS results indicated that UST improved the wettability of the BG particles, leading to a more homogeneous dispersion compared to the unprocessed sample (Figure 2A,D). The rolling process further enhanced this dispersion, as illustrated in Figure 2G. The morphology of the BG particles underwent a significant transformation into a spherical shape, likely due to deformation and breakdown of BG nanoparticle aggregates during the rolling process. The shape of the reinforcing particles has been shown to influence the biodegradation of Mg/HA nanocomposites [47]. Similarly, in this study, the UST Rolled MMNC exhibited the lowest corrosion current density (1.55 ± 0.67 $\mu A/cm^2$), corrosion rate (0.03 ± 0.02 mmpy), and polarization resistance (21,579.54 ± 11,839.93 $Ohm{\cdot}cm^2$). However, the immersion results indicated an increased corrosion rate, necessitating a prolonged EIS test to elucidate the underlying reasons. One potential explanation could be the higher density of grain boundaries, which may have facilitated passivation during the initial stages of corrosion, as measured by electrochemical methods. Additionally, the presence of significant amounts of oxide could have contributed to the disintegration of the matrix, while solution penetration may have diminished corrosion resistance. Oxidation may have occurred during the heat treatment before rolling, as evidenced by SEM/EDS (Figure 2I). The UST Rolled MMNC had the highest O content of 54.4 ± 0.7 wt.%. The homogeneous dispersion of the BG particles also increased yield strength, following the order UST Rolled MMNC > UST MMNC > unprocessed MMNC. In metal matrix composites, particle strengthening is a mechanism that enhances yield strength [48]. Furthermore, the shape, size, and quantity of the particles can significantly influence mechanical properties, such as crack initiation within the matrix. Smaller particles with a narrow size distribution tend to reduce the likelihood of crack formation.

Hot rolling also offers additional benefits, such as increased pressure and temperature, enhancing the nucleation and growth of intermetallic phases. This increases yield stress and ultimate compressive strength (UCS) in the UST Rolled MMNC while reducing elongation at break. However, the formation of intermetallic phases can increase the risk of micro-galvanic pitting corrosion due to variations in electrochemical corrosion potentials relative to the Mg matrix. The segregation of intermetallics at grain boundaries, as in Figure 1, can lead to pitting corrosion along these boundaries. This can result in intergranular corrosion, pit-to-crack transition, and early material failure. Another major drawback of the hot rolling process is the introduction of stress-induced twinning and microcracks. The roughness from twinning at the micron scale increases the surface area, which can significantly enhance pitting corrosion [49]. Likewise, microcracks formed during the significant plastic deformation during hot rolling may act as initiation sites for further pitting corrosion.

The corrosion behavior observed during electrochemical corrosion testing can be attributed to the microstructural changes induced by the UST rolling process. Specifically, the increased compressive residual stress and dislocation density contribute to enhanced recrystallization and grain refinement [50,51], while also promoting the closure of matrix-reinforcement interfacial gaps. High dislocation levels and subtle defects introduced by the rolling process can create diffusion pathways for elements to reach the metal surface, forming a protective oxide layer that enhances corrosion resistance [51]. Corrosion rates calculated via PDP analysis are supported by a study from Deng et al., examining the corrosion behavior of Mg alloys. They concluded that hot rolling significantly reduced the corrosion rate by up to three times, with the effect being temperature-dependent [52].

Generally, the charge transfer resistance (R_3 in Table 2) indicates the dissolution rate of Mg, and higher values equate to slower rates of dissolution, specifically lower corrosion rates [53]. When analyzed, the charge transfer resistance values (R_3) were highest in the WE43 and UST Rolled MMNCs, at 12,873.20 $\pm$ 1074.57 Ohm·cm^2 and 20,213.85 $\pm$ 13,413.07 Ohm·cm^2, respectively. These data demonstrate over a two-fold difference from the UST MMNC, Mg, and unprocessed MMNCs compared to WE43 and a four-fold difference compared to the UST Rolled MMNCs. These conclusions further underscore the significant impact of integrated UST and rolling processes on the corrosion resistance of MMNC. It is also important to note that the electrochemical behavior of the MMNCs is expected to change as the surfaces of the samples are oxidized and corroded. Due to this expectation, further study into the electrochemical behavior of already-immersed samples should be conducted to validate existing EIS and immersion discrepancies, as well as to study the possible beneficial effects of creating a passivation layer via immersion prior to electrochemical studies.

Immersion testing was used to confirm the corrosion rate of MMNC calculated from electrochemical testing. However, extended immersion testing, compared to electrochemical testing, may accelerate pit formation and the transition from pits to cracks due to micro-galvanic corrosion between the matrix, intermetallics, and nano-reinforcements, potentially leading to significantly higher corrosion rates [54]. The corrosion of the Mg matrix is evident from the formation of small bubbles on the sample surfaces as immersion time increases. While similar bubbles are observed on electrochemical corrosion samples, they appear at a much smaller scale and in fewer quantities. Variation may also be due to the creation of a passivation layer during the initial stages of the electrochemical corrosion test, prior to the EIS and PDP measurements.

The study of Mg-based alloys and composites is well established, with extensive research exploring a variety of testing media for these samples. Initially, solutions of NaCl had been used according to ASTM standards, but due to the significant differences in ionic compositions, physiological comparisons between these results and in vivo studies showed significant differences in the corrosion behavior and resulting corrosion products [55]. Due to these differences, immersion tests have evolved to include specific valid compositions such as Hanks Balanced Salt Solutions (HBSS), simulated body fluid (SBF), and Dulbecco's Phosphate-Buffered Saline (DPBS). Each of these fluids have various ionic profiles, specially formulated to imitate some aspect of physiological solutions.

Mena-Morcillo and Veleva recently studied the effect that varying the immersion media had on the corrosion rate of AZ31 and AZ91 [56]. They observed a significant variation in corrosion rates between HBSS, SBF, and Ringer's solution. Specifically, HBSS resulted in a mass loss rate approximately three times lower than that observed in SBF. Additionally, Mei et al. discussed criteria for selecting the appropriate immersion media in their 2020 review [55]. They concluded that simulated body fluids like SBF and HBSS were suitable for open immersion tests due to the lack of sugars and other elements that would cause microbial contaminations to tamper with the data. They also concluded that the corrosion rate of Mg samples in HBSS-like media would be "revelated to a lesser extent" when compared to NaCl solutions due to the various ionic interactions between the Mg matrix and the ionic solution. However, the human body has a complex ionically stable and pH-stable condition that significantly contributes to the behavior of the implants in situ.

It is widely reported that static immersion tests, like the one we conducted, often show higher corrosion rates compared to in vivo studies [57–59]. This discrepancy arises because in vivo specimens experience a constantly changing environment due to the circulatory system, which maintains a homeostatic ion exchange. Additionally, the presence of enzymes and proteins in vivo can contribute to the formation of protective layers.

Although UST and integrated UST rolling processes are effective in the microstructural refinement of MMNC, further optimization can be achieved in future regarding the ultrasound intensity, duration, amplitude, and hot rolling temperature to further reduce grain sizes, de-agglomerate and disperse nano-reinforcements, and reduce intermetallic and twining formation. Further optimization of MMNC composition and processing may enhance microstructure, corrosion resistance, and mechanical properties in future studies. In future studies, the formation of corrosion products will be investigated to better understand the corrosion mechanisms and the effects of biomineralization on corrosion rates. Other severe plastic deformation methods can also be used to reduce grain sizes, thereby elucidating the effects of processing on microstructure and resulting corrosion and mechanical properties.

5. Conclusions

The fabrication of a novel MMNC composition using ultrasonic melt processing and hot rolling resulted in significant increases in the yield stress and ultimate compressive strength of samples while decreasing the elongation at break. Likewise, ultrasonication independently increased the corrosion resistance of the samples when immersed in a static HBSS solution and improved the corrosion behavior, as seen in the electrochemical analysis. The sonication of the melt did not significantly change the mechanical properties of the MMNC but did result in lower corrosion rates. Further processing of samples with hot rolling increased the corrosion rate, but it remained lower than the unprocessed and UST-processed MMNCs.

Author Contributions: Conceptualization, A.L., P.S. and M.R.; methodology, A.L., S.B., P.S. and M.R.; software, A.L. and S.B.; validation, A.L., S.B., P.S. and M.R.; formal analysis, A.L., S.B. and M.R.; investigation, A.L., S.B., P.S. and M.R.; resources, M.R.; data curation, A.L., S.B. and M.R.; writing—original draft preparation, A.L. and S.B.; writing—review and editing, P.S. and M.R.; visualization, M.R.; supervision, M.R.; project administration, M.R.; funding acquisition, M.R. All authors have read and agreed to the published version of the manuscript.

Funding: This material is based upon work supported by the National Science Foundation under Grant No. CMMI 2142610.

Data Availability Statement: The raw data supporting the conclusions of this article will be made available by the authors on request.

Acknowledgments: We thank Sun Latt for his assistance with nanoparticle synthesis and Valeria Cardenas for her help with TEM imaging. We also thank Melanie Coathup for providing the grinder used for sample polishing, the Burnett School of Biomedical Sciences for access to microscopy facilities, and the Advanced Materials Processing and Analysis Center at UCF for electron microscopy support. ChatGPT and Grammarly were used for sentence editing and paraphrasing.

Conflicts of Interest: The authors declare no conflicts of interest.

References

1. Paiva, J.C.C.; Oliveira, L.; Vaz, M.F.; Costa-De-Oliveira, S. Biodegradable Bone Implants as a New Hope to Reduce Device-Associated Infections—A Systematic Review. *Bioengineering* **2022**, *9*, 409. [CrossRef] [PubMed]
2. Prasad, A. Bioabsorbable polymeric materials for biofilms and other biomedical applications: Recent and future trends. *Mater. Today Proc.* **2021**, *44*, 2447–2453. [CrossRef]
3. Hayes, J.S.; Richards, R. The use of titanium and stainless steel in fracture fixation. *Expert Rev. Med. Devices* **2010**, *7*, 843–853. [CrossRef] [PubMed]
4. Disegi, J.; Eschbach, L. Stainless steel in bone surgery. *Injury* **2000**, *31*, D2–D6. [CrossRef]
5. Moghaddam, N.S.; Jahadakbar, A.; Amerinatanzi, A.; Skoracki, R.; Miller, M.; Dean, D.; Elahinia, M. Fixation release and the bone bandaid: A new bone fixation device paradigm. *Bioengineering* **2017**, *4*, 5. [CrossRef]
6. Xiu, P.; Jia, Z.; Lv, J.; Yin, C.; Cheng, Y.; Zhang, K.; Song, C.; Leng, H.; Zheng, Y.; Cai, H.; et al. Tailored surface treatment of 3D printed porous Ti6Al4V by microarc oxidation for enhanced osseointegration via optimized bone in-growth patterns and interlocked bone/implant interface. *ACS Appl. Mater. Interfaces* **2016**, *8*, 17964–17975. [CrossRef]
7. Moghaddam, N.S.; Andani, M.T.; Amerinatanzi, A.; Haberland, C.; Huff, S.; Miller, M.; Elahinia, M.; Dean, D. Metals for bone implants: Safety, design, and efficacy. *Biomanuf. Rev.* **2016**, *1*, 1.
8. Savarino, L.; Maci, G.S.; Greco, M.; Baldini, N.; Giunti, A. Metal ion release from fracture fixation devices: A potential marker of implant failure. *J. Biomed. Mater. Res. Part B Appl. Biomater. Off. J. Soc. Biomater. Jpn. Soc. Biomater. Aust. Soc. Biomater. Korean Soc. Biomater.* **2008**, *86*, 389–395. [CrossRef]

9. Blumenthal, N.C.; Posner, A.S.; Cosma, V.; Gross, U. The effect of glass–ceramic bone implant materials on the in vitro formation of hydroxyapatite. *J. Biomed. Mater. Res.* **1988**, *22*, 1033–1041. [CrossRef]
10. Ibrahim, H.; Esfahani, S.N.; Poorganji, B.; Dean, D.; Elahinia, M. Resorbable bone fixation alloys, forming, and post-fabrication treatments. *Mater. Sci. Eng. C* **2017**, *70*, 870–888. [CrossRef]
11. Reyes, C.D.; Petrie, T.A.; Burns, K.L.; Schwartz, Z.; García, A.J. Biomolecular surface coating to enhance orthopaedic tissue healing and integration. *Biomaterials* **2007**, *28*, 3228–3235. [CrossRef] [PubMed]
12. Andersson, T.; Agholme, F.; Aspenberg, P.; Tengvall, P. Surface immobilized zoledronate improves screw fixation in rat bone: A new method for the coating of metal implants. *J. Mater. Sci. Mater. Med.* **2010**, *21*, 3029–3037. [CrossRef] [PubMed]
13. Savaedi, Z.; Mirzadeh, H.; Aghdam, R.M.; Mahmudi, R. Effect of grain size on the mechanical properties and bio-corrosion resistance of pure magnesium. *J. Mater. Res. Technol.* **2022**, *19*, 3100–3109. [CrossRef]
14. Ma, N.; Peng, Q.; Li, X.; Li, H.; Zhang, J.; Tian, Y. Influence of scandium on corrosion properties and electrochemical behaviour of mg alloys in different media. *Int. J. Electrochem. Sci.* **2012**, *7*, 8020–8034. [CrossRef]
15. Mushahary, D.; Sravanthi, R.; Li, Y.; Kumar, M.J.; Harishankar, N.; Hodgson, P.D.; Wen, C.; Pande, G. Zirconium, calcium, and strontium contents in magnesium based biodegradable alloys modulate the efficiency of implant-induced osseointegration. *Int. J. Nanomed.* **2013**, *8*, 2887–2902.
16. Thakur, B.; Barve, S.; Pesode, P. Investigation on mechanical properties of AZ31B magnesium alloy manufactured by stir casting process. *J. Mech. Behav. Biomed. Mater.* **2023**, *138*, 105641. [CrossRef]
17. Sathishkumar, P.; Deepakaravind, V.; Gopal, P.; Azhagiri, P. Analysis the mechanical properties and material characterization on magnesium metal matrix nano composites through stir casting process. *Mater. Today Proc.* **2021**, *46*, 7436–7441. [CrossRef]
18. Ponappa, K.; Aravindan, S.; Rao, P. Magnesium Metal-Matrix Composites–Types and Fabrication Approaches. In *Advances in Corrosion Control of Magnesium and Its Alloys*; CRC Press: Boca Raton, FL, USA, 2023; pp. 23–34.
19. Arora, G.S.; Saxena, K.K.; Mohammed, K.A.; Prakash, C.; Dixit, S. Manufacturing techniques for Mg-Based metal matrix composite with different reinforcements. *Crystals* **2022**, *12*, 945. [CrossRef]
20. Yin, Z.; Le, Q.; Chen, X.; Jia, Y. The grain refinement of Mg alloy subjected to dual-frequency ultrasonic melt treatment: A physical and numerical simulation. *J. Mater. Res. Technol.* **2022**, *21*, 1554–1569. [CrossRef]
21. Hu, W.; Le, Q.; Liao, Q.; Wang, T. Effects of Ultrasonic Treatment on Grain Refinement and Gas Removal in Magnesium Alloys. *Crystals* **2024**, *14*, 237. [CrossRef]
22. Eskin, G. Cavitation mechanism of ultrasonic melt degassing. *Ultrason. Sonochem.* **1995**, *2*, S137–S141. [CrossRef]
23. Lebon, G.B.; Tzanakis, I.; Pericleous, K.; Eskin, D.; Grant, P.S. Ultrasonic liquid metal processing: The essential role of cavitation bubbles in controlling acoustic streaming. *Ultrason. Sonochem.* **2019**, *55*, 243–255. [CrossRef] [PubMed]
24. Fatemi-Varzaneh, S.; Zarei-Hanzaki, A.; Haghshenas, M. The room temperature mechanical properties of hot-rolled AZ31 magnesium alloy. *J. Alloys Compd.* **2009**, *475*, 126–130. [CrossRef]
25. Cao, F.; Shi, Z.; Song, G.-L.; Liu, M.; Dargusch, M.S.; Atrens, A. Influence of hot rolling on the corrosion behavior of several Mg–X alloys. *Corros. Sci.* **2015**, *90*, 176–191. [CrossRef]
26. He, F.; Lu, T.; Fang, X.; Li, Y.; Zuo, F.; Deng, X.; Ye, J. Effects of strontium amount on the mechanical strength and cell-biological performance of magnesium-strontium phosphate bioceramics for bone regeneration. *Mater. Sci. Eng. C* **2020**, *112*, 110892. [CrossRef]
27. Bornapour, M.; Celikin, M.; Cerruti, M.; Pekguleryuz, M. Magnesium implant alloy with low levels of strontium and calcium: The third element effect and phase selection improve bio-corrosion resistance and mechanical performance. *Mater. Sci. Eng. C* **2014**, *35*, 267–282. [CrossRef]
28. Bian, D.; Chu, X.; Xiao, J.; Tong, Z.; Huang, H.; Jia, Q.; Liu, J.; Li, W.; Yu, H.; He, Y.; et al. Design of single-phased magnesium alloys with typically high solubility rare earth elements for biomedical applications: Concept and proof. *Bioact. Mater.* **2023**, *22*, 180–200. [CrossRef]
29. Liu, J.; Lin, Y.; Bian, D.; Wang, M.; Lin, Z.; Chu, X.; Li, W.; Liu, Y.; Shen, Z.; Liu, Y.; et al. In vitro and in vivo studies of Mg-30Sc alloys with different phase structure for potential usage within bone. *Acta Biomater.* **2019**, *98*, 50–66. [CrossRef]
30. Sheweita, S.; Khoshhal, K. Calcium metabolism and oxidative stress in bone fractures: Role of antioxidants. *Curr. Drug Metab.* **2007**, *8*, 519–525. [CrossRef]
31. Yoshizawa, S.; Brown, A.; Barchowsky, A.; Sfeir, C. Magnesium ion stimulation of bone marrow stromal cells enhances osteogenic activity, simulating the effect of magnesium alloy degradation. *Acta Biomater.* **2014**, *10*, 2834–2842. [CrossRef]
32. Hing, K.A.; Revell, P.A.; Smith, N.; Buckland, T. Effect of silicon level on rate, quality and progression of bone healing within silicate-substituted porous hydroxyapatite scaffolds. *Biomaterials* **2006**, *27*, 5014–5026. [CrossRef] [PubMed]
33. *ASTM E112–24*; Standard Test Methods for Detemining Average Grain Size. ASTM: West Conshohocken, PA, USA, 2024.
34. *ASTM E9*; Standard Test Methods of Compression Testing of Metallic Materials at Room Temperature. ASTM: West Conshohocken, PA, USA, 2019.
35. *ASTM G59–97*; Standard Test Method for Conducting Potentiodynamic Polarization Resistance Measurements. ASTM: West Conshohocken, PA, USA, 2023.
36. *ASTM G31*; Standard Guide for Laboratory Immersion Corrosion Testing of Metals. ASTM: West Conshohocken, PA, USA, 2021.
37. Lei, Y.; Zhan, M.; Xin, H.; Ma, L.; Yuan, Y.; Zhang, H.; Zheng, Z. Comparison of the Strain Rate Sensitivity in AZ31 and WE43 Magnesium Alloys under Different Loading Conditions. *Crystals* **2023**, *13*, 554. [CrossRef]
38. Elias, A.C. *Principles and Presentation of Corrosion*; Jones, D.A., Ed.; Prentice Hall: Upper Saddle River, NJ, USA, 1996.

39. Feliu, S., Jr. Electrochemical Impedance Spectroscopy for the Measurement of the Corrosion Rate of Magnesium Alloys: Brief Review and Challenges. *Metals* **2020**, *10*, 775. [CrossRef]
40. Kirkland, N.; Birbilis, N.; Staiger, M. Assessing the corrosion of biodegradable magnesium implants: A critical review of current methodologies and their limitations. *Acta Biomater.* **2012**, *8*, 925–936. [CrossRef]
41. Bagha, P.S.; Paternoster, C.; Khakbiz, M.; Sheibani, S.; Gholami, N.; Mantovani, D. Surface Modification of an Absorbable Bimodal Fe-Mn-Ag Alloy by Nitrogen Plasma Immersion Ion Implantation. *Materials* **2023**, *16*, 1048. [CrossRef]
42. Priyadarshi, A.; Khavari, M.; Subroto, T.; Prentice, P.; Pericleous, K.; Eskin, D.; Durodola, J.; Tzanakis, I. Mechanisms of ultrasonic de-agglomeration of oxides through in-situ high-speed observations and acoustic measurements. *Ultrason. Sonochemistry* **2021**, *79*, 105792. [CrossRef]
43. Kudryashova, O.; Vorozhtsov, S.; Khrustalyov, A.; Stepkina, M. Ultrasonic dispersion of agglomerated particles in metal melt. In *AIP Conference Proceedings*; AIP Publishing: Melville, NY, USA, 2016.
44. Saha, P.; Roy, M.; Datta, M.K.; Lee, B.; Kumta, P.N. Effects of grain refinement on the biocorrosion and in vitro bioactivity of magnesium. *Mater. Sci. Eng. C* **2015**, *57*, 294–303. [CrossRef]
45. Aung, N.N.; Zhou, W. Effect of grain size and twins on corrosion behaviour of AZ31B magnesium alloy. *Corros. Sci.* **2010**, *52*, 589–594. [CrossRef]
46. Emadi, P.; Andilab, B.; Ravindran, C. Effects of sonication amplitude on the microstructure and mechanical properties of AZ91E magnesium alloy. *J. Magnes. Alloys* **2022**, *10*, 3397–3405. [CrossRef]
47. Razavi, M.; Huang, Y. Effect of hydroxyapatite (HA) nanoparticles shape on biodegradation of Mg/HA nanocomposites processed by high shear solidification/equal channel angular extrusion route. *Mater. Lett.* **2020**, *267*, 127541. [CrossRef]
48. Yang, Z.; Fan, J.; Liu, Y.; Nie, J.; Yang, Z.; Kang, Y. Effect of the particle size and matrix strength on strengthening and damage process of the particle reinforced metal matrix composites. *Materials* **2021**, *14*, 675. [CrossRef] [PubMed]
49. Gerashi, E.; Alizadeh, R.; Langdon, T.G. Effect of crystallographic texture and twinning on the corrosion behavior of Mg alloys: A review. *J. Magnes. Alloys* **2022**, *10*, 313–325. [CrossRef]
50. Chen, H.; He, Z.; Lu, L. Correlation of surface features with corrosion behaviors of interstitial free steel processed by temper rolling. *J. Mater. Sci. Technol.* **2020**, *36*, 37–44. [CrossRef]
51. Xu, H.; Tian, T.; Hua, B.; Zhan, W.; Niu, L.; Han, B.; Zhang, Q. Effect of in-situ rolling and heat treatment on microstructure, mechanical and corrosion properties of wire-arc additively manufactured 316L stainless steel. *J. Mater. Res. Technol.* **2023**, *27*, 3349–3361. [CrossRef]
52. Deng, B.; Dai, Y.; Lin, J.; Zhang, D. Effect of rolling treatment on microstructure, mechanical properties, and corrosion properties of WE43 alloy. *Materials* **2022**, *15*, 3985. [CrossRef]
53. Cao, C.-N.; Zhang, J.-Q. *An Introduction to Electrochemical Impedance Spectroscopy*; Science Press: Beijing, China, 2002.
54. Han, L.; Zhang, Z.; Dai, J.; Li, X.; Bai, J.; Huang, Z.; Guo, C.; Xue, F.; Chu, C. In vitro bio-corrosion behaviors of biodegradable AZ31B magnesium alloy under static stresses of different forms and magnitudes. *J. Magnes. Alloys* **2023**, *11*, 1043–1056. [CrossRef]
55. Mei, D.; Lamaka, S.V.; Lu, X.; Zheludkevich, M.L. Selecting medium for corrosion testing of bioabsorbable magnesium and other metals—A critical review. *Corros. Sci.* **2020**, *171*, 108722. [CrossRef]
56. Mena-Morcillo, E.; Veleva, L. Degradation of AZ31 and AZ91 magnesium alloys in different physiological media: Effect of surface layer stability on electrochemical behaviour. *J. Magnes. Alloys* **2020**, *8*, 667–675. [CrossRef]
57. Sanchez, A.H.M.; Luthringer, B.J.; Feyerabend, F.; Willumeit, R. Mg and Mg alloys: How comparable are in vitro and in vivo corrosion rates? A review. *Acta Biomater.* **2015**, *13*, 16–31. [CrossRef]
58. Myrissa, A.; Agha, N.A.; Lu, Y.; Martinelli, E.; Eichler, J.; Szakács, G.; Kleinhans, C.; Willumeit-Römer, R.; Schäfer, U.; Weinberg, A.-M. In vitro and in vivo comparison of binary Mg alloys and pure Mg. *Mater. Sci. Eng. C* **2016**, *61*, 865–874. [CrossRef]
59. Walker, J.; Shadanbaz, S.; Kirkland, N.T.; Stace, E.; Woodfield, T.; Staiger, M.P.; Dias, G.J. Magnesium alloys: Predicting in vivo corrosion with in vitro immersion testing. *J. Biomed. Mater. Res. Part B Appl. Biomater.* **2012**, *100B*, 1134–1141. [CrossRef]

metals

MDPI

Article

Corrosion Behavior and Mechanical Properties of Zn–Ti Alloys as Biodegradable Materials

Alexandra-Tamara Șutic [1], Romeu Chelariu [1,*], Ramona Cimpoeșu [1], Ana-Maria Roman [1], Bogdan Istrate [2], Viorel Goanță [2], Marcelin Benchea [2], Mihaela Moscu [3,*], Adrian Alexandru [1], Nicanor Cimpoeşu [1] and Georgeta Zegan [4]

1 Department of Materials Science, Faculty of Materials Science and Engineering, "Gheorghe Asachi" Technical University of Iasi, 41 Dimitrie Mangeron Blvd., 700050 Iasi, Romania; alexandra-tamara.sutic@student.tuiasi.ro (A.-T.Ș.); ramona.cimpoesu@academic.tuiasi.ro (R.C.); ana-maria.roman@academic.tuiasi.ro (A.-M.R.); adrian.alexandru@academic.tuiasi.ro (A.A.); nicanor.cimpoesu@academic.tuiasi.ro (N.C.)

2 Faculty of Mechanical Engineering, "Gheorghe Asachi" Technical University of Iasi, 43 Dimitrie Mangeron Blvd., 700050 Iasi, Romania; bogdan.istrate@academic.tuiasi.ro (B.I.); viorel.goanta@academic.tuiasi.ro (V.G.); marcelin.benchea@academic.tuiasi.ro (M.B.)

3 Department of Morpho-Functional Sciences I, Faculty of Medicine, "Grigore T. Popa" University of Medicine and Pharmacy, 16 University Street, 700115 Iasi, Romania

4 Department of Surgical, Faculty of Dental Medicine, "Grigore T. Popa" University of Medicine and Pharmacy, 16 University Street, 700115 Iasi, Romania; georgeta.zegan@umfiasi.ro

* Correspondence: romeu.chelariu@academic.tuiasi.ro (R.C.); mihaela.moscu@umfiasi.ro (M.M.)

Abstract: The influence of the chemical composition and structural state of Zn–Ti alloys on corrosion behaviour and mechanical properties was studied. Zn-based alloys were investigated, more precisely, pure technical Zn and Zn with 0.10, 0.25 and 1.00 wt.% Ti. The microstructure and chemical composition of these materials were analysed using light optical microscopy (LOM), scanning electron microscopy (SEM), energy dispersive X-ray spectroscopy (EDS) and X-ray diffraction (XRD). The chemical composition of the alloys and the surface after immersion were analysed using an EDS detector from Bruker. The alloys' electro-chemical corrosion resistance was further investigated through linear (LP) and cyclic (CP) potentiometry and open-circuit potential (OCP) analysis. A tensile/compression equipment (Instron 34SC-5) was used to determine the compression behaviour. UMT testing equipment was used to determine microhardness (by Rockwell indentation) and COF vs. length. For percentages higher than 0.25 wt.% Ti, the formation of a primary $TiZn_{16}$ intermetallic compound in the (α-Zn + $TiZn_{16}$) eutectic matrix was observed, a slight influence of $TiZn_{16}$ on the Zn corrosion resistance results, and a greater influence on the mechanical properties was confirmed.

Keywords: Zn–Ti biodegradable alloy; corrosion resistance; degradation rate

1. Introduction

There are three main classes of biodegradable metals: Mg-, Fe- and Zn-based materials [1,2]. Although Mg-based materials have compatible mechanical properties with bone tissue and high osteopromotive and biocompatible behaviour during degradation, their biomedical applications are limited by their high degradation rate, which can cause both local gas accumulation and loss of mechanical integrity before the surrounding bone is fully healed [3]. Currently, Mg-based biodegradable materials are being used in the manufacture of commercial coronary stents and bone screws and pins [1,4,5]. For Fe-based materials, best machinability, excellent mechanical properties and cost-effectiveness are the main advantages as biodegradable materials; however, they have a slow degradation rate in the bone environment, and therefore, research has mainly focused on cardiovascular applications [1–4,6]. So far, there are no commercially available Fe-based materials on the global market, and their development is expected to require more time as their performance depends on the complexity of their interactions with the physiological environment [6]. Investigations on Zn-based materials for biomedical applications have only recently emerged as a result of the favourable biodegradability characteristics in physiological environments, which are intermediate between those of Mg and Fe biodegradable metals [2,4]. Although the role of Zn in structural, catalytic and signalling processes is thoroughly proven and the human body has developed its own homeostatic capacity to maintain tight regulation of zinc levels in relation to its essentiality/toxicity duality [7,8], both deficiency and excess of Zn levels are associated with many disease conditions in the human body [9–11]. Despite its physiological relevance, Zn is not usually considered a first choice as a biodegradable medical material, corresponding to less than 100 publications in the last decade [3]. Even under these conditions, Zn-based materials have a great potential for clinical applications, such as biodegradable vascular scaffolds and bone implants, when considering their favourable association of mechanical properties, degradation behaviour and biocompatibility [4,12]. However, no clinical use of Zn-based orthopaedic devices has been reported [2].

Chen et al. [13] summarised both the issues and the perspectives of Zn-based materials for orthopaedic internal fixation implants. The study showed that biodegradable Zn-based materials have great potential for application as temporary implants for internal osteosynthesis due to their versatility, which can be modified and controlled by both chemical composition and processing, mechanical properties, degradation behaviour, biocompatibility and bacteriostatic activity. Furthermore, it has been shown that there are still important concerns, such as the correlation of the degradation rate with the tissue healing and the toxic dose of Zn^{2+} level, the stress corrosion and corrosion fatigue behaviour, the high longitudinal elastic modulus compared to the bone tissue and the creep effect at physiological human body temperature [7,13].

During the decades of Zn-based biodegradable material development, various strategies have been employed to improve their performance, but all have resulted in modifications of the microstructural features. Alloying has been widely used, namely for binary, tertiary, quaternary and quintenary alloys, using different chemical elements [14–21]. By alloying and casting, the main microstructural features consist of a dendritic Zn-based solid solution as the primary phase (matrix) and secondary phases, often intermetallic compounds, formed by various reactions [22]. Thermomechanical processing can provide a further improvement in performance by altering the microstructural features (grain size, texture, phase volume fraction ratio, phase morphology, etc.), resulting in enhanced mechanical properties, degradation behaviour and biocompatibility [16–21].

As far as the binary Zn–Ti system has been studied, only a few studies have been conducted towards its use as a biodegradable material. In his master's thesis, Yin [23] extensively investigated some Zn–Ti alloys with Ti = 0.01, 0.1 and 0.03 wt.% as biodegradable materials for stents. The study showed the best combination of tensile properties for the hot-extruded Zn-0.1 wt.% Ti alloy without testing the corrosion and degradation behaviour, the friction and wear performance or the biocompatibility of these Zn–Ti alloys. Wang et al. [24] extended their investigations on some Zn–Ti alloys (Ti contents of 0.05, 0.1, 0.2 and 0.3 wt.%, cast + hot-rolled processing) by a comprehensive evaluation of their corrosion and degradation properties, friction and wear behaviour and in vitro cytotoxicity, motivated by the consideration that a high Ti ion concentration was toxic to osteoblast-like $SaOS_2$ cells, and the threshold of the Ti ion concentration was 15.5 μg/L. It was concluded that the microstructure was formed from an α-Zn matrix and a TiZn16 intermetallic phase, with increased Ti content causing microstructural modifications and improved mechanical properties, higher corrosion current densities and higher corrosion rates, improved wear behaviour and no cytotoxicity to MG-63 cells for extracts with Ti ion concentrations $\leq$ 25%.

In this work, some cast binary Zn–xTi alloys with specific contents of x = 0.00, 0.10, 0.25 and 1.00 wt.% have been studied regarding the relationship of their chemical composition and processing with their microstructural features and, hence, with their mechanical properties, degradation behaviour and frictional behaviour in order to determine their availability as biodegradable materials for temporary internal osteosynthesis implants.

Although various Zn alloy systems have been investigated, in the present study, the binary Zn–Ti alloy system was chosen in consideration of (i) the availability of the Zn–Ti phase diagram, (ii) the design of the cast Zn–Ti alloy and (iii) the potential of Ti as an alloying element to induce changes in microstructural features and, hence, significant changes in mechanical properties, degradation rate and biocompatibility.

2. Materials and Methods

2.1. Materials

A two-step laboratory-scale processing pathway was used: The first was the synthesis of a Zn-5 wt.% Ti master alloy (Zn-5Tima) from ingots of commercially pure Zn (cp-Zn, 99.99 wt.%, each impurity < 0.005 wt.%) and Ti sponge (Tisp, 99.90 wt.%, each impurity < 0.02 wt.%), and the second was the synthesis of the targeted Zn–xTi alloys, x = 0.00 wt.%, 0.10 wt.%, 0.25 wt.% and 1.00 wt.%, using cp-Zn and necessary quantities from Zn-5Tima. To perform melting and alloying, both for the master alloy and targeted alloys, an induction furnace with an argon protective atmosphere and a silicon carbide crucible were used (medium frequency static generator, 20 kW/8 kHz/300 V, Electrotehnica SA Bucharest, București, Romania). The melted material (casting temperature approximately 600 °C) was poured into a metal mould preheated at 225 °C (cooling rate approximately 3–4 °C/s until the melt was completely solidified), then held in the mould for 10 min and finally extracted and air cooled. The primary ingots of the Zn–xTi alloys had approximately 500 g and dimensions of 25 mm × 40 mm × 70 mm.

The primary ingots were subjected to mechanical and, where necessary, electroerosion processing to produce the final samples. After removing the upper part of the primary ingots, secondary ingots measuring 25 mm × 25 mm × 70 mm were processed and used to produce the four semi-finished samples (12 mm × 12 mm × 70 mm). The samples were used to characterise the microstructure, mechanical properties and the corrosion, wear and friction behaviour. Samples measuring 12 mm × 12 mm × 5 mm were prepared for microstructural characterisation. For the mechanical compression tests (10 mm × 12 mm) and for corrosion and degradation behaviour (10 mm × 5 mm), the samples were produced by electroerosion cutting and turning. A further set of specimens was prepared for micro-indentation testing to investigate the friction characteristics. To determine the chemical compositions of the investigated materials, the energy dispersive spectroscopy (EDS) method was used and performed on the grinded and polished samples (see Section 2.2) using an X-ray energy dispersive detector (energy dispersive spectroscopy (EDS), Bruker, X-Flash 6-10, Billerica, MA, USA, working mode in Esprit 2.2 for spectrum acquisition: automatic mode and Mapping).

2.2. Microstructure Analysis

To observe and analyse microstructure features, the samples were subjected to conventional preparation methods, mechanical grinding (SiC disks, P80-P2400 grit), mechanical polishing (alumina suspension, 1.00 μm, 0.30 μm, 0.04 μm) and an etchant containing 50 g CrO_3, 4 g Na_2SO_4 and 1000 mL H_2O, either alone or, where it was necessary, in combination with an etchant containing 5 g $FeCl_3$, 10 mL HCl and 240 mL alcohol.

Optical microscopy (OM) was performed using an ISM-M1000 Namicon (INSIZE, Suzhou, China) trinocular inverted mass metallographic microscope equipped with a MotiCam camera (10+, 10.0 MP) specialised in microscopic analysis and Motic Images Plus 3.0 (×86) software (version 3.0.12.41, Motic, China Group Co., Ltd. 2015, Hong Kong, China). For scanning electron microscopy (SEM), a VegaTescan LMH II (TESCAN, Brno-Kohoutovice, Czech Republic) microscope was used with a cathode supply voltage of 30 kV, secondary electron detector and working distance of 15.5 mm.

The X-ray diffraction analysis (XRD) was performed using an Xpert PRO MPD 3060 diffractometer from PANalytical (Almelo, The Netherlands) with a Cu X-ray tube ($K\alpha$ = 1.54051 Å), 2θ = 25°–75°, step size = 0.13, time/step: 99.45 s and scan speed of 0.033667°/s with a number of 3808 steps. The XRD patterns were analysed and phases indexed using MATCH! Software version 3.16 Build 288 (Crystal Impact, Bonn, Germany). The samples for X-ray diffraction analysis (XRD) were not etched.

2.3. Mechanical Testing

Compressive testing was performed on an INSTRON 8801 testing machine (Norwood, MA, USA) (maximum load 100 kN). The compression samples were strained with a traverse speed (TS) = -0.68×10^{-3} mm/s up to a strain of 0.85. To evaluate the surface hardness, the microindentation tests on a CETR-UMT-2 tribometer were performed (Campbell, Camden, NJ, USA) using a Rockwell diamond tip (Milwaukee, WI, USA) (radius 200 μm and angle 120°, and a vertical load of 10 N). The results were interpreted using the Test Viewer software (version 2.16).

2.4. Corrosion and Degradation Behavior

2.4.1. Immersion Testing

The samples were immersed in a NaCl solution (9 mg/mL) in a thermostatically controlled enclosure at 37 °C (20 mL/cm^2 ratio). For each immersion interval (3, 7 and 14 days), the degradation rate (DR) based on the mass loss was obtained for all Zn–xTi alloys. The samples were weighed using an AS220 Partner analytical balance (RADWAG Balances & Scales, Radom, Poland). The samples were subjected to ultrasonic cleaning in technical alcohol using an ultrasonic equipment (PRO 50 ASonic, Ultrasonic Cleaner, Beijing, China) for 60 min. The DRs [mm/y] were obtained according to ASTM G31-21 using the following equation [25], where W = mass loss [g]; A = sample area [cm^2]; t = time [h] and ρ = density [g/cm^3]:

$$DR = \frac{8.76 \times 10^4 \times W}{A \times t \times \rho} \quad (1)$$

In addition, the pH of the NaCl solution was registered for the first 72 h with a Hanna HI98191 (Darmstadt, Germany) pH meter equipped with a HI72911B titanium-bodied pH electrode and a temperature sensor that were inserted into the sample container during the experiment. The pH calibration was performed regularly at five points with pre-programed standard buffer solutions with pH of 4.01; 6.86; 7.00; 9.18 and 10.01, achieving a pH accuracy of ±0.002 to ±0.001. The data were acquired minute-by-minute using the Hanna HI-92000 software (version 5.0.38) with a USB connection.

2.4.2. Electrochemical Testing

A VoltaLab-21 potentiometer (Radiometer, Copenhagen, Denmark) was used to determine the corrosion resistance by analysing linear and cyclic potentiometry in NaCl as the electrolyte solution. The acquisition and processing of data were performed using the Volta Master 4 software (version 6.0.25130). A three-electrode cell was used to analyse the corrosion behaviour. A sample with an exposed surface of 0.63 cm^2 was used as the working electrode. A saturated calomel electrode (SCE) was used as a reference, and a platinum electrode was used as an auxiliary.

The solution was stirred with a magnetically driven, Teflon-coated stirring bar to remove hydrogen. Potentiodynamic polarisation curves were obtained at a scanning rate of 1 mV/s, and cyclic voltammetry curves were obtained at a scan rate of 10 mV/s. For linear potentiometry, scanning was performed in the range 500–1200 mV vs. FREE. The open circuit potential (OCP) was registered for 10 min before the electrochemical tests.

2.5. Friction Behavior Test

To evaluate friction behaviour, microscratch tests were performed using a CETR-UMT-2 tribometer (Campbell, Camden, NJ, USA). The tests involved a sharp stylus to produce a controlled and continuous microscratch mark on the material surface. Additionally, the AE signals provided information on the load-displacement characteristics during the microscratch test, helping to identify damage initiation and failure mechanisms. Two types of microscratch tests were performed: one with a constant load and the other with a linearly increasing force. The tests were performed with vertical forces of 10 N and 19 N, moving the table 10 mm in 60 s at a rate of 0.167 mm/s. A widia cutting blade with a tip radius of 0.4 mm was used for the microscratch tests.

2.6. Statistical Analysis

For the mechanical, immersion corrosion and friction testing, the tests were replicated three times under the same testing conditions for all Zn–xTi alloys. To demonstrate the effect of the Ti content on the properties of the Zn–xTi alloys, the experimental data were statistically analysed using either one-way repeated measures ANOVA (OWR-ANOVA) or two-way repeated measures ANOVA (TWR-ANOVA), performed by Tukey's test at a significance level of 0.05.

All experimental activities of this study have been conducted in accordance with the principles of occupational health and safety, the use of compliant instrumentation and machinery and proper personal protective equipment [26].

3. Results

3.1. Chemical Composition

The chemical compositions of the investigated Zn-based materials were determined using EDS and are shown in Table 1.

Table 1. Chemical compositions of the Zn–xTi biodegradable materials.

Element Average (St. Dev.) (wt.%)	Zn–xTi Alloys			
	Zn–0.00Ti	Zn–0.10Ti	Zn–0.25Ti	Zn–1.00Ti
Zn	**99.99 (0.001)**	**99.814 (0.028)**	**99.663 (0.037)**	**99.007 (0.013)**
Ti	**-**	**0.113 (0.025)**	**0.259 (0.037)**	**0.910 (0.011)**
Al	0.003 (0.002)	0.015 (0.002)	0.014 (0.004)	0.016 (0.002)
Ni	-	0.016 (0.002)	0.013 (0.001)	0.019 (0.003)
Si	-	0.013 (0.001)	0.019 (0.003)	0.013 (0.001)
Cu	0.002 (0.001)	0.015 (0.002)	0015 (0.002)	0.014 (0.004)
Fe	0.004 (0.001)	0.014 (0.004)	0.016 (0.002)	0.019 (0.003)
Sn	0.002 (0.001)	-	-	-

3.2. Microstructure

3.2.1. Zn–Ti Phase Diagram and Alloy Design

It is well-known that, by adding small quantities of Ti into Zn-based alloys, a grain refinement effect and a change in the eutectic morphology are produced in correlation with the cooling rate of the melt [27–35]. In addition, the grain size can be controlled by the cooling rate of the melt, as Leone et al. [30] reported. For example, at 0.004–0.009 wt.% Ti for all cooling rates, a completely columnar microstructure was observed, but for 0.1 wt.% Ti and a 0.5 °C/s cooling rate, a small columnar microstructure was observed in contrast with a completely grain-refined microstructure observed at a cooling rate greater than 5 °C/s. To limit the grain-refinement effect determined by either the cooling rate or between the Ti wt.% content and the cooling rate, a 3–4 °C/s range value for the last was maintained for all Zn–xTi alloys. In contrast with other Zn–X phase diagrams, e.g., Zn–Al, the Zn–Ti phase diagram is still in dispute [35], especially the crystal structures of the low melting temperature compounds that are formed at the Zn-rich zone of the Zn–Ti phase diagram [25,36].

Ti has a very low solubility in Zn at room temperature, approximately 0.0004–0.0005 wt.% (0.00055–0.00068 at.%) Ti [31,37,38], and at 0.160 wt.% (0.118 at.%) Ti content, a eutectic reaction has been reported [24,27–36]:

$$L \rightarrow \alpha\text{-Zn} + TiZn_{16};\ T = 418.6\ ^\circ C/691.75\ K;\ Ti = 0.160\ wt.\%\ (0.118\ at.\%) \quad (2)$$

The Ti content corresponding to the eutectic reaction is dependent on the cooling rate, with a 0.18 wt.% (0.25 at.%) value at a very low crystallisation rate [37]. Ti solubility in Zn depends on the temperature, thus at 300 °C, it varies in the 0.007–0.015 wt.% range (0.0095–0.0204 at.%), and at 400 °C, it is approximately 0.02 wt.% (0.027 at.%) [38].

The TiZn16 intermetallic compound formed with an α-Zn eutectic, shown above, has an approximate 4.39 wt.% (5.90 at.%) Ti content [32,34]. According to the Ti–Zn phase diagram [28,31,34,39] and the chemical compositions of the Zn–xTi alloys (Figure 1), the main microstructural features for the four Zn–xTi alloys are as follows: (i) Zn–0.00Ti contains only α-Zn phase; (ii) Zn–0.10Ti, a hypoeutectic alloy, forms from α-Zn phase as a dendritic matrix and (α-Zn + $TiZn_{16}$) eutectic; (iii) Zn–0.25Ti, a hyper-eutectic alloy very close to the eutectic Ti content, contains an α-Zn + $TiZn_{16}$ eutectic and primary TiZn16 (presumably with a low volume fraction); and (iv) Zn–1.00Ti, a hyper-eutectic alloy with a Ti content distant from the eutectic concentration, contains an α-Zn + $TiZn_{16}$ eutectic and primary $TiZn_{16}$ (presumably with a higher volume fraction and modified morphology).

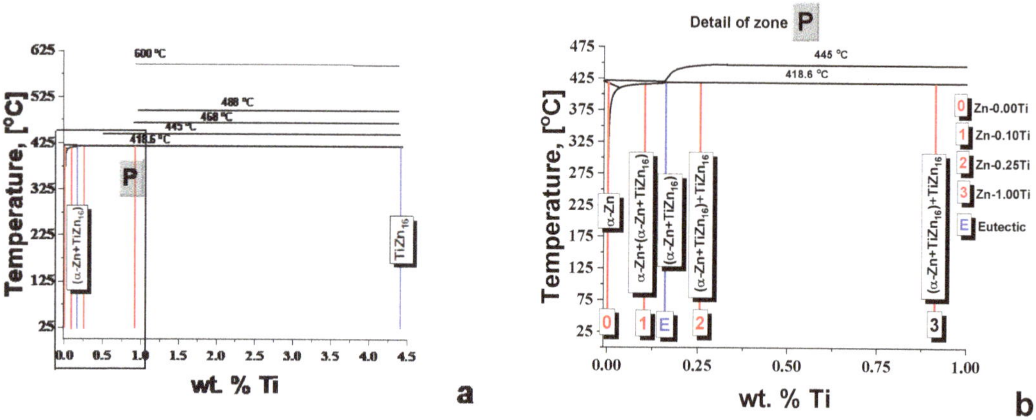

Figure 1. Details of the Zn–Ti phase diagram for wt.% Ti intervals of interest (adapted from Refs. [28,31,34,39]): (**a**) 0.00–4.5 wt.% Ti corresponding to TiZn16 content area; (**b**) the wt.% Ti content for the Zn–xTi alloys.

3.2.2. XRD Analysis

The experimental XRD patterns are shown in Figure 2. According to the Zn–Ti phase diagram (Figure 1) [28,31,34,39], the phases formed are α-Zn (space group $P6_3/mmc$ (194)) and TiZn16 intermetallic compound (space group Cmcm (63)) [34].

To index the diffractograms, the Match! software as well as the XRD standard patterns from the Crystallography Open Database (COD) and the literature data were used [31,36,40–42]. The lattice parameters for the two phases are as follows: (i) α-Zn, a = 2.6648 Å, b = 4.4967 Å (COD Entry 96-900-8523); (ii) $TiZn_{16}$, a = 7.7200 Å, b = 11.4490 Å, c = 11.7550 Å (COD Entry 96-210-6429). Match! software provides phase composition based on the experimental XRD pattern. Thus, the volume percentage of the $TiZn_{16}$ phase was (i) 2.7% for Zn–0.10Ti; (ii) 7.6% for Zn–0.25Ti and (iii) 20.3% for Zn–1.00Ti.

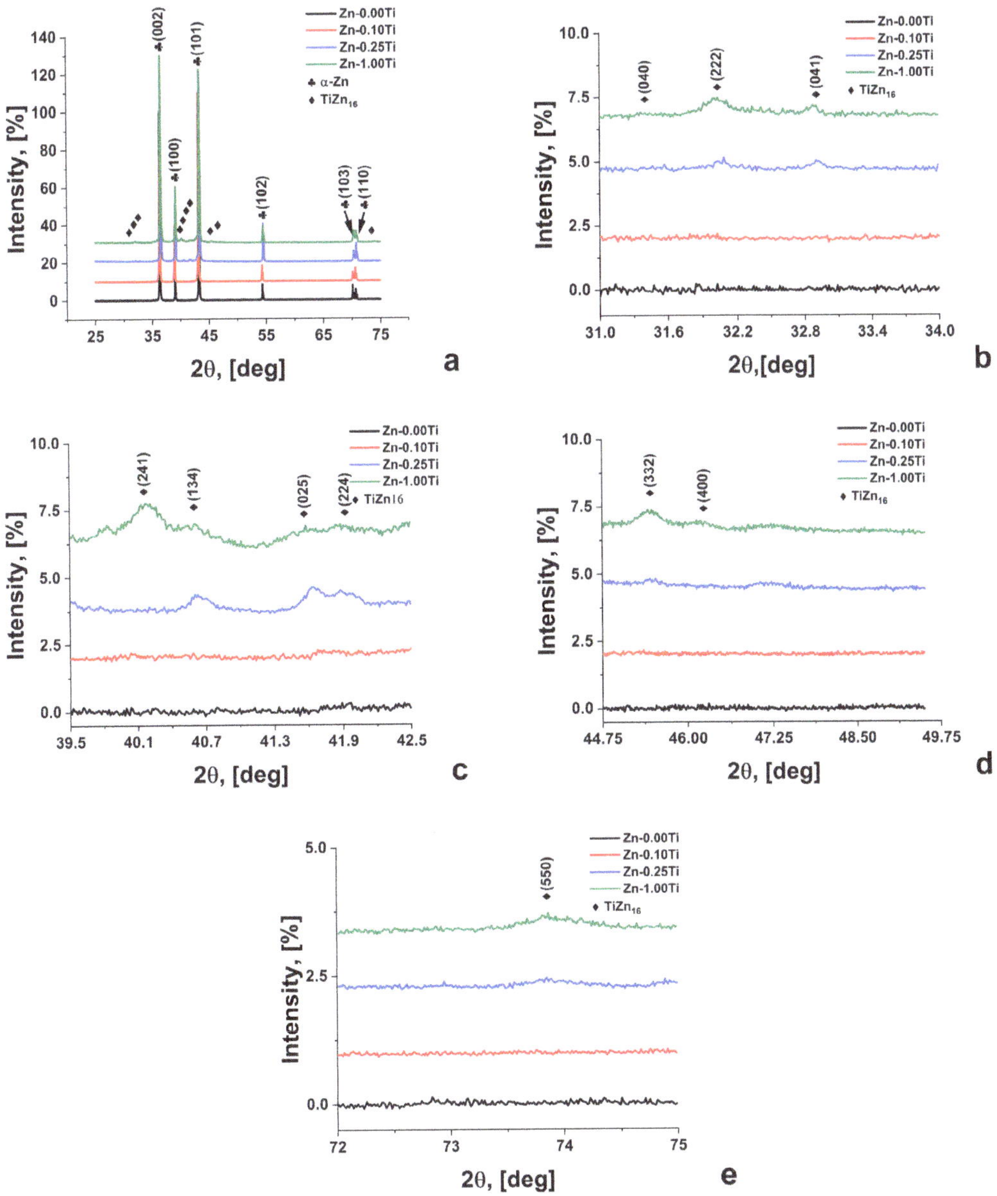

Figure 2. XRD patterns of Zn–xTi alloys (2θ): (**a**) 25°–75° for α-Zn; (**b**) 31°–34° for $TiZn_{16}$; (**c**) 39.5°–42.5° for $TiZn_{16}$; (**d**) 44.75°–49.75° for $TiZn_{16}$; (**e**) 72°–75° for $TiZn_{16}$.

3.2.3. Optical Microscopy

Figure 3a–d show the optical microstructure (OM) of the Zn–xTi alloys. The as-cast Zn–0.00Ti contains large grains (above 500 μm), as shown in Figure 3a.

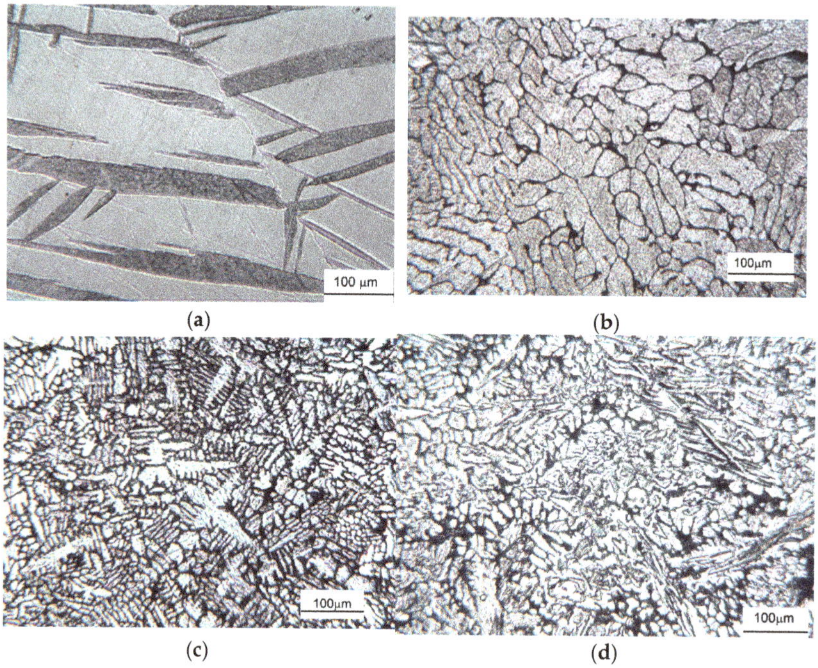

Figure 3. Optical micrographs of the experimental alloy microstructures: (**a**) Zn–0.00Ti; (**b**) Zn–0.10Ti; (**c**) Zn–0.25Ti; (**d**) Zn–1.00Ti.

In Figure 3b, the microstructure of the as-cast Zn–0.10Ti is characterised by an α-Zn dendritic morphology, with the average grain size decreasing to approximately 200 (±50) µm. Also, the dendritic arm size in relation with the relatively low solidification rate and low Ti content has values in the 20–70 µm range with a shape ratio between 0.3 and 0.5. As it is observed from the Zn–Ti binary phase diagram (Figure 1), the Zn–0.1 wt.% Ti alloy is a hypoeutectic alloy. It can be considered that the Zn–0.10Ti alloy comprises primary Zn dendrites and a eutectic of Zn and intermetallic phase mixtures ($TiZn_{16}$) placed in the interdendritic space of primary α-Zn as it is confirmed by the OM from Figure 3b.

Analysing the microstructure from Figure 3c, the OM of the as-cast Zn–0.25Ti shows a decreasing grain size compared to the as-cast pure Zn and Zn–0.10Ti. The average grain size decreases to approximately 100 (±27) µm with the Ti content increasing from 0.10 to 0.25 wt.%. Given the low solidification rate and higher Ti content above but close to the (α-Zn + $TiZn_{16}$) eutectic Ti content (approximately 0.16 wt.% Ti), Zn–0.25Ti can be considered as a hypereutectic alloy. The (α-Zn + $TiZn_{16}$) eutectic morphology has, as the main feature, the relative coarse lamella of α-Zn phase in eutectic and small and rare primary $TiZn_{16}$.

Zn–1.00Ti is also a hypereutectic alloy, but the Ti content is much higher than the eutectic. The eutectic morphology is different compared to Zn–0.25Ti, with both eutectic regions having finer or larger lamellae of α-Zn, especially the larger primary $TiZn_{16}$.

3.2.4. SEM Microscopy

The SEM images of the Zn–xTi alloy microstructures are shown in Figure 4a–d, and the same features were observed in the OM images.

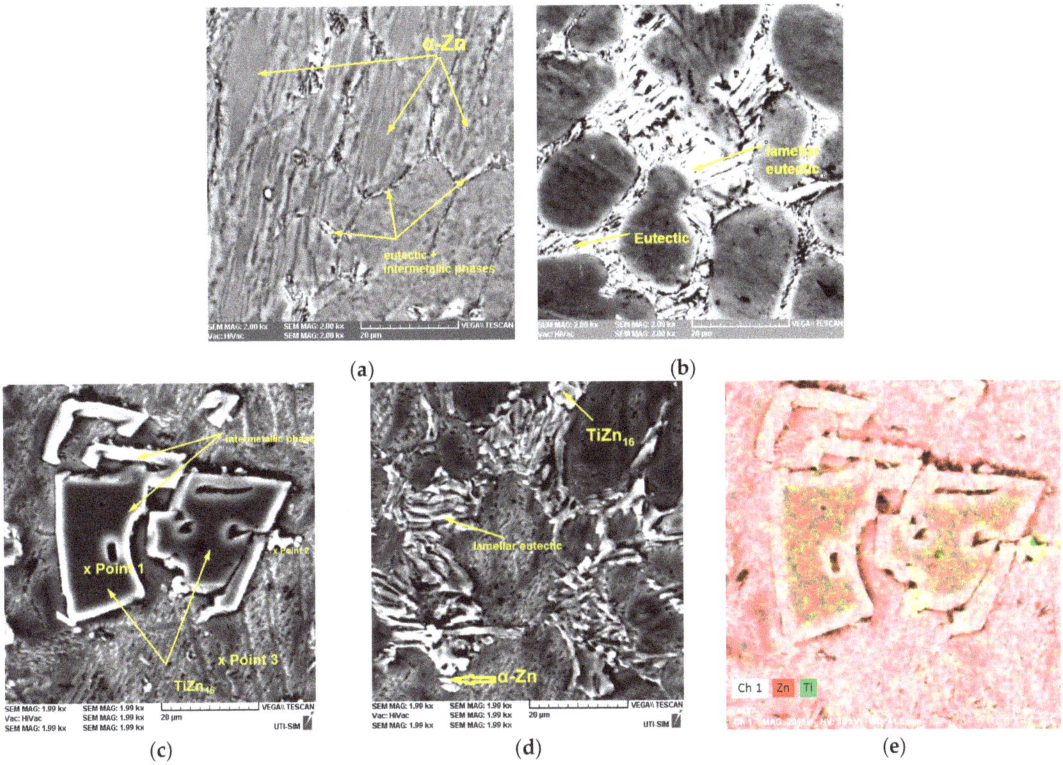

Figure 4. Microstructural and chemical insights of Zn–Ti alloys (**a**) The Zn–0.10Ti alloy structure; (**b**) Zn–0.25Ti; (**c**) Zn–1.00Ti-area1; (**d**) Zn–1.00Ti-area2 and (**e**) the elemental distribution of Zn and Ti on the area given in (**c**).

Supplementary, EDS was performed on three different points (Figure 4c), with 90 nm diameter spots on different structural entities presenting the formation of two different TiZn intermetallic compounds, points 1 and 2 spread in the eutectic matrix and point 3. The energy spectrum obtained from a 4 mm^2 area confirmed the presence of Zn and Ti elements at different energies.

The Zn and Ti elemental distributions, shown in Figure 4e, confirmed the formation of two TiZn intermetallic compounds. In point 1, a compound with 94.97 (93.26) wt.(at.)% Zn and 5.03 (6.74) wt.(at.)% Ti was observed. The chemical composition corresponds to the $TiZn_{16}$ phase. In point 2, the chemical composition has an average value of approximately 82.85 (77.95) wt.(at.)% Zn and 17.15 (22.04) wt.(at.)% Ti. According to Vassilev et al., the chemical composition of the compound identified in point 2 is close to the $TiZn_3$ compound [33], although not identified by XRD analysis, presumably on account of its low volume fraction.

3.3. Mechanical Properties

3.3.1. Compression

The stress–strain compression curves were analysed and interpreted according to ASTM E9-09 [43]. Figure 5 shows representative stress–strain curves recorded for all the alloys investigated and the compression data derived from the stress–strain curves shown in the comparative plots.

The average (st. dev.) values of the yield strength at 0.2% offset ($YS_{0.2}$) were 111.49 (6.08) MPa, 170.26 (2.48) MPa, 226.13 (10.20) MPa and 248.02 (6.32) MPa for Zn–

0.00Ti, Zn–0.10Ti, Zn–0.25Ti and Zn–1.00Ti. For the same alloys, the average (st. dev.) values of the compressive strength at 50% strain were 361.40 (8.10) MPa, 484.32 (20.14) MPa, 520.55 (19.20) MPa and 524.23 (23.17) MPa. OWR-ANOVA statistical analysis showed that, for both mechanical properties, there were significant differences between the mean values for Zn–0.00Ti, Zn–0.10Ti, Zn–0.25Ti or Zn–1.00Ti. Although the average values for Zn–0.25Ti are lower than those for Zn–1.00Ti alloys, statistically there are no significant differences.

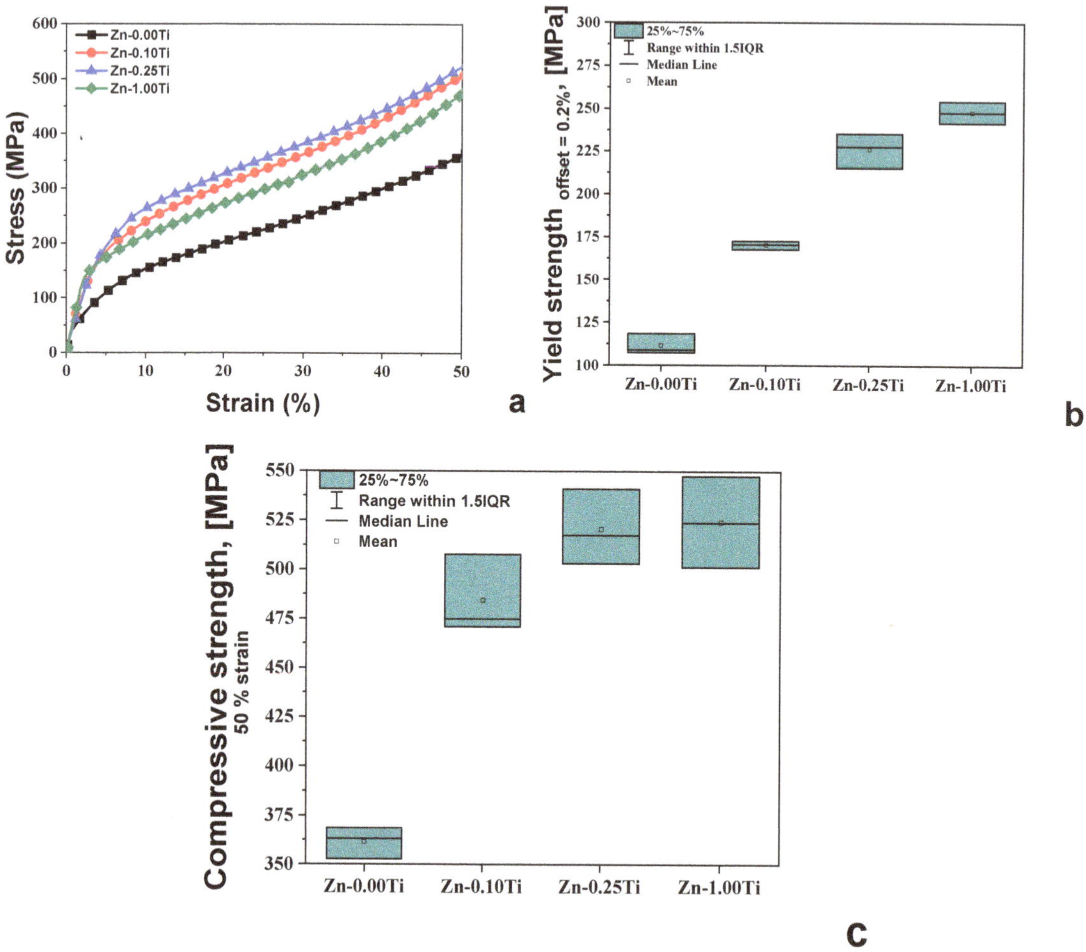

Figure 5. Representative compression curves and data: (**a**) compression stress–strain representative curves for all Zn–xTi alloys; (**b**) yield strength (offset = 0.20%); (**c**) compression strength (50% strain).

3.3.2. Microhardness

The results of the microindentation tests are graphically shown in Figure 6. The average (st. dev.) values of the microhardness were 0.328 (0.037) GPa, 0.531 (0.020) GPa, 0.619 (0.034) GPa and 0.694 (0.023) GPa. Despite the average values increasing with increasing Ti content, the OWR-ANOVA statistical analysis showed that there are significant differences between the average microhardness values of Zn–0.00Ti compared to all other alloys and of Zn–0.10Ti compared to Zn–0.00Ti and Zn–1.00Ti.

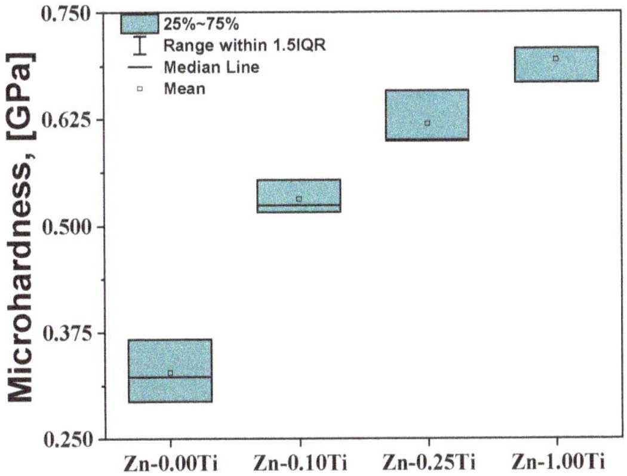

Figure 6. Microhardness results of Zn–xTi alloys.

3.3.3. Friction Properties

Table 2 shows the values of applied force (F) and the mean (st. dev.) values of friction coefficient (COF) and friction force (Ff). The TWR-ANOVA statistical analysis showed that there were significant differences between the average values of COF corresponding to Zn–0.00Ti and Zn–1.00Ti for 10 N applied force, and Zn–0.00Ti, Zn–0.10Ti, Zn–0.25Ti and Zn–1.00Ti, the exception being Zn–0.25Ti and Zn–1.00Ti for 19 N applied force.

Table 2. The average values of the coefficient of friction and force friction.

Alloy	Applied Force, F [N]	COF Average (St. Dev.)	Ff, [N] Average (St. Dev.)
Zn–0.00Ti	10 19	0.578 (0.327) 0.804 (0.401)	3.96 (3.28) 9.82 (7.67)
Zn–0.10Ti	10 19	0.422 (0.270) 0.726 (0.393)	2.91 (2.92) 9.16 (7.53)
Zn–0.25Ti	10 19	0.316 (0.193) 0.661 (0.423)	2.12 (2.18) 8.67 (7.9)
Zn–1.00Ti	10 19	0.245 (0.154) 0.622 (0.417)	1.65 (1.76) 8.23 (7.69)

Using the COF values from Table 2 and the TWR-ANOVA results, Figure 7 graphically shows the dependence of COF on Ti content and applied force.

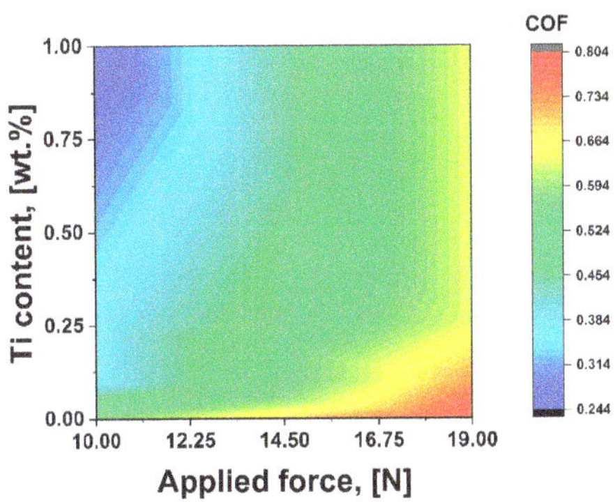

Figure 7. COF vs. applied force and Ti content.

3.4. Degradation Behavior

The corrosion resistance of the alloys was evaluated through immersion and electrochemical tests.

3.4.1. Immersion

When interacting with an electrolyte solution, metallic materials, particularly alloys, undergo an exchange of electrons and ions, causing changes in the chemistry of the metal/electrolyte interface. The chemical processes at the metal/electrolyte interface produce acid/base radicals that lead to passivation of the metal substrate or changes in electrolyte pH, whether the reaction products are soluble or not. For the present study, the pH of the electrochemical solution was recorded during the immersion tests, and its evolution is shown in Figure 8. In addition, the chemical compositions of the corroded surfaces of the Zn–xTi alloys after the immersion tests are shown in Table 3. These measurements were used to investigate the nature of the corrosion products. The degradation rates obtained for the three different immersion intervals are also given in Table 4.

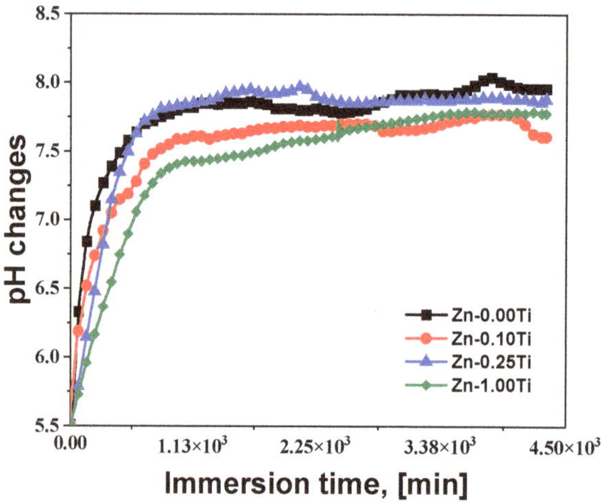

Figure 8. The representative pH variation plots recorded during the immersion tests for 72 h.

Table 3. The chemical composition of the corroded surfaces for the tested alloys after the immersion intervals.

Time/Samples/Elements		Zn	Ti	O	Na	C	Cl
		at.%	at.%	at.%	at.%	at.%	at.%
3 days	Zn–0.00Ti	24.4	-	43.6	14.7	16.5	0.6
	Zn–0.10Ti	19.5	0.10	46.8	9.9	22.9	0.8
	Zn–0.25Ti	29.6	0.10	37.8	16.5	10.2	5.7
	Zn–1.00Ti	19.4	0.40	46.8	9.2	23.7	0.5
7 days	Zn–0.00Ti	32.8	-	33.8	18.7	13.5	1.2
	Zn–0.10Ti	28.1	0.10	37.9	17.6	15.9	0.5
	Zn–0.25Ti	28.2	0.20	37.4	17.6	16.3	0.4
	Zn–1.00Ti	23.5	0.30	40.6	14.5	20.3	0.8
14 days	Zn–0.00Ti	19.7	-	45.5	11.5	22.4	0.9
	Zn–0.10Ti	28.2	0.02	38.3	21.6	11.3	0.6
	Zn–0.25Ti	25.5	0.03	41.04	15.7	17.2	0.5
	Zn–1.00Ti	25.6	0.2	38.1	18.3	17.7	0.2
St. Dev. %		0.11	0.01	0.61	0.18	0.44	0.06

Table 4. Degradation rates determined for all investigated Zn–xTi alloys.

Alloys/Immersion Time	Degradation Rates (μm/y)		
	3 Days	7 Days	14 Days
Zn–0.00Ti	268 (67)	163 (53)	13 (3)
Zn–0.10Ti	226 (11)	179 (2)	128 (5)
Zn–0.25Ti	217 (37)	102 (1)	96 (2)
Zn–1.00Ti	166 (17)	152 (17)	159 (7)

The TWR-ANOVA statistical analysis showed that, for all Zn–xTi alloys, significant differences of DR mean values were found depending on the immersion time. Ti content caused a significant difference after 14 days immersion for Zn–0.10Ti and Zn–1.00Ti. From the DR and TWR-ANOVA mean values, the dependence of DR on Ti content and immersion time is plotted in Figure 9.

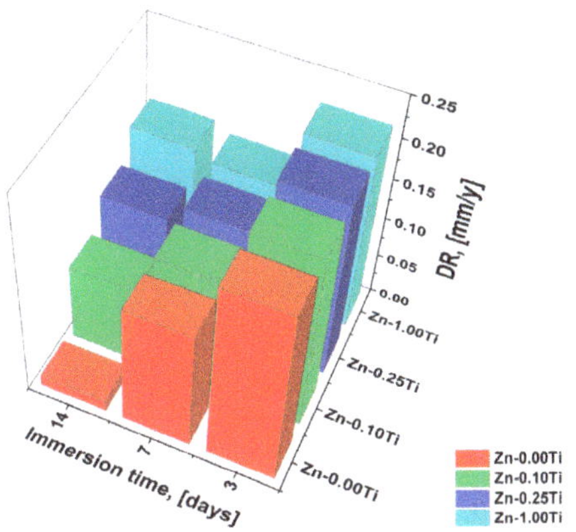

Figure 9. Degradation rate (DR) variation vs. wt.% Ti and immersion time.

The corroded surfaces of the investigated alloys were analysed by SEM after the three immersion intervals and ultrasonic cleaning to remove the unstable compounds. The results are shown in Figure 10. In the same figure, the distribution of the main elements identified on the surface, Zn, O, Cl, Na and Ti, are also illustrated.

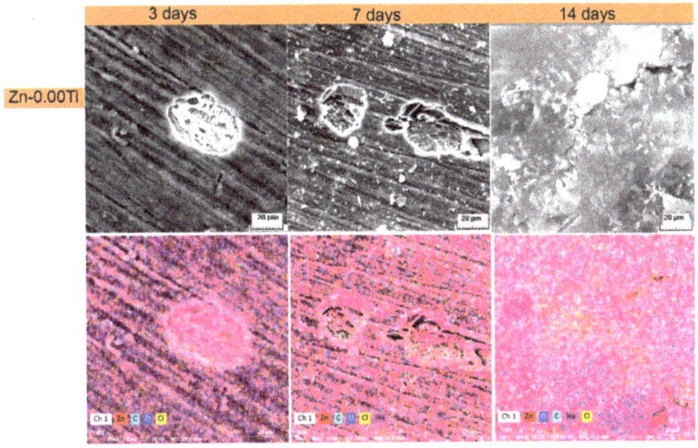

Figure 10. *Cont.*

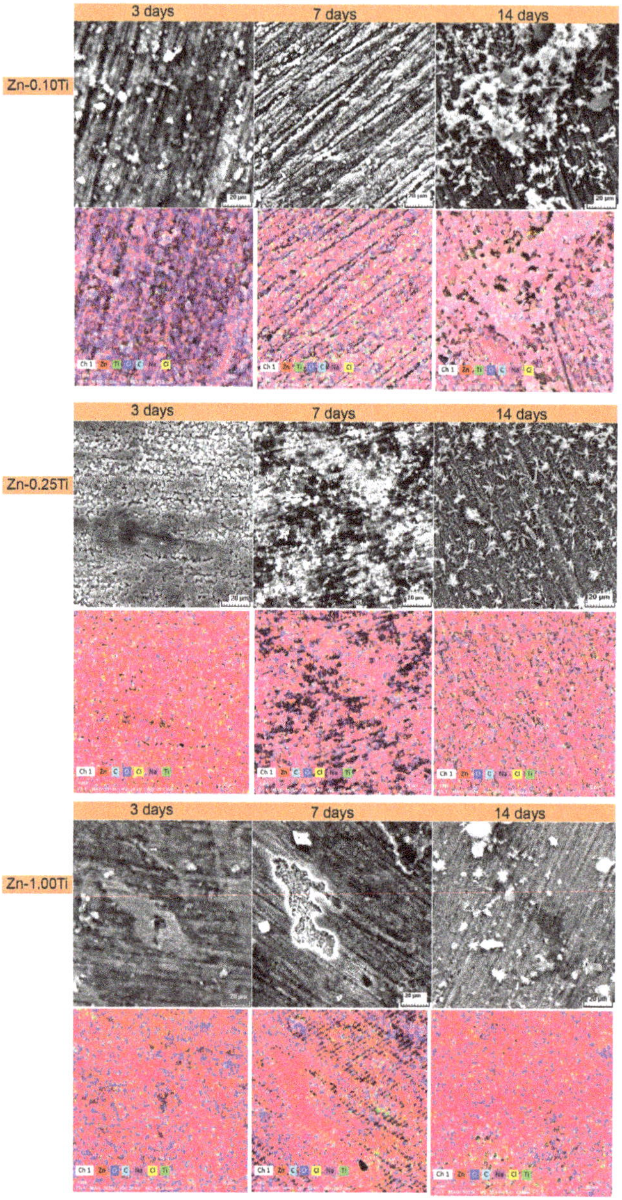

Figure 10. SEM and EDS images of the samples after immersion in NaCl (9 mg) for 3, 7 and 14 days.

3.4.2. Electrochemical Tests

Electrochemical experiments on the open-circuit potential (OCP), linear polarisation and cyclic voltammetry were performed to further investigate the influence of the Ti addition to Zn (Figure 11). Using Tafel extrapolation (Figure 11a), I_{corr} has been evaluated, and degradation (corrosion) rates (DR) have been determined (Table 5).

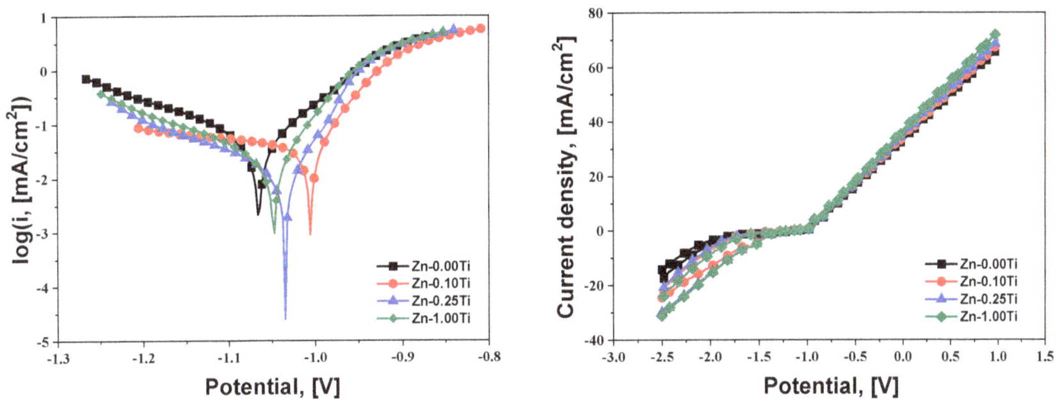

(**a**) (**b**)

Figure 11. Electro-corrosion results of the experimental alloys in NaCl: (**a**) Tafel diagrams; (**b**) voltammograms with cyclic curves.

Table 5. Electrochemical corrosion parameters (in 0.9% NaCl saline solution).

Alloy	Corrosion Process Parameters							
	OCP, [mV]	E (I = 0), [mV]	I_{corr}, [mA/cm^2]	Rp, [Ohm/cm^2]	DR, [μm/y]	β_c, [mV/dec]	β_a, [mV/dec]	OCP, [mV]
Zn–0.00Ti	−1010	−1065.8	0.0625	468.95	934.2	−203.6	93.6	−1060
Zn–0.10Ti	−935	−1005.6	0.0334	645.31	500.0	−496.8	51.8	−1115
Zn–0.25Ti	−1020	−1035.3	0.0162	1430.00	241.9	−197.8	49.7	−1100
Zn–1.00Ti	−990	−1048.4	0.0156	906.85	233.5	−146.5	50.1	−1050

Experimental electrochemical test results show that the Zn–1.00Ti alloy had the lowest corrosion rate, 233.5 μm/y and current density I_{corr} 0.0156 mA/cm^2. Linear polarisation curves show no significant changes in the anodic branch caused by the addition of Ti. Cyclic voltammograms show similar characteristics for all Zn–xTi alloys. Furthermore, a passive region between −1.4 and −0.9 V was observed for all alloys.

4. Discussion

The alloying of Zn with Ti resulted in different microstructural features corresponding to the low-Ti region of the binary Zn–Ti phase diagram, affecting the properties of the Zn–xTi alloys, as can be observed from the results on mechanical properties and corrosion behaviour presented above. As the Ti content increased, a different eutectic morphology was observed, and the volume fraction of the TiZn16 intermetallic compound increased continuously (Figure 4a–d). As reported in the literature, the increasing Ti addition caused the modification of the cast cp-Zn to a very coarse grain size (above 1 mm) [44,45], to a hypoeutectic and then two hypereutectic alloys with the formation of primary $TiZn_{16}$ intermetallic phase [28,39,44] and finer eutectic. As the percentage of Ti increases, the appearance of the lamellar eutectic is observed (Figure 4b) as well as its interruption for higher percentages of Ti (Figure 4d).

The EDS investigation confirmed that primary $TiZn_{16}$ intermetallic phases were formed in the as-cast Zn–0.25Ti and Zn–1.00Ti alloys (Figure 4d). The chemical composition of the $TiZn_{16}$ intermetallic compound is approximately 94.97 (93.26) wt.(at.)% Zn and 5.03 (6.74) wt.(at.)% Ti (point 1 from the SEM image in Figure 4c), corresponding to an approximate mole ratio for Zn:Ti of 16:1. As shown in the Zn–Ti phase diagram (Figure 1), it is considered that $TiZn_{16}$ is the intermetallic compound formed in the Zn–0.25Ti alloy and Zn–1.00Ti. These results are in agreement with a previous study, which demonstrated the presence of the primary Zn–Ti intermetallic phase in as-cast Zn–Ti alloys (0.16, 0.18, 0.23 and 0.34 wt.% Ti) [27]. By increasing the amount of Ti added to Zn from 0.25 to 1 wt.%, a new intermetallic compound was identified by EDS analysis (point 2 in Figure 4c), $TiZn_3$, but without validation on the XRD pattern [33,34].

The higher values of mechanical properties were obtained as the Ti content increased from 0.00 to 1.00 wt.%. However, the increase between 0.25 wt.% and 1.00 wt.% is greater than between 0.00 wt.% and 0.25 wt.%; a smaller increase was observed in the 0.25–1.00 wt.% Ti range. A similar compressive strength value (Figure 5) was measured for a Zn–1 wt.% Ti alloy (approximately 265 MPa) and reported in the literature [46].

The average hardness values of the alloys, based on the results shown in Figure 6, increased upon the addition of a small amount of Ti to cp-Zn. Two main reasons could be identified for the increase in hardness of Zn–xTi compared to the cast cp-Zn. The first is related to the Ti addition resulting in a grain refinement process through the appearance of a eutectic and intermetallic phases at the grain boundaries [47,48]. Secondly, at a higher addition than 0.25 wt.% Ti, the formation of the $TiZn_{16}$ primary intermetallic compound enhanced the hardening of Zn–xTi alloys.

The variation tendency of the friction coefficient is similar to the microhardness but inversely proportional. The friction coefficient of cast cp-Zn decreased by Ti addition. By increasing the Ti content to 1 wt.%, the friction coefficient decreases by 50% at low loads (10 N) and by approximately 25% at higher loads (19 N).

The hydroxide-based compounds formed from the start of the alloy-solution interaction resulted in a constant increase in the pH of the solution. The larger pH variations at the beginning of the immersion, namely the first 5–8 h, can be attributed to a more pronounced electron/ion exchange and the formation of corrosion compounds. In all cases, the pH of the electrolyte solution increases from approximately 5 to 7.5–8.0 as a result of a much more frequent exchange of ions between the immersed metallic material and the solution. After the first hours, a lower intensity in the pH variation is attributed to the stabilisation of the corrosion process, as the alloy is protected by a corrosion compound layer formed during this interval. This is supported by the Pourbaix diagram of Zn. There is also a strong correlation between the electrochemical behaviour associated with Zn corrosion and the pH of the electrolyte. The Pourbaix diagram of Zn in an aqueous environment [49] shows that the Zn can be oxidized to dissolved forms of Zn^{2+}, $Zn(OH)_2$, $HZnO^{2-}$ and ZnO_2^{2-} with increasing pH. The diagram shows the stability of $Zn(OH)_2$ or ZnO compounds in the range of 6–8 pH.

After the immersion tests, the Zn–xTi alloyed samples showed a higher percentage of oxides formed on the surface after 3 or 7 days of immersion and a lower percentage after 14 days. On the surface of all samples, regardless of the immersion time, Na and Cl were identified as salts formed from the interaction between the alloy and the electrolyte solution or deposited from the NaCl solution (Table 3 and Figure 10). The DRs of the alloys, determined by the mass losses resulting after the three immersion intervals (Equation (1)), show a reduction in the corrosion rate caused by the increase in the oxidized layer on the surface, which correlates with the increase in O wt.% (Figure 9 and Tables 3 and 4). This pattern, observed in all cases, is inferior for the Zn–1.00Ti alloy, where the determined degradation rates are more similar for all immersion intervals (Table 4 and Figure 9), attributed to the influence of titanium on the corrosion rate of pure zinc through the formation of a eutectic between α-Zn and the TiZn16 intermetallic compound. The degradation rate results are consistent with those reported by Liu et al. (2019) [50] for pure Zn. In addition to the constituent elements Zn and Ti, the oxygen responsible for the oxide layer formation and the other elements identified resulted from compounds that remained on the surface, mostly in the corrosion traces that emerged and were not removed by ultrasonic cleaning. The corrosion and degradation of the alloys occurred progressively by the repeated removal of layers that reached an advanced degree of oxidation, losing their stability and mechanical integrity and entering the electrolyte solution as reactive compounds [51,52]. Furthermore, the addition of at least 1% Ti decreased the DR due to the titanium oxide formed on the surface in contrast to the degradation rate of Zn–0.00Ti (as-cast cp-Zn) (Table 4).

Zinc and zinc-based alloys are very active metals in ionic media, especially in the presence of chloride ions.

Tafel slopes provide information on the reaction mechanism (Figure 11a). In this case, the small value of the anodic slope indicates the following reaction:

$$(Zn \rightarrow Zn^{2+} + 2e^{-}) \quad (3)$$

The above active reaction is controlled only by concentration, while the cathodic reaction is slower and controls the overall corrosion rate. The control of activation is determined by the electron transfer rate from the anode to the cathode (Figure 12).

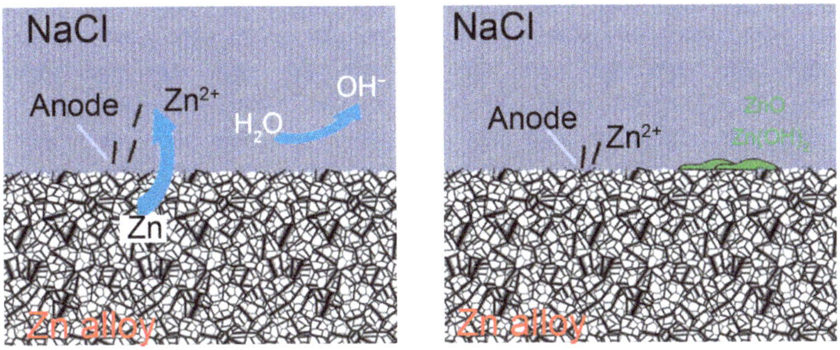

Figure 12. Schematic representation of the corrosion mechanism of zinc alloy surfaces in NaCl solution.

In the presence of dissolved oxygen at the cathode, the following Reaction (4) occurs:

$$\tfrac{1}{2}O_2 + H_2O + 2e^{-} \rightarrow 2OH^{-} \quad (4)$$

Followed by the Reactions (5) and (6):

$$Zn^{2+} + 2OH^{-} = Zn(OH)_2 \quad (5)$$

$$Zn(OH)_2 \rightarrow ZnO + H_2O \quad (6)$$

Corrosion potential (E), corrosion current density, (I_{corr}), anodic Tafel slope (β_a) and cathodic Tafel slope (β_c) are parameters calculated by the Tafel extrapolation (Table 5). The corrosion rate varies between 233.5 [μm/y] for Zn–1.00Ti and 934.2 [μm/y] for Zn–0.00Ti. It decreases with increasing Ti percentage, which means that the newly formed intermetallic compound $ZnTi_{16}$ increases corrosion resistance. The corrosion current density (I_{corr}) also decreases with increasing percentage of Ti from 0.0625 [mA/cm^2] to 0.0156 [mA/cm^2]. This is believed to be caused by the stability of the corrosion product layers. In the NaCl solution, the corrosion potential of Zn alloy is approximately 1.00 V lower than the corrosion potential reported in potentiodynamic tests of Zn alloys in SBF [53].

The cyclic voltammograms of the four alloys shown in Figure 11b are all similar in patterning. For all curves, the linear current–voltage dependence starts at an overpotential of −1000 mV. It can be observed that the reverse branch (cathodic curve) overlaps almost perfectly with the direct branch (anodic curve). Therefore, the generalised corrosion caused by the overpotential does not significantly alter the active surface nor causes passivation through the deposition of reaction products. It may be explained by the formation of OH^{-} ions, which leads to an increase in the pH; consequently, the formation of zinc hydroxide chloride on the surface alloys confirms the results shown in Figure 8.

5. Conclusions

In this study, titanium was used as the alloying element to design the Zn–xTi (x = 0.1, 0.25 and 1.0 wt.%) series of alloys. Their microstructure, mechanical properties and corrosion behaviour have been systematically analysed. The main conclusions are as follows:

- For contents higher than 0.25 wt.% Ti, the formation of a primary TiZn16 intermetallic compound in the (α-Zn + $TiZn_{16}$) eutectic matrix is observed both by X-ray diffraction and by structural and chemical analyses. Alloying with Ti leads to refinement of α-Zn dendrites for hypoeutectic alloys and refinement of eutectic grains for hypereutectic alloys. These changes in microstructural features led to an improvement in the mechanical properties of Zn–xTi alloys.
- After the interaction between the alloys and the saline solution, the pH increased after the first hours (10–15 h) and then remained stable between 7.5 and 8, but different values of the degradation rate were observed, decreasing for Zn–0.00Ti (as-cast cp-Zn) and remaining almost constant for Zn–1.00Ti.
- Electrochemical tests performed following immersion, i.e., at a pH of approximately 5.5–6.0, showed that the degradation rate decreased as the Ti content increased.

Author Contributions: Conceptualization, R.C. (Romeu Chelariu), N.C. and A.-T.Ş.; methodology, R.C. (Ramona Cimpoeşu), B.I. and A.-M.R.; software, M.M.; validation, R.C. (Romeu Chelariu), G.Z. and A.A.; formal analysis, V.G., B.I. and M.B.; investigation, R.C. (Ramona Cimpoeşu), N.C., A.-M.R., M.M. and M.B.; resources, A.-T.Ş.; data curation, V.G.; writing—original draft preparation, R.C. (Romeu Chelariu), N.C. and A.-T.Ş.; writing—review and editing, M.M., B.I. and V.G.; visualization, R.C. (Ramona Cimpoeşu) and B.I.; supervision, G.Z., N.C. and M.M.; project administration, N.C., M.B. and A.A. All authors have read and agreed to the published version of the manuscript.

Funding: This work was supported by a National Research Grants of the TUIASI, project number GNaC2023_285/2024.

Data Availability Statement: The original contributions presented in the study are included in the article, further inquiries can be directed to the corresponding author.

Conflicts of Interest: The authors declare no conflict of interest.

References

1. Hermawan, H. Updates on the research and development of absorbable metals for biomedical applications. *Prog. Biomater.* **2018**, *7*, 93–110. [CrossRef] [PubMed]
2. Li, C.; Guo, C.; Fitzpatrick, V.; Ibrahim, A.; Zwierstra, M.J.; Hanna, P.; Lechtig, A.; Nazarian, A.; Lin, S.J.; Kaplan, D.L. Design of biodegradable, implantable devices towards clinical translation. *Nat. Rev. Mater.* **2020**, *5*, 61–81. [CrossRef]
3. Khan, A.R.; Grewal, N.S.; Zhou, C.; Yuan, K.; Zhang, H.J.; Jun, Z. Recent advances in biodegradable metals for implant applications: Exploring in vivo and in vitro responses. *Results Eng.* **2023**, *20*, 101526. [CrossRef]
4. Yang, H.; Lin, W.; Zheng, Y. Advances and perspective on the translational medicine of biodegradable metals. *Biomater. Transl.* **2021**, *2*, 177–187. [PubMed]
5. Wanga, J.; Doua, J.; Wanga, Z.; Hua, C.; Yub, H.; Chen, C. Research progress of biodegradable magnesium-based biomedical materials: A review. *J. Alloys Compd.* **2022**, *923*, 166377. [CrossRef]
6. Rabeeh, V.P.M.; Hanas, T. Progress in manufacturing and processing of degradable Fe-based implants: A review. *Prog. Biomater.* **2022**, *11*, 163–191. [CrossRef]
7. Nriagu, J. Zn toxicity in humans. In *Encyclopedia of Environmental Health*; Elsevier: Amsterdam, The Netherlands, 2011; pp. 801–807.
8. Plum, L.M.; Rink, L.; Haase, H. The Essential Toxin: Impact of Zinc on Human Health. *Int. J. Environ. Res. Public Health* **2010**, *7*, 1342–1365. [CrossRef] [PubMed]
9. Sangeetha, V.J.; Dutta, S.; Moses, J.A.; Anandharamakrishnan, C. Zinc nutrition and human health: Overview and implications. *eFood* **2022**, *3*, e17. [CrossRef]
10. Chasapis, C.T.; Loutsidou, A.C.; Spiliopoulou, C.A.; Stefanidou, M.E. Zinc and human health: An update. *Arch. Toxicol.* **2012**, *86*, 521–534. [CrossRef]
11. Jomova, K.; Makova, M.; Alomar, S.Y.; Alwasel, S.H.; Nepovimova, E.; Kuca, K.; Rhodes, C.J.; Valko, M. Essential metals in health and disease. *Chem. Biol. Interact.* **2022**, *367*, 110173. [CrossRef]
12. Kim, T.; Wang See, C.; Li, X.; Zhu, D. Orthopedic implants and devices for bone fractures and defects: Past, present and perspective. *Eng. Regen.* **2020**, *1*, 6–18. [CrossRef]

13. Chen, K.; Gu, X.; Zheng, Y. Feasibility, challenges and future prospects of biodegradable zinc alloys as orthopedic internal fixation implants. *Smart Mater. Manuf.* **2024**, *2*, 100042. [CrossRef]
14. Wang, X.; Lu, H.; Li, X.; Li, L.; Zheng, Y. Effect of cooling rate and composition on microstructures and properties of Zn-Mg alloys. *Trans. Nonferrous Met. Soc. China* **2007**, *17*, s122–s125.
15. Vojtech, D.; Kubásek, J.; Šerák, J.; Novák, P. Mechanical and corrosion properties of newly developed biodegradable Zn-based alloys for bone fixation. *Acta Biomater.* **2011**, *7*, 3515–3522. [CrossRef]
16. Zhuo, X.; Wu, Y.; Ju, J.; Liu, H.; Jiang, J.; Hu, Z.; Bai, J.; Xue, F. Recent progress of novel biodegradable zinc alloys: From the perspective of strengthening and toughening. *J. Mater. Res. Technol.* **2022**, *17*, 244–269. [CrossRef]
17. Liu, Y.; Du, T.; Qiao, A.; Mu, Y.; Yang, H. Zinc-Based Biodegradable Materials for Orthopaedic Internal Fixation. *J. Funct. Biomater.* **2022**, *13*, 164. [CrossRef]
18. Kong, L.; Heydari, Z.; Lami, G.H.; Saberi, A.; Baltatu, M.S.; Vizureanu, P. A Comprehensive Review of the Current Research Status of Biodegradable Zinc Alloys and Composites for Biomedical Applications. *Materials* **2023**, *16*, 4797. [CrossRef]
19. Hussain, M.; Ullah, S.; Raza, M.R.; Abbas, N.; Ali, A. Recent Developments in Zn-Based Biodegradable Materials for Biomedical Applications. *J. Funct. Biomater.* **2023**, *14*, 1. [CrossRef]
20. Shishir, R.; Nasiruddin, U.; Manojkumar, P.; Lokeshkumar, E.; Ponnilavan, V.; Parfenov, E.V.; Valiev, R.Z.; Rameshbabu, N. Zinc as a potential bio-degradable temporary implant material. *Mater. Today Proc.* 2023, *in press*. [CrossRef]
21. Yang, H.; Jia, B.; Zhang, Z.; Qu, X.; Li, G.; Lin, W.; Zhu, D.; Dai, K.; Zheng, Y. Alloying design of biodegradable zinc as promising bone implants for load-bearing applications. *Nat. Commun.* **2020**, *11*, 401. [CrossRef] [PubMed]
22. Baker, H. Alloy Phase Diagrams. In *ASM Handbook*; ASM International: Materials Park, OH, USA, 1992; Volume 3.
23. Yin, Z.Y. Microstructural Evolution and Mechanical Properties of Zn–Ti Alloys for Biodegradable Stent Applications. Master's Thesis, Michigan Technological University, Houghton, MI, USA, 2017.
24. Wang, K.; Tong, X.; Lin, J.; Wei, A.; Li, Y.; Dargusch, M.; Wen, C. Binary Zn–Ti alloys for orthopedic applications: Corrosion and degradation behaviors, friction and wear performance, and cytotoxicity. *J. Mater. Sci. Technol.* **2021**, *74*, 216–229. [CrossRef]
25. *ASTM G31-21*; Standard Guide for Laboratory Immersion Corrosion Testing of Metals. ASTM International: West Conshohocken, PA, USA, 2021.
26. Bernevig-Sava, M.A.; Darabont, D.C.; Lohan, M.; Mihalache, E.; Bejinariu, C. Selection and verification of personal protective equipment in the context of current legal requirements. *Qual.-Access Success* **2019**, *20*, 109.
27. Spittle, J.A. The Effects of Composition and Cooling Rate on the as-Cast Microstructures of Zn-Ti Alloys. *Metallography* **1972**, *5*, 423–447. [CrossRef]
28. Goto, S.; Esashi, K.; Koda, S.; Morozumi, S. Structure-Controlling of Zn-Ti Hyper-Eutectic Alloys by Unidirectionally Solidification. *J. Jpn. I Met. Mat.* **1973**, *37*, 466–473.
29. Spittle, J.A. Lamellar and Rod Morphologies in the Zn-TiZn15 Eutectic. *Metallography* **1973**, *6*, 115–121. [CrossRef]
30. Leone, G.L.; Kerr, H.W. Grain structures and coupled growth in Zn-Ti alloys. *J. Cryst. Growth* **1976**, *32*, 111–116. [CrossRef]
31. Murray, J.L. The Ti-Zn (Titanium-Zinc) System. *Bull. Alloy Phase Diagr.* **1984**, *5*, 52–56. [CrossRef]
32. Vassilev, G.P.; Liu, X.J.; Ishida, K. Reaction kinetics and phase diagram studies in the Ti–Zn system. *J. Alloys Compd.* **2004**, *375*, 162–170. [CrossRef]
33. Vassilev, G.P. Contribution to the zinc-rich side of the Ti–Zn system. *Z. Metallkd.* **2004**, *95*, 813–817. [CrossRef]
34. Okamoto, H. Ti-Zn (Titanium-Zinc). *J. Phase Equilib. Diff.* **2008**, *29*, 211–212. [CrossRef]
35. Yang, S.; Su, X.; Wang, J.; Yin, F.; Li, Z.; Tu, H.; Peng, H. The Zn-rich corner of the Zn–Al–Ti system at 723 K. *J. Alloys Compd.* **2010**, *499*, 194–199. [CrossRef]
36. Chen, X.; Jeitschko, W.; Danebrock, M.E.; Evers, B.H.; Wagner, K. Preparation, Properties, and Crystal Structures of Ti_3Zn_{22} and $TiZn_{16}$. *J. Solid State Chem.* **1995**, *118*, 219–226. [CrossRef]
37. Boczkal, G.; Mikułowski, B.; Wolczynski, W. Oscillatory structure of the Zn-Cu-Ti single crystals. *Mater. Sci. Forum.* **2010**, *649*, 113–118. [CrossRef]
38. Boczkal, G. Second phase morphology in the Zn-$Ti_{0.1}$-$Cu_{0.1}$ single crystals obtained at different growth rates. *Arch. Metall. Mater.* **2012**, *57*, 479–484. [CrossRef]
39. Boczkal, G. Structure and properties of Zn-$Ti_{0.2}$-$Cu_{0.15}$ single crystal containing eutectic precipitates. *Arch. Metall. Mater.* **2013**, *58*, 1019–1022. [CrossRef]
40. Jette, E.R.; Foote, F. Precision determination of lattice constants. *J. Chem. Phys.* **1935**, *3*, 605–616. [CrossRef]
41. Farrar, R.A.; King, H.W. Axial ratios and solubility limits of H.C.P. eta and epsilon phases in the systems Cd–Mn and Zn–Mn. *Metallography* **1970**, *3*, 61–70. [CrossRef]
42. Saillard, M.; Develey, G.; Becle, C. The Structure of $TiZn_{16}$. *Acta Cryst. B* **1981**, *37*, 224–226. [CrossRef]
43. *ASTM E9-09*; Standard Test Methods of Compression Testing of Metallic Materials at Room Temperature. ASTM International: West Conshohocken, PA, USA, 2018.
44. Mostaed, E.; Sikora-Jasinska, M.; Mostaed, A.; Loffredo, S.; Demir, A.G.; Previtali, B.; Mantovani, D.; Beanland, R.; Vedani, M. Novel Zn-based alloys for biodegradable stent applications: Design, development and in vitro degradation. *J. Mech. Behav. Biomed. Mater.* **2016**, *60*, 581–602. [CrossRef] [PubMed]

45. Sikora-Jasinska, M.; Mostaed, E.; Mostaed, A.; Beanland, R.; Mantovani, D.; Vedani, M. Fabrication, mechanical properties and in vitro degradation behavior of newly developed Zn Ag alloys for degradable implant applications. *Mater. Sci. Eng. C* **2017**, *77*, 1170–1181. [CrossRef]
46. Zhang, S.; Yuan, P.; Wang, X.; Wang, T.; Zhao, L.; Cui, C. Fabrication and Properties of Zn-3Mg-1Ti Alloy as a Potential Biodegradable Implant Material. *Materials* **2022**, *15*, 940. [CrossRef]
47. Manasijević, D.; Balanović, L.; Marković, I.; Minić, D.; Premović, M.; Đorđević, A.; Gorgievski, M.; Stamenković, U. Microstructure and thermal properties of the Bi–Ag alloys. *J. Therm. Anal. Calorim.* **2022**, *147*, 1965–1972. [CrossRef]
48. Anderson, E.A.; Boyle, E.J.; Ramsey, P.W. Rolled Zinc-titanium Alloys. *Trans. Am. Inst. Min. Metall. Eng.* **1944**, *156*, 278–287.
49. Guo, X.; He, G. Opportunities and challenges of zinc anodes in rechargeable aqueous batteries. *J. Mater. Chem. A* **2023**, *11*, 11987–12001. [CrossRef]
50. Liu, X.; Yang, H.; Xiong, P.; Li, W.; Huang, H.H.; Zheng, Y. Comparative studies of Tris-HCl, HEPES and $NaHCO_3/CO_2$ buffer systems on the biodegradation behaviour of pure Zn in NaCl and SBF solutions. *Corros. Sci.* **2019**, *157*, 205–219. [CrossRef]
51. Panaghie, C.; Zegan, G.; Sodor, A.; Cimpoeșu, N.; Lohan, N.M.; Istrate, B.; Roman, A.M.; Ioanid, N. Analysis of Degradation Products of Biodegradable ZnMgY Alloy. *Materials* **2023**, *16*, 3092. [CrossRef]
52. Panaghie, C.; Cimpoeșu, R.; Istrate, B.; Cimpoeșu, N.; Bernevig, M.A.; Zegan, G.; Roman, A.M.; Chelariu, R.; Sodor, A. New Zn3Mg-xy alloys: Characteristics, microstructural evolution and corrosion behavior. *Materials* **2021**, *14*, 2505. [CrossRef] [PubMed]
53. Gerhátová, Ž.; Babincová, P.; Drienovský, M.; Pašák, M.; Cernicková, I.; Duriška, L.; Havlík, R.; Palcut, M. Microstructure and Corrosion Behavior of Sn–Zn Alloys. *Materials* **2022**, *15*, 7210. [CrossRef] [PubMed]

metals MDPI

Article

A First-Time Investigation into Ecofriendly and Biocompatible Mg-Se Binary System for a Greener Earth

Michael Johanes, Vasuudhaa Sonawane and Manoj Gupta *

Department of Mechanical Engineering, National University of Singapore, 9 Engineering Drive 1, Singapore 117575, Singapore
* Correspondence: mpegm@nus.edu.sg

Abstract: In this study, the Mg-15Se binary system was, for the first time, investigated and synthesized using the powder metallurgy (PM) method, including microwave sintering and hot extrusion. The resulting material was shown to possess visible pores with a porosity of 2.91%, higher than other Mg materials synthesized using this method in the literature. Despite this, the material not only exhibited a comparable corrosion response with pure Mg but also a significantly superior mechanical response (76% greater damping capacity, 57% increase in hardness, and increases of 21%, 50%, and 51% for compressive yield strength, ultimate compressive strength, and fracture strain, respectively). Thus, this not only opens the door for future work concerning the addition of medicinal Se to nutritional Mg element and the optimization of process parameters but also could potentially be making inroads into the biomedical field with the use of selenium as a biomedical-oriented alloying element.

Keywords: magnesium; selenium; mechanical properties; corrosion resistance; lightweight materials; powder metallurgy

1. Introduction

Magnesium as a material has immense potential within the modern context; it is not only very abundant within the Earth's crust but also one of the lightest structural metals in existence [1]. This, along with world Mg primary production exceeding one million tons [2], presents a compelling case for the development of Mg materials to not only leverage its present quantities but also its qualities, such as low density for use in weight-saving engineering applications and biocompatibility for temporary implant applications.

In the past, studies on the property enhancement of Mg materials were conducted utilizing micro- and nano-sized reinforcements and alloying elements [3–7]. These studies have shown that from the viewpoint of Mg as a structural material in engineering applications, the addition of alloying elements and/or reinforcements provided significant improvements in the hardness, load bearing, and deformability of Mg materials and thus their viability as engineering materials.

Another potential application of Mg materials is in the medical sector, specifically as implants. Contemporary applications of metals in the biomedical field take the form of fasteners and supports such as plates, screws, pins, etc. [8], with materials such as steel and titanium having widespread use [9]. In this field, a different set of challenges presents itself, ranging from the need to match the property of human bone as closely as possible to avoid stress-shielding [10], minimizing the typically high corrosion rates in the human body environment [11], and overcoming concerns regarding toxicity or lack of biological incompatibility of the very materials used as implants or fasteners in the case of steels (via their alloying elements, e.g., nickel and chromium) [12]. As a metal which is not only biocompatible with, but also potentially bioresorbable within, the human body [13,14], magnesium is potentially well suited for such applications.

Selenium is a metalloid with several key functions within the human body as a biocompatible and essential trace metal, ranging from probing biological functions (biosensing) and catalytic functions involving enzyme production to the regulation of hormone metabolism and antioxidants [15,16]. Furthermore, there is also potential for its use in the treatment of cancer (specifically clear cell renal cell carcinoma, ccRCC), provided that optimal dosage and sequence of administration are applied [17]. In addition, its anticancer properties are also derived from multiple mechanisms such as countering heavy metal toxicity, antioxidative properties, as well as the ability to inhibit the angiogenesis, migration, and invasion of certain types of cancer cells [18,19]. Furthermore, the co-supplementation of Mg and low-dose Se has shown promise in reducing liver total cholesterol levels in animal trials [20].

As an additive to biomedical materials, selenium also poses several challenges, including toxicity when ingested above a certain threshold [21]. Despite these challenges, the importance of selenium within the human body makes it an attractive frontier for the development of biomedical-oriented materials.

With these in mind, this work sets out to develop and investigate the prospect of selenium as an alloying element using magnesium as the base metal in light of the positive properties of both elements. As of writing, this study represents the first work of any kind concerning the addition of selenium as an alloying element in magnesium, thus providing an important basis on which further works and studies may be designed and conducted.

2. Materials and Methods

2.1. Materials

Table 1 outlines the raw materials used in this work. Figure 1 shows micrographs of the raw Mg (Merck Group, Darmstadt, Germany) and Se powders (Alfa Aesar GmbH & Co KG, Haverhill, MA, USA) used, as well as that of the blended Mg-15Se powder, showing their different sizes and morphologies.

Table 1. Raw materials used in this work.

Raw Material	Supplier	Purity	Density (g/cc)
Magnesium Powder, 60–300 μm	Merck	≥98.5%	1.738
Selenium Powder, 200 mesh (~74 μm)	Alfa Aesar	99.999%	4.28

The Mg and Se powders appear to be thoroughly mixed, with the larger Mg powder being more prominent in the foreground against the smaller Se particles in the background. The two powders were analyzed for size distribution. The results are presented in Table 2 and Figure 2.

Table 2. Experimental average sizes of Mg and Se powders used in this work.

Raw Material	Average Powder Diameter (μm)
Mg powder	254 ± 63
Se powder	28 ± 11

It can be seen that while the Mg powder (with its average diameter of 254 μm) conformed to the size indicated by the supplier, the Se powder was actually found to be finer in size (28 μm average diameter) compared to the information supplied (74 μm).

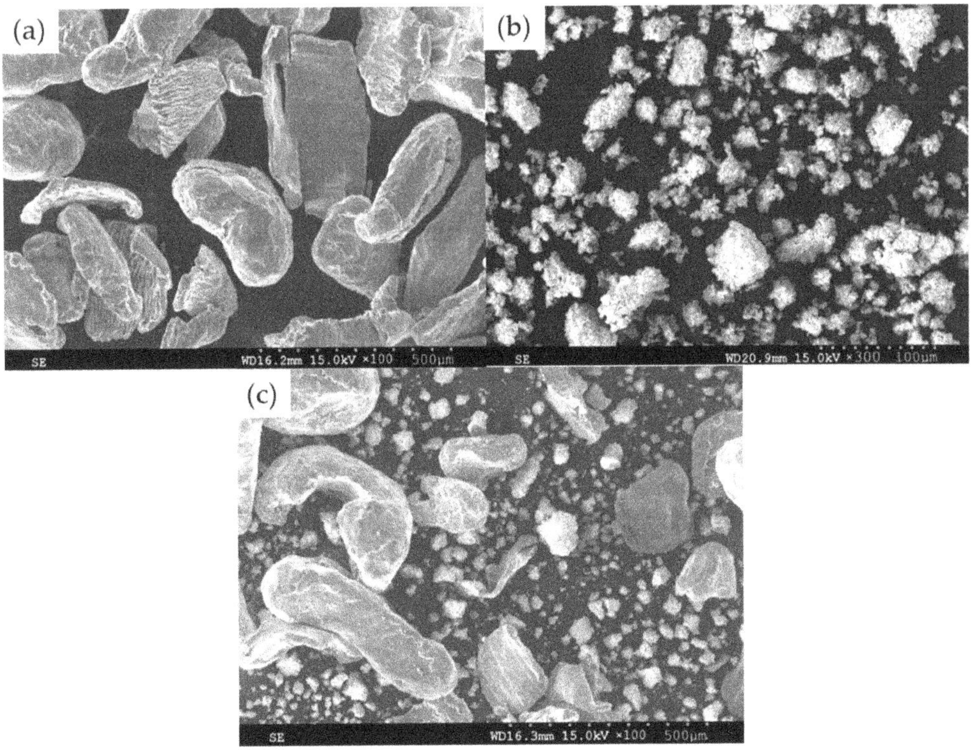

Figure 1. Micrographs of powder materials used in this work: (**a**) Mg, (**b**) Se, and (**c**) Mg-15Se after blending.

2.2. Synthesis

Mg-15Se (85 wt.% Mg and 15 wt.% Se) was synthesized using powder metallurgy by mixing the two powders in a sealed vessel using an Inversina 2 L tumbler mixer (Bioengineering AG, Wald, Switzerland) at a mixing speed of 50 rpm for a duration of 1 h. The resulting powder mixture was then compacted in a hydraulic press to a pressure of 1000 psi (6.89 MPa) with a 60 s holding time to generate a billet of 35.5 mm diameter and 45 mm height.

The billet was then sprayed with a layer of colloidal graphite and subjected to microwave sintering using a Sharp R898C(S) microwave oven rated at 900 W to a temperature of 200 °C, following which it was heated at 400 °C for 60 min and then subjected to direct hot extrusion through an 8 mm diameter die at 350 °C (corresponding to an extrusion ratio of 20.25).

2.3. Materials Characterisation

2.3.1. Microstructure

Flat and parallel samples were fine-finished using 4000 grit sandpaper, followed by polishing using alumina suspension to 0.05-micron size with Deionized (DI) water as the polishing lubricant. A Hitachi S-4300 Field Emission Scanning Electron Microscope (FESEM, Hitachi, Ltd., Tokyo, Japan) was used to obtain scanning electron micrographs, and the accompanying energy dispersive X-ray (EDX) analysis capabilities also provided material composition analysis results.

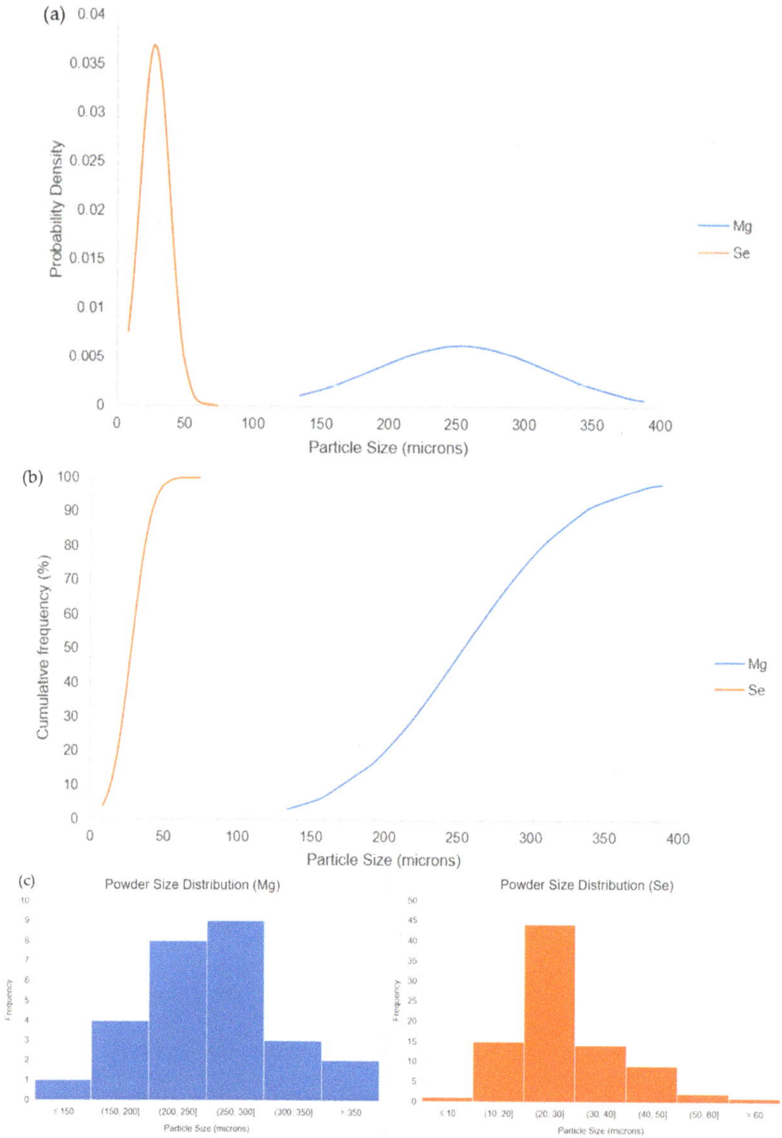

Figure 2. Distribution of powder sizes of Mg and Se powders used in this work in (**a**) normal distribution form, (**b**) cumulative frequency form, and (**c**) histogram form.

Grain size characterization was conducted by etching polished samples using a solution of 4.2 g citric acid, 10 mL ethylene glycol, and 100 mL DI water for a duration of 34 s. Grain images were taken using a Leica DM2500 optical microscope (Leica Microsystems (SEA) Pte Ltd., Singapore), and MATLAB (version R2013b, Natick, MA, USA) was used to conduct image analysis and grain quantification in accordance with standard ASTM E112-13(2021).

2.3.2. X-ray Diffraction

X-ray diffraction was conducted using a Shimadzu XRD-6000 X-ray diffractometer (Shimadzu Corporation, Kyoto, Japan) with Cu-Kα radiation (λ = 1.54056 Å) and scanning parameters of speed, 2° per minute, and scanning angle 2θ in the range of 10° to 80°.

2.3.3. Density and Porosity

The density of the compacted Mg-15Se billet (both green and sintered) was calculated assuming a perfect cylinder and by mass measurement using a GH-252 electronic scale (AND Company, Limited, Tokyo, Japan).

The experimental density of the extruded Mg-15Se material was determined by the Archimedes principle, using an AD-1653 Density Determination Kit (AND Company, Limited, Tokyo, Japan) mounted on a GH-252 electronic scale (AND Company, Limited, Tokyo, Japan). A minimum of 5 samples' readings were taken.

In addition, a Hitachi S-4300 Field Emission Scanning Electron Microscope (FESEM, Hitachi, Ltd., Tokyo, Japan) was utilized to observe the presence of pores on the surface. To calculate the experimental porosity of the materials, a MATLAB program (Version R2013b, The Mathworks Inc., Natick, MA, USA) was used to measure the area fraction of the pores by dividing the total area of the detected pores by the total area of the representative sample images.

2.3.4. Damping Analysis

Rods of 50 mm length were cut and subjected to impulse excitation in conjunction with Response Frequency Damping Analyzer (RFDA, version 8.1.2) software (IMCE, Genk, Belgium). The resulting recorded vibration signals from the rod were then analyzed to obtain the damping properties (attenuation coefficient, damping capacity, and elastic modulus) of the material.

2.3.5. Mechanical Properties

A Shimadzu HMV-2 hardness tester (Shimadzu Corporation, Kyoto, Japan) was used to perform microhardness characterization in accordance with the procedures outlined in ASTM E-384 using a diamond indenter with phase angle of 136° with a load of 25 g force at a dwell time of 15 s. In total, 15 hardness measurements were taken across 1 sample surface.

Flat and parallel samples with an L/D ratio of 1 (8 mm diameter and height) were subjected to compressive load testing to failure using an MTS E-44 compressive testing machine (MTS Systems, Eden Prairie, MN, USA) with a strain rate of 0.5%/min (8.3×10^{-5} s^{-1}) in accordance with the ASTM standard E9-09. A minimum of 3 representative samples were compressed to failure to obtain the compressive properties.

Compressive fractography was also conducted using an FESEM to observe and investigate the fracture surface.

2.3.6. Thermal Properties

Using a Shimadzu DTG-60H Thermogravimetric Analyzer (Shimadzu Corporation, Kyoto, Japan), samples of approximately 2 mm × 2 mm × 2 mm size were subjected to an environment of 30 °C to 1400 °C at a rate of 10 °C per minute in purified air (50 mL/min flow rate) to determine the material's thermal response. The temperature reading immediately prior to a sudden spike or increase in temperature and followed by recovery, corresponding to material ignition, was taken as the ignition temperature.

Samples of the same dimensions were also subjected to an environment of 30 °C to 600 °C at a rate of 5 °C per minute in argon gas of a 25 mL/min flow rate using a Shimadzu DSC-60 Digital Scanning Calorimeter (Shimadzu Corporation, Kyoto, Japan) to measure the thermal flow of the sample with increasing sample temperature.

The coefficient of thermal expansion (CTE) was investigated on samples of 8 mm diameter and 5 mm length using a TMA PT1000 thermo-mechanical analyzer (Linseis

Messgeraete GmbH, Selb, Germany) at 50 °C to 400 °C at a 5 °C/min ramp rate in argon gas of a 0.1 L/min flow rate.

2.3.7. Corrosion Response

Samples of approximately 1.5 mm thickness were subjected to corrosion testing by immersion in PBS (Phosphate-Buffered Saline, Thermo Fisher Scientific Inc., Waltham, MA, USA) solution at a temperature of 37 °C for a duration of 28 days or until sample disintegration, whichever was earlier. Weight loss data were obtained in 24 h intervals by the removal of corrosion products, which was achieved by immersing the corroded samples in a solution consisting of 1.5 g $AgNO_3$ and 15 g of CrO_3 in 100 mL of DI water. This was followed by cleaning with DI water, and the samples were then weighed. The following formula was used to calculate the corrosion rate using the weight loss method [22]:

$$\text{Corrosion rate} = \frac{87.6 \times \text{Weight loss (mg)}}{\text{Experimental Density } (g/cm^3) \times \text{Surface Area } (cm^2) \times \text{Immersion Time (hours)}}$$

3. Results

3.1. Synthesis

The material was synthesized successfully with a solid, continuous extrudate generated with minimal surface-level cracks. However, it was observed that the sintering setup as well as the furnace used for billet soaking immediately prior to extrusion were stained red, which implies the loss of selenium content despite the presence of a colloidal graphite layer applied to the billet prior to both sintering and hot extrusion. This is likely due to the fact that the sintering temperature and extrusion temperature conditions were near and above the melting point of selenium (220 °C), respectively.

Furthermore, it was found that the resulting extruded rod has visible pores, as seen in Figure 3.

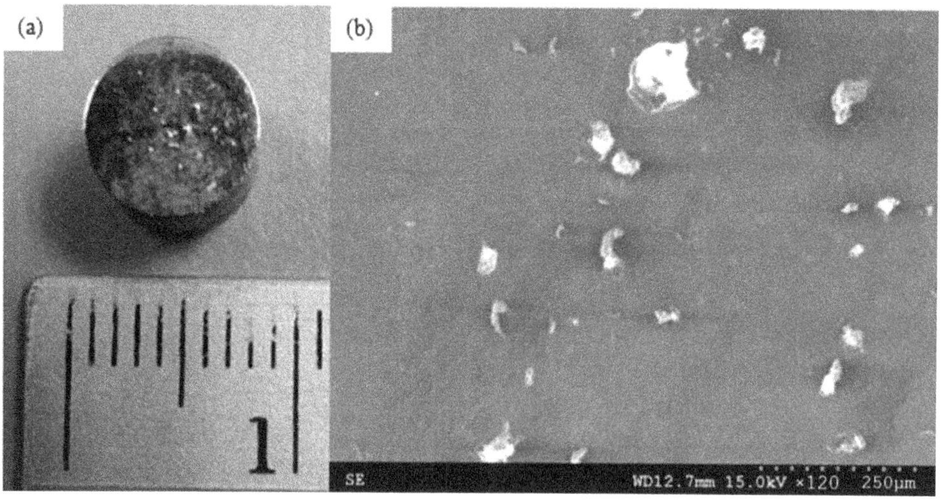

Figure 3. Cross-section of Mg-15Se showing visible pores on (**a**) macro-scale and (**b**) under SEM.

3.2. Microstructure

Figure 4 shows the microstructure of Mg-15Se, showing the presence of further small pores as well as swirls of white/bright particles against the Mg matrix.

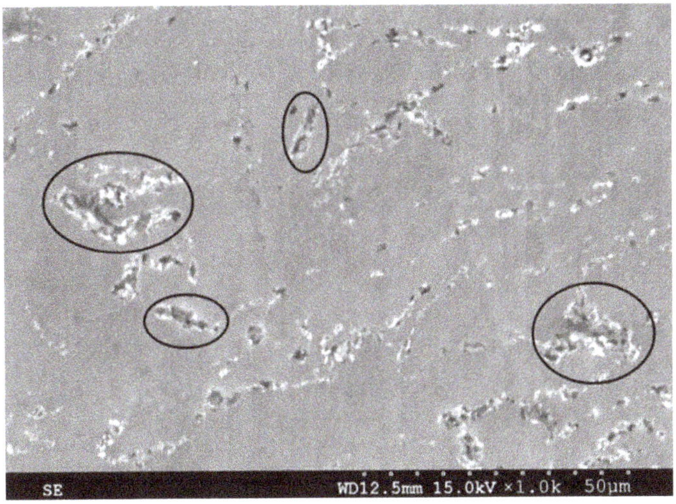

Figure 4. Scanning electron micrograph of Mg-15Se, with some pores circled in black.

Figure 5 shows the mapping results of Mg-15Se and selected regions for EDS analysis within the microstructure, while Table 3 shows EDS results indicating the detected elements corresponding to the various selected spectrum locations, with Se-containing regions being primarily constrained to areas with bright particles or pores, while the matrix area is predominantly Mg with some Se content distributed within.

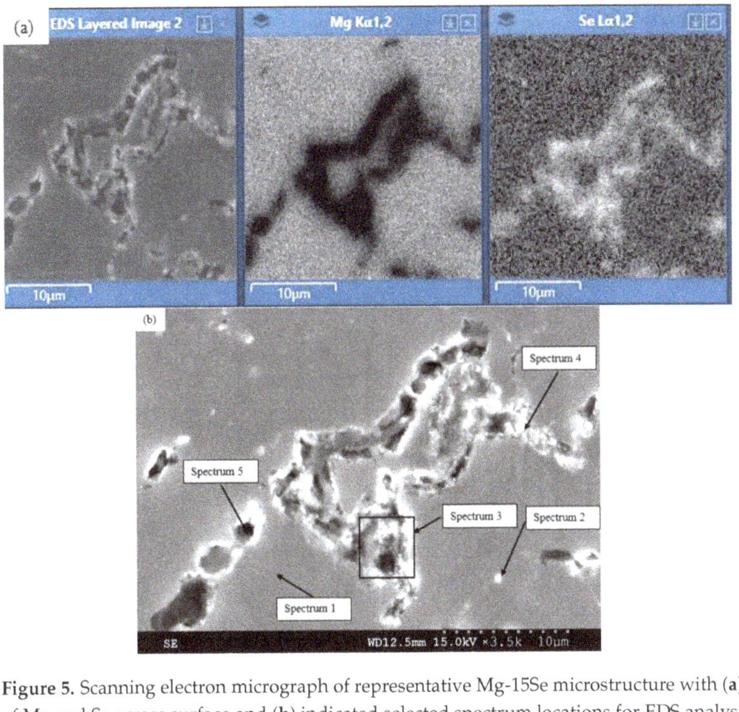

Figure 5. Scanning electron micrograph of representative Mg-15Se microstructure with (**a**) mapping of Mg and Se across surface and (**b**) indicated selected spectrum locations for EDS analysis.

Table 3. Tabulated EDS results of selected Mg-15Se spectrum regions.

Spectrum	Detected Element (wt. %)	
	Mg	Se
1	99.4	0.6
2	78.9	21.1
3	53.1	46.9
4	69.1	30.9
5	79.4	20.6

The grain diameter of Mg-15Se was observed to be much lower than pure Mg, as seen in Table 4. A representative micrograph of the grains can be seen in Figure 6.

Table 4. Grain size characterization results of Mg-15Se.

Material	Average Grain Diameter (μm)
Pure Mg [23]	34 ± 2
Mg-15Se	9 ± 3 (↓74%)

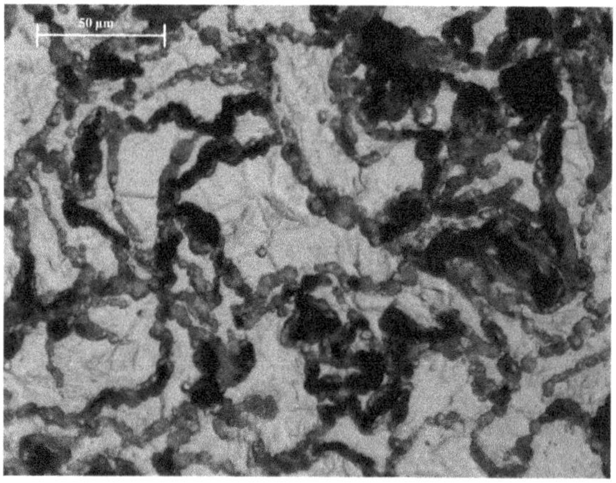

Figure 6. Optical of Mg-15Se material studied in this work showing the grain morphology.

3.3. *X-ray Diffraction*

The X-ray diffractogram of Mg-15Se studied in this work is shown in Figure 7. The diffractogram displayed several peaks belonging to Mg, Se, MgO, and MgSe. This was checked against the Powder Diffraction File (PDF-4+ 2023) [24,25] with card numbers 00-004-0770 (Mg), 00-051-1389 (Se), 00-004-0829 (MgO), and 01-073-6986 (MgSe). The result implies that during the synthesis, MgSe was formed.

The intensities of the Mg-15Se peaks at 2θ = 32°, 34°, and 36° (corresponding to the 10-10 prism, 0002 basal, and 10-11 pyramidal planes of magnesium, respectively) are compared with that of pure Mg, as seen in Table 5. Mg-15Se exhibited weaker prism and pyramidal texture. Mg-15Se shows a maximum peak intensity at 34°, exhibiting strong basal texture perpendicular to the extrusion direction.

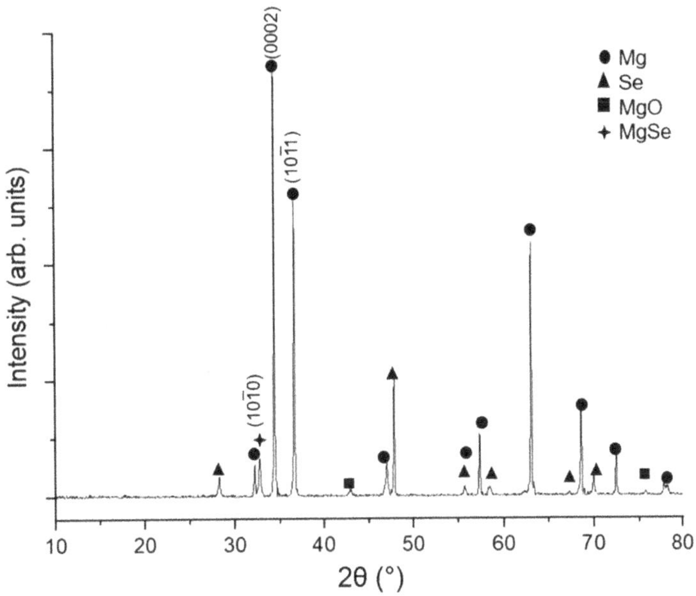

Figure 7. X-ray diffractogram of Mg-15Se material studied in this work.

Table 5. X-ray diffraction results of Mg-15Se with pure Mg as a reference.

Material	Plane	I/I_{max}
Pure Mg [26]	10-10 Prism	0.147
	0002 Basal	1
	10-11 Pyramidal	0.794
Mg-15Se	10-10 Prism	0.074
	0002 Basal	1
	10-11 Pyramidal	0.692

3.4. Density and Porosity

The green and sintered densities of the Mg-15Se material have been outlined in Table 6, showing the very slight decrease in density just prior to extrusion.

Table 6. Green and sintered densities of Mg-15Se compacted billets.

Material	Theoretical Density (g/cm^3)	Experimental Density (g/cm^3)
Mg-15Se, green compact	1.9079	1.6962
Mg-15Se, microwave sintered compact	1.9079	1.6956

The pore area fraction of the extruded material was found to be 2.91% using image analysis. Thus, the theoretical density calculated using the rule of mixtures would have to be adjusted to account for this reduction in material across a given volume. Table 7 outlines this process:

Table 7. Adjustment of Mg-15Se theoretical density accounting for observed porosity within the material.

Material	Initial Theoretical Density (g/cm^3)	Adjusted Theoretical Density (g/cm^3)	Porosity (%)
Mg-15Se	1.908	1.854	2.91 *

* Porosity calculated using image analysis.

Based on the results shown in Table 7, which account for the pore area fraction of the Mg-15Se material, it is then possible to calculate the retained Se content within the material using the rule of mixtures, which was found to be approximately 12.5%. This is outlined in Table 8.

Table 8. Theoretical and measured experimental density of Mg-15Se in this work.

Material	Experimental Density (g/cm^3)	Retained Se Content (%)	Porosity (%)
Pure Mg [22]	1.736	-	0.21
Mg-15Se	1.823	12.5	2.91 *

* Porosity calculated using image analysis.

3.5. Damping Analysis

The vibration response curve plotted as part of the damping analysis in Figure 8 was prescribed an exponential best-fit curve with a general form of Ae^{-bt}, where b is the derived attenuation coefficient. The results indicated that Mg-15Se exhibited a superior damping capacity, as outlined in Table 9, with only a minor reduction in Young's modulus compared to pure Mg.

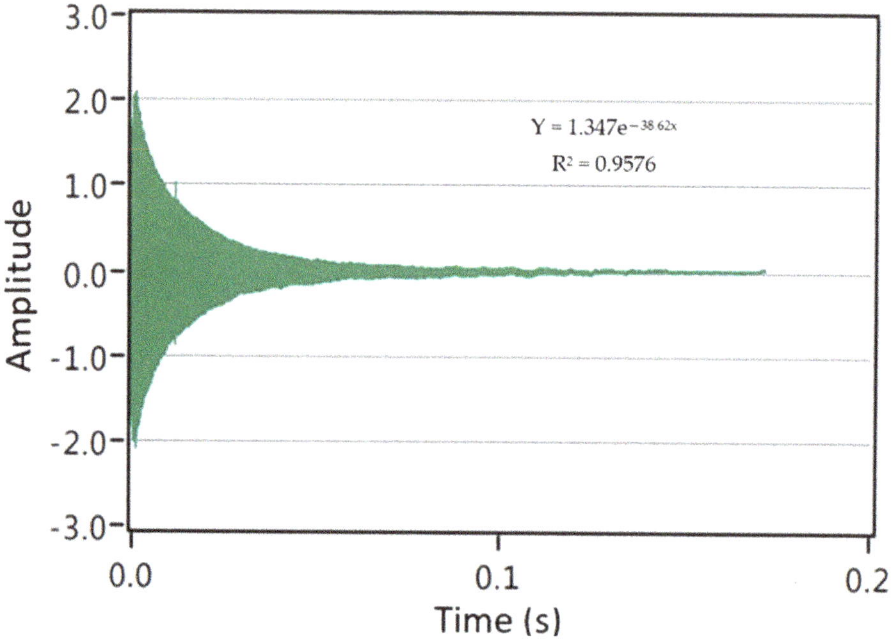

Figure 8. Vibration response curve of Mg-15Se with best-fit curve equation.

Table 9. Damping properties and Young's modulus of samples.

Material	Attenuation Coefficient	Damping Capacity	Young's Modulus (GPa)
Pure Mg [23]	-	0.000656	44.7
Mg-15Se	38.62	0.001155 (↑76%)	42.9 (↓4%)

3.6. Mechanical Properties

3.6.1. Hardness

The microhardness values of Mg-15Se studied in this work are presented in Table 10, showing an increase in hardness relative to pure Mg.

Table 10. Average microhardness values of Mg-15Se compared to pure Mg.

Material	Average Microhardness (HV)
Pure Mg [23]	46 ± 3
Mg-15Se	72 ± 5 (↑57%)

3.6.2. Compressive Properties

The compressive test results of Mg-15Se are compiled in Table 11, with the stress–strain curve outlined in Figure 9. Significant increases in yield strength, ultimate compressive strength, fracture strain, and work of fracture were all observed with respect to the Mg-15Se material over pure Mg.

Table 11. Compressive properties of Mg-15Se studied in this work with Pure Mg as a reference.

Material	0.2% Compressive Yield Strength (MPa)	Ultimate Compressive Strength (MPa)	Fracture Strain (%)	Energy Absorbed (MJ/m^3)
Pure Mg [23]	72 ± 5	174 ± 7	16 ± 2	23 ± 2
Mg-15Se	87 ± 3 (↑21%)	263 ± 12 (↑51%)	24 ± 2 (↑50%)	39 ± 6 (↑70%)

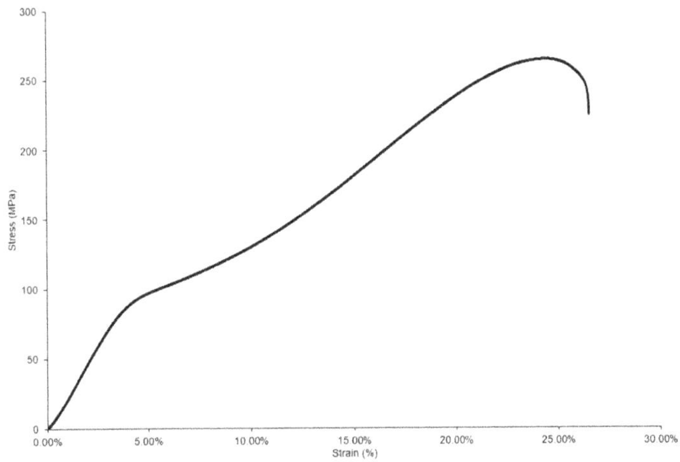

Figure 9. Representative stress–strain curve of Mg-15Se material.

Figure 10 shows the resulting microstructure of the fractured samples, showing the 45-degree fracture angle and the resulting shear band morphology, as well as the presence of cracks on the fractographs.

Figure 10. Fractography of Mg-15Se in this work: (**a**) macro-scale photograph showing fracture line and (**b**,**c**) micro-scale fracture surfaces showing shear bands and cracks respectively.

3.7. *Thermal Properties*

Table 12 shows the ignition temperature results of the TGA testing. The results are compared with pure Mg, and it was found that there was a 64 °C increase in ignition temperature.

Table 12. Ignition temperature of Mg-15Se with pure Mg as a reference.

Material	Ignition Temperature (°C)
Pure Mg [26]	581
Mg-15Se	645.5 ± 1.5 (↑ 11.1%)

Figure 11 displays the DSC response curve of Mg-15Se.

Table 13 shows the coefficient of the thermal expansion of Mg-15Se, which is slightly reduced when compared to pure Mg.

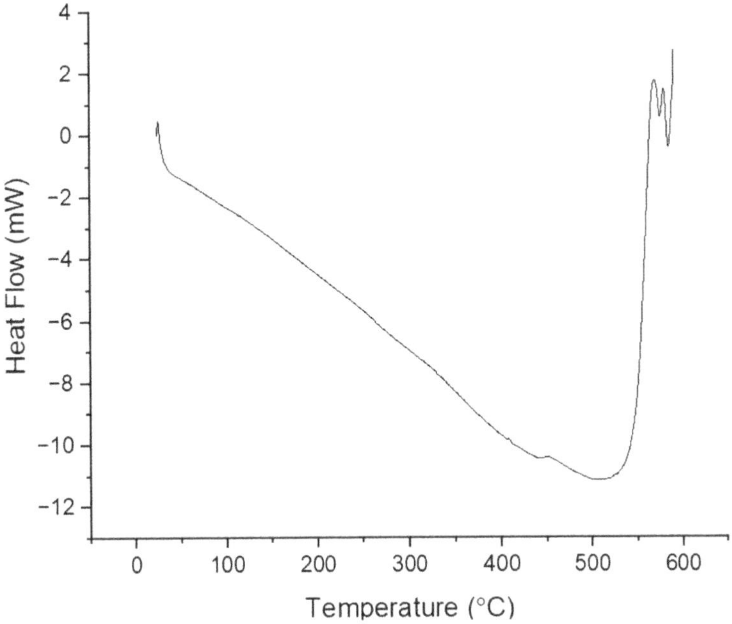

Figure 11. Plot of heat flow vs. temperature of Mg-15Se.

Table 13. Coefficient of thermal expansion of Mg-15Se with pure Mg as a reference.

Material	Average Coefficient of Thermal Expansion ($\times 10^{-6}$/K)
Pure Mg [26]	25.3
Mg-15Se	24.2 ± 0.5 (↓4.8%)

3.8. Corrosion Response

Figure 12 shows the corrosion rates of Mg-15Se with exposure duration. It was found that the samples would last, on average, 16 days in the corrosive environment before disintegrating. Table 14 shows the average corrosion rate of Mg-15Se.

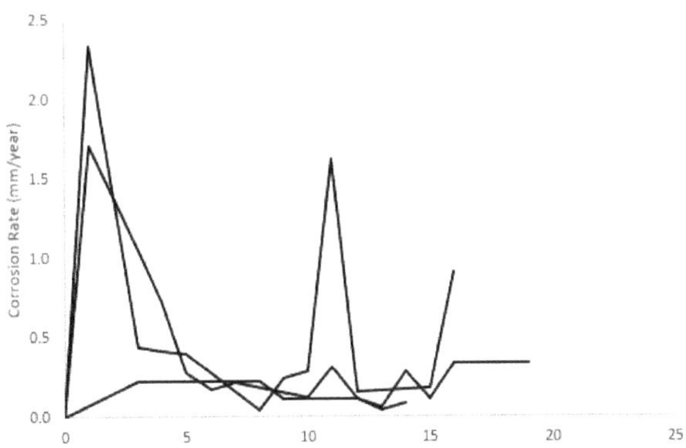

Figure 12. Corrosion rates of Mg-15Se in this work, with each line representing one sample.

Table 14. Corrosion rate of Mg-15Se with other pure Mg materials from the literature as a reference.

Material	Average Corrosion Rate (mm/Year)
Pure Mg, extruded and cold drawn [27]	2.2
Pure Mg, extruded [27] [28]	1.2
Pure Mg, porous (scaffolds) [29]	0.5–3.7
AZ91, powder metallurgy [27,30]	1.6
AZ91-0.5CNT (vol.%), powder metallurgy [30]	12.5
AZ91-0.5Graphene (vol.%), powder metallurgy [30]	3.5
AZ91-0.5Fullerine (vol.%), powder metallurgy [30]	2.2
Mg-15Se (current study)	1.5 ± 1.1

4. Discussion

4.1. Synthesis

While Mg-15Se was successfully synthesized, there was Se loss, which likely occurred during exposure to elevated temperatures in processing, specifically during pre-extrusion heating to 400 °C, which is beyond the melting point of selenium. In addition, the presence of porosity of nearly 3% despite sintering and an extrusion ratio of 20.25 implies that the high Se volume fraction contributes to increased porosity, suggesting that the synthesis parameters could be further optimized for future works, including more refined blending for increased homogeneity, as well as a reduction in Se loss during the extrusion stage, where the temperature was beyond the melting point of selenium.

4.2. Microstructure

The resulting pores and observation that selenium was detected mainly near/at the pores indicate the loss of selenium primary during the soaking/extrusion stage. This increased porosity was accompanied by a matrix, which appears to be continuous, implying that the matrix portion of the material is otherwise bonded well.

The significant degree of grain refinement (by 74% compared to pure Mg) can be attributed to Particle-Stimulated Nucleation (PSN) on the part of micro-scale selenium used as the alloying element and MgSe phase [31,32]. As selenium peak was observed in XRD studies (Figure 4), the present material can be classified as an alloy–composite. Note that the Mg-Se binary phase diagram has not yet been investigated and does not exist in the open literature.

4.3. X-ray Diffraction

The X-ray results indicate that apart from Mg and Se, small amounts of MgO and MgSe were also detected. Due to the blending of powder as well as compaction of the billet taking place without the use of protective/inert gas, some oxidation of the magnesium powder resulting in MgO can be expected as a result of the processing method. For MgSe, this intermetallic phase was previously found to be present at room temperature in a past work by Broch [33]. Given that the powder blending was conducted, as well as pre-extrusion heating, at 400 °C (beyond the melting point of selenium of 221 °C), it is likely that this intermetallic phase was present across the surface of the material after precipitation, though it is difficult to ascertain/isolate this phase specifically using SEM or EDS characterization.

With regard to the texture, the strong basal texture which was observed in the longitudinal direction is consistent with that of pure Mg and its alloys [34], unlike with the addition of nano-scale ceramics, as previously investigated by Parande et al. [26], suggesting that the addition of Se alters the texture of Mg in a different way compared to other composite reinforcements.

4.4. Density and Porosity

It was found that the retained Se content within the material was approximately 12.5%. This was derived from the observed pore area fraction, which was 2.91%, which equates to the true porosity of the material and that above pure Mg (0.21%). A possible reason for this, apart from the use of Se as an alloying element, is the lower sintering temperature; the material in this work was sintered to just 200 °C (due to concerns of Se melting) compared to the reference material, which was sintered to 640 °C. A higher sintering temperature/duration was previously observed to have resulted in narrower and smaller pores in both Mg and non-Mg materials [35–37], likely as a result of more interconnected particles as sintering temperature increases. Another reason for the high porosity of the Mg-Se alloy can be attributed to the partial leaching/melting of selenium during the extrusion step, as indicated in Section 4.2.

4.5. Damping Analysis

The damping analysis indicates that Mg-15Se possesses superior damping properties when compared to pure Mg, notably damping capacity. In addition, Young's modulus was also slightly decreased to 42.9 GPa, which can be attributed to the presence of selenium, which has a lower Young's modulus of 10 GPa [38]. This represents an improvement from a biomedical standpoint; the human cortical bone has a Young's modulus of 19 to 22.5 GPa [39,40], and thus Mg-15Se is closer to the human bone in this property (property-matching).

With regard to the improvement in damping capacity, this can be attributed to the increased presence of pores within the material as a natural consequence of the powder metallurgy method used; these act as sites of energy conversion (via deformation and displacement) from kinetic energy (from vibrations, etc.) to thermal energy, leading to their dissipation. This beneficial effect of pores and voids has been observed in past works involving the intentional synthesis of porous Mg materials [41,42].

4.6. Mechanical Properties

4.6.1. Hardness

The increase in room temperature hardness of Mg-15Se can be explained by the Hall–Petch relationship [43], indicated by the much smaller average grain diameter when compared to pure Mg. This decrease in grain diameter can be attributed to the presence of secondary phases such as Se, MgSe, and MgO in the matrix (Figure 4). Further, the secondary phases with different intrinsic CTE when compared to Mg lead to an increase in dislocation density, thus making localized deformation more difficult.

4.6.2. Compressive Properties

Substantial increases in compressive yield strength (21%), ultimate compressive strength (51%), fracture strain (50%), and work of fracture (70%) were observed, which can be attributed to the strengthening mechanisms associated with the presence of secondary phases as well as grain refinement, as observed from grain size characterization in accordance with the Hall–Petch relationship [44–47]. Furthermore, as mentioned earlier in Section 4.2, grain refinement was also achieved through PSN, the benefits of which were sufficient to enhance the overall compressive properties, including fracture strain despite the higher relative strength of the basal texture exhibited by Mg-15Se from X-ray analysis.

The resulting fracture surface was found to be In line with that of other similarly processed Mg materials [23], though with deeper cracks on the fracture surface, which are similar to those found in porous Mg-SiO_2 nanocomposites [35], suggesting a possible link between the presence of these cracks and the relatively high porosity of the Mg-15Se studied.

4.7. Thermal Properties

The ignition temperature of Mg-15Se was observed to be 11% higher than pure Mg at 645 °C. While increased ignition resistance is not significant in the context of biomedical applications, this may provide some potential for the further use of selenium in the alloying of other Mg materials as ignition resistance has historically been a compromise in Mg materials.

The DSC results indicate a strongly exothermic thermal flow starting at approximately 540 °C and peaking just above 550 °C and remaining almost entirely exothermic past that temperature. This can be attributed to the formation of MgSe in an inert environment [48], which was estimated by Mills [49,50] to have an enthalpy of formation of −293 kJ/mol.

Mg-15Se was found to have a slightly lower CTE compared to pure Mg despite selenium having a higher CTE of 37×10^{-6}/K [47]. This can be attributed to the presence of secondary phases (Figure 4) which outplayed the role of higher porosity. The results thus indicate the capability of Se addition to increase the dimensional stability of magnesium.

4.8. Corrosion Response

The corrosion rates of Mg-15Se in this work were comparable with those of pure Mg in the past literature using a variety of processing methods (extrusion, casting, etc.), as well as being more corrosion-resistant than AZ91 alloy with carbon addition using powder metallurgy [27–30]. This is significant as the pure Mg materials were processed by both liquid-to-solid (casting) processing methods, as well as solid-to-solid (scaffold) processing methods, highlighting the feasibility of powder metallurgy, while the Mg-15Se material is resistant to corrosion relative to carbon-reinforced AZ91.

The corrosion rate exhibited by Mg-15Se despite the presence of pores and a pore area fraction (in turn, porosity) of nearly 3% can be explained by the Pilling–Bedworth Ratio (PBR) of selenium (1.69) [51], which is within the range required for protective oxide layers. This would explain the brief spikes/increases in corrosion rates with immersion duration, but this would then decrease to consistently low values as the Se becomes the next to be subject to corrosion, lowering the corrosion rate for that specific time interval.

This highlights the potential for applications of selenium addition for use in corrosive environments, which has historically been a challenge for Mg-based materials. The high corrosion resistance of selenium, implied by its PBR of 1.69, also synergizes with a potential use of Se-containing Mg materials in the biomedical field, where the increased corrosion resistance conferred would result in a longer useful lifespan of any Mg-based implant. In addition, it was demonstrated that mechanical enhancements were exhibited without compromising corrosion response.

5. Conclusions

In this study, Mg-15Se was successfully synthesized with powder metallurgy and, for the first time, investigated for the effect of selenium addition as an alloying element on Mg, with the following conclusions being observed:

1. The resulting Mg-15Se material had a higher porosity (2.91%) when compared to pure magnesium (0.21%). This can be attributed to the partial loss of selenium primarily during the extrusion stage.
2. The resulting Mg-15Se material exhibited a relatively stronger basal texture than that of pure Mg, and it also underwent significant grain refinement (74% smaller grain diameter than pure Mg).
3. Mg-15Se had a superior damping capacity to pure Mg (76% increase).
4. Mg-15Se had superior hardness (57% increase) as well as compression properties (21% increase for 0.2% yield strength, at least 50% increases for ultimate compressive strength, failure strain, and energy absorbed).
5. The average corrosion rate for Mg-15Se is comparable with extruded pure Mg from multiple works, despite its high porosity, due to the high PBR of Se at 1.69, showcasing the lack of compromise in corrosion resistance whilst possessing a superior mechanical response.
6. Thermal analysis also indicated that Mg-15Se is more ignition-resistant than pure Mg despite its much higher porosity.
7. In this study, it was discovered that Se loss occurred during the extrusion stage; additional steps to reduce Se loss by use of a lower temperature in processing or

further protective/containment measures is possible, but this would have to be studied to see if there is an adverse effect on the final material.

In essence, considering the potential of Se as a biomedical-oriented element, the prospects of subsequent studies go beyond merely structural applications; as the use of Mg materials in the biomedical field increases, selenium can be utilized as an alloying element without undue concerns of degradation in neither mechanical nor corrosion response. In addition, further refinements for Mg-Se materials (in biomedical contexts, e.g., bioresorbable implants) would include controlling the corrosion rate of the material as well as the rate of Se release/dosage to ensure both the good functioning/structural integrity of the implant as well as the nutritional/supplemental nature of Se within medically safe ranges.

Author Contributions: Conceptualization, M.G.; methodology, M.J. and M.G.; validation, M.J.; formal analysis, M.J. and V.S.; investigation, V.S.; resources, M.J. and M.G.; data curation, M.J. and V.S.; writing—original draft preparation, M.J.; writing—review and editing, M.J. and M.G.; visualization, M.J. and V.S.; supervision, M.G.; project administration, M.J. and M.G. All authors have read and agreed to the published version of the manuscript.

Funding: This research received no external funding.

Data Availability Statement: The original contributions presented in the study are included in the article, further inquiries can be directed to the corresponding author.

Acknowledgments: The authors acknowledge Juraimi Bin Madon for assistance with extrusion and Ng Hong Wei for assistance with TGA, DSC, and CTE testing.

Conflicts of Interest: The authors declare no conflicts of interest.

References

1. Meenashisundaram, G.K.; Gupta, M. Emerging Environment Friendly, Magnesium-Based Composite Technology for Present and Future Generations. *JOM* **2016**, *68*, 1890–1901. [CrossRef]
2. U.S. Geological Survey. Mineral commodity summaries 2023. In *Mineral Commodity Summaries*; U.S. Geological Survey: Reston, VA, USA, 2023; Report. Available online: http://pubs.er.usgs.gov/publication/mcs2023 (accessed on 3 January 2024).
3. Gupta, S.; Parande, G.; Tun, K.S.; Gupta, M. Enhancing the Physical, Thermal, and Mechanical Responses of a Mg/2 wt.%CeO_2 Nanocomposite Using Deep Cryogenic Treatment. *Metals* **2023**, *13*, 660. [CrossRef]
4. Garcés, G.; Pérez, P.; Adeva, P. Effect of the extrusion texture on the mechanical behaviour of Mg–SiCp composites. *Scr. Mater.* **2005**, *52*, 615–619. [CrossRef]
5. Garcés, G.; Rodríguez, M.; Pérez, P.; Adeva, P. Effect of volume fraction and particle size on the microstructure and plastic deformation of Mg–Y_2O_3 composites. *Mater. Sci. Eng. A* **2006**, *419*, 357–364. [CrossRef]
6. Umeda, J.; Kawakami, M.; Kondoh, K.; Ayman, E.L.S.; Imai, H. Microstructural and mechanical properties of titanium particulate reinforced magnesium composite materials. *Mater. Chem. Phys.* **2010**, *123*, 649–657. [CrossRef]
7. Goh, C.S.; Wei, J.; Lee, L.C.; Gupta, M. Development of novel carbon nanotube reinforced magnesium nanocomposites using the powder metallurgy technique. *Nanotechnology* **2006**, *17*, 7–12. [CrossRef]
8. Jin, W.; Chu, P.K. Orthopedic Implants. In *Encyclopedia of Biomedical Engineering*; Narayan, R., Ed.; Elsevier: Oxford, UK, 2019; pp. 425–439.
9. Chouirfa, H.; Bouloussa, H.; Migonney, V.; Falentin-Daudré, C. Review of titanium surface modification techniques and coatings for antibacterial applications. *Acta Biomater.* **2019**, *83*, 37–54. [CrossRef] [PubMed]
10. Sadati, M.; Ghofrani, S.; Mehrizi, A.A. Investigation of porous cells interface on elastic property of orthopedic implants: Numerical and experimental studies. *J. Mech. Behav. Biomed. Mater.* **2021**, *120*, 104595. [CrossRef]
11. Das, P.; Kumar, T.S.S.; Sahu, K.; Gollapudi, S. Corrosion, Stress Corrosion Cracking and Corrosion Fatigue Behavior of Magnesium Alloy Bioimplants. *Corros. Rev.* **2022**, *40*, 289–333. [CrossRef]
12. Pacheco, K.A. Allergy to Surgical Implants. *Clin. Rev. Allergy Immunol.* **2019**, *56*, 72–85. [CrossRef]
13. Charyeva, O.; Dakischew, O.; Sommer, U.; Heiss, C.; Schnettler, R.; Lips, K.S. Biocompatibility of magnesium implants in primary human reaming debris-derived cells stem cells in vitro. *J. Orthop. Traumatol.* **2016**, *17*, 63–73. [CrossRef]
14. Sandler, G.; Nguyen, L.; Lam, L.; Manglick, M.P.; Soundappan, S.S.V.; Holland, A.J.A. Trampoline Trauma in Children: Is It Preventable? *Pediatr. Emerg. Care* **2011**, *27*, 1052–1056. [CrossRef] [PubMed]
15. Wang, J.; Chen, M.; Zhang, Z.; Ma, L.; Chen, T. Selenium: From fluorescent probes to biomedical application. *Coord. Chem. Rev.* **2023**, *493*, 215278. [CrossRef]
16. Huang, Y.; Su, E.; Ren, J.; Qu, X. The recent biological applications of selenium-based nanomaterials. *Nano Today* **2021**, *38*, 101205. [CrossRef]

17. Rataan, A.O.; Geary, S.M.; Zakharia, Y.; Rustum, Y.M.; Salem, A.K. Potential Role of Selenium in the Treatment of Cancer and Viral Infections. *Int. J. Mol. Sci.* **2022**, *23*, 2215. (In English) [CrossRef]
18. Kuršvietienė, L.; Mongirdienė, A.; Bernatonienė, J.; Šulinskienė, J.; Stanevičienė, I. Selenium Anticancer Properties and Impact on Cellular Redox Status. *Antioxidants* **2020**, *9*, 80. (In English) [CrossRef]
19. Radomska, D.; Czarnomysy, R.; Radomski, D.; Bielawski, K. Selenium Compounds as Novel Potential Anticancer Agents. *Int. J. Mol. Sci.* **2021**, *22*, 1009. (In English) [CrossRef]
20. Zhang, Q.; Qian, Z.-Y.; Zhou, P.-H.; Zhou, X.-l.; Zhang, D.-L.; He, N.; Zhang, J.; Liu, Y.-H.; Gu, Q. Effects of oral selenium and magnesium co-supplementation on lipid metabolism, antioxidative status, histopathological lesions, and related gene expression in rats fed a high-fat diet. *Lipids Health Dis.* **2018**, *17*, 165. [CrossRef]
21. Ferro, C.; Florindo, H.F.; Santos, H.A. Selenium Nanoparticles for Biomedical Applications: From Development and Characterization to Therapeutics. *Adv. Healthc. Mater.* **2021**, *10*, 2100598. [CrossRef]
22. Singh, I.B.; Singh, M.; Das, S. A comparative corrosion behavior of Mg, AZ31 and AZ91 alloys in 3.5% NaCl solution. *J. Magnes. Alloys* **2015**, *3*, 142–148. [CrossRef]
23. Parande, G.; Manakari, V.; Meenashisundaram, G.K.; Gupta, M. Enhancing the hardness/compression/damping response of magnesium by reinforcing with biocompatible silica nanoparticulates. *Int. J. Mater. Res.* **2016**, *107*, 1091–1099. [CrossRef]
24. Gates-Rector, S.; Blanton, T. The Powder Diffraction File: A quality materials characterization database. *Powder Diffr.* **2019**, *34*, 352–360. [CrossRef]
25. Gates-Rector, S.; Blanton, T. PDF-4+ 2023. Available online: https://www.cambridge.org/core/journals/powder-diffraction/article/powder-diffraction-file-a-quality-materials-characterization-database/3C1FDF449C7A82A8E6FBC8F76342187A (accessed on 19 December 2023).
26. Parande, G.; Manakari, V.; Meenashisundaram, G.K.; Gupta, M. Enhancing the tensile and ignition response of monolithic magnesium by reinforcing with silica nanoparticulates. *J. Mater. Res.* **2017**, *32*, 2169–2178. [CrossRef]
27. Atrens, A.; Shi, Z.; Mehreen, S.U.; Johnston, S.; Song, G.; Chen, X.; Pan, F. Review of Mg alloy corrosion rates. *J. Magnes. Alloys* **2020**, *8*, 989–998. [CrossRef]
28. Dvorský, D.; Kubásek, J.; Voňavková, I.; Vojtěch, D. Structure, mechanical and corrosion properties of extruded Mg-Nd-Zn, Mg-Y-Zn and Mg-Y-Nd alloys. *Mater. Sci. Technol.* **2019**, *35*, 520–529. [CrossRef]
29. Jia, G.; Chen, C.; Zhang, J.; Wang, Y.; Yue, R.; Luthringer-Feyerabend, B.J.C.; Willumeit-Roemer, R.; Zhang, H.; Xiong, M.; Huang, H.; et al. In vitro degradation behavior of Mg scaffolds with three-dimensional interconnected porous structures for bone tissue engineering. *Corros. Sci.* **2018**, *144*, 301–312. [CrossRef]
30. Turan, M.E.; Sun, Y.; Aydin, F.; Zengin, H.; Turen, Y.; Ahlatci, H. Effects of carbonaceous reinforcements on microstructure and corrosion properties of magnesium matrix composites. *Mater. Chem. Phys.* **2018**, *218*, 182–188. [CrossRef]
31. Doherty, R.D.; Hughes, D.A.; Humphreys, F.J.; Jonas, J.J.; Jensen, D.J.; Kassner, M.E.; King, W.E.; McNelley, T.R.; McQueen, H.J.; Rollett, A.D. Current issues in recrystallization: A review. *Mater. Sci. Eng. A* **1997**, *238*, 219–274. [CrossRef]
32. Van Aken, D.C.; Krajewski, P.E.; Vyletel, G.M.; Allison, J.E.; Jones, J.W. Recrystallization and grain growth phenomena in a particle-reinforced aluminum composite. *Metall. Mater. Trans. A* **1995**, *26*, 1395–1405. [CrossRef]
33. Nayeb-Hashemi, A.A.; Clark, J.B. The Mg−Se (Magnesium-Selenium) system. *J. Phase Equilibria* **1987**, *8*, 118–119. [CrossRef]
34. Nie, J.F.; Shin, K.S.; Zeng, Z.R. Microstructure, Deformation, and Property of Wrought Magnesium Alloys. *Metall. Mater. Trans. A* **2020**, *51*, 6045–6109. [CrossRef]
35. Johanes, M.; Gupta, M. An Investigation into the Potential of Turning Induced Deformation Technique for Developing Porous Magnesium and Mg-SiO_2 Nanocomposite. *Materials* **2023**, *16*, 2463. [CrossRef]
36. Avci, U.; Güleç, A. Effect of Different Sintering Temperatures on Microstructure and Mechanical Properties for Pure Al Material Produced by Powder Metallurgy. *El-Cezeri J. Sci. Eng.* **2021**, *8*, 462–470.
37. Wahi, A.; Muhamad, N.; Sulong, A.B.; Ahmad, R.N. Effect of Sintering Temperature on Density, Hardness and Strength of MIM $Co_{30}Cr_6Mo$ Biomedical Alloy. *J. Jpn. Soc. Powder Powder Metall.* **2016**, *63*, 434–437. [CrossRef]
38. American Elements. Selenium Metal. American Elements. Available online: https://www.americanelements.com/selenium-metal-7782-49-2. (accessed on 3 January 2024).
39. Mirzaali, M.J.; Schwiedrzik, J.J.; Thaiwichai, S.; Best, J.P.; Michler, J.; Zysset, P.K.; Wolfram, U. Mechanical properties of cortical bone and their relationships with age, gender, composition and microindentation properties in the elderly. *Bone* **2016**, *93*, 196–211. [CrossRef]
40. Rho, J.-Y.; Tsui, T.Y.; Pharr, G.M. Elastic properties of human cortical and trabecular lamellar bone measured by nanoindentation. *Biomaterials* **1997**, *18*, 1325–1330. [CrossRef]
41. Xie, Z.-K.; Tane, M.; Hyun, S.-K.; Okuda, Y.; Nakajima, H. Vibration–damping capacity of lotus-type porous magnesium. *Mater. Sci. Eng. A* **2006**, *417*, 129–133. [CrossRef]
42. Li, B.; Lavernia, E.J. Spray Forming of MMCs. In *Comprehensive Composite Materials*, 1st ed.; Kelly, A., Zweben, C., Eds.; Elsevier: Amsterdam, The Netherlands, 2000; Volume 3, pp. 617–653.
43. Taha, A.S.; Hammad, F.H. Application of the Hall-Petch Relation to Microhardness Measurements on Al, Cu, Al-MD 105, and Al-Cu Alloys. *Phys. Status Solidi (A)* **1990**, *119*, 455–462. [CrossRef]

44. Meenashisundaram, G.K.; Nai, M.H.; Gupta, M. Effects of Primary Processing Techniques and Significance of Hall-Petch Strengthening on the Mechanical Response of Magnesium Matrix Composites Containing TiO_2 Nanoparticulates. *Nanomaterials* **2015**, *5*, 1256–1283. [CrossRef]
45. Yu, H.; Xin, Y.; Wang, M.; Liu, Q. Hall-Petch relationship in Mg alloys: A review. *J. Mater. Sci. Technol.* **2018**, *34*, 248–256. [CrossRef]
46. Wang, Y.; Choo, H. Influence of texture on Hall–Petch relationships in an Mg alloy. *Acta Mater.* **2014**, *81*, 83–97. [CrossRef]
47. Ross, R.B. Selenium Se. In *Metallic Materials Specification Handbook*; Ross, R.B., Ed.; Springer US: Boston, MA, USA, 1992; pp. 282–283.
48. ChemicalBook Inc. Magnesium Selenide. ChemicalBook. Available online: https://www.chemicalbook.com/ChemicalProductProperty_EN_CB6937531.htm (accessed on 26 December 2023).
49. Olin, Å.; Noläng, B.I.; Öhman, L.; Osadchii, E.G.; Rosén, E. *Chemical Thermodynamics of Selenium*; Elsevier: Amsterdam, The Netherlands, 2005.
50. Mills, K.C. *Thermodynamic Data for Inorganic Sulphides, Selenides and Tellurides*; John Wiley & Sons, Inc: Hoboken, NJ, USA, 1974.
51. Syvertsen, M. Oxide Skin Strength on Molten AA5XXX Aluminum Alloy—Effect of Beryllium and Alternatives. In *Light Metals 2017*; Springer International Publishing: Cham, Switzerland, 2017; pp. 1451–1455.

metals

MDPI

Article

Dependence of Electrochemical Characteristics of a Biodegradable Fe-30Mn-5Si wt.% Alloy on Compressive Deformation in a Wide Temperature Range

Pulat Kadirov [1,*], Yury Pustov [2], Yulia Zhukova [3], Maria Karavaeva [1], Vadim Sheremetyev [1], Andrey Korotitskiy [4], Alexandra Baranova [1] and Sergey Prokoshkin [1]

1 Metal Forming Department, National University of Science and Technology "MISIS", 119049 Moscow, Russia; karavaeva.ma@yandex.ru (M.K.); sheremetyev@misis.ru (V.S.); baranova.al.pavlovna@yandex.ru (A.B.); prokoshkin@tmo.misis.ru (S.P.)

2 Department of Steel Metallurgy, New Production Technologies and Protection of Metals, National University of Science and Technology "MISIS", 119049 Moscow, Russia; pustov@misis.ru

3 Center of Nanomaterials and Nanotechnologies, National University of Science and Technology "MISIS", 119049 Moscow, Russia; zhukova@misis.ru

4 Ultrafine-Grained Metallic Materials Laboratory, National University of Science and Technology "MISIS", 119049 Moscow, Russia; akorotitskiy@gmail.com

* Correspondence: pulat_1993-2009@mail.ru

Abstract: Fe-30Mn-5Si alloy subjected to a compression test at various deformation temperatures ranging from 350 to 900 °C with a strain rate of 1 s^{-1} are studied. It was found that the Fe-30Mn-5Si alloy exhibits high resistance to the dynamic recrystallization process in a whole studied range of deformation temperatures. There are no differences in structure formation in the zone of action of tangential tensile stresses and peripheral and central zones of localized compressive stresses. The room-temperature X-ray diffraction study shows the presence of a single-phase state (FCC γ-austenite) after deformation temperature range from 350 to 700 °C and a two-phase state (FCC γ-austenite + HCP ε-martensite) after deformation test at 900 °C. The presence of a two-phase state provides a higher rate of biodegradation compared with a single-phase state. The changes in the biodegradation rate dependence on the structure change with an increase in the deformation temperature are explained. Favorable temperature regimes for subsequent thermomechanical processing are proposed based on the relationship between structure formation and biodegradation rate to obtain semi-products from the Fe-30Mn-5Si alloy.

Keywords: biodegradable shape memory alloys; electrochemical behavior; thermomechanical treatment; rheological behavior; grain structure; phase composition

1. Introduction

In the past two decades, there has been a growing interest in biodegradable metallic materials for use as temporary devices in traumatology, osteosynthesis, orthopedics, and cardiovascular surgery [1–7]. Magnesium (Mg), zinc (Zn), iron (Fe), and some of their alloys are the most favorable candidates for these applications [8–14]. Among these biodegradable alloys, Fe-based alloys appear to be the most promising ones due to their excellent combination of mechanical properties, similar to stainless steels [15], sufficient biocompatibility [16,17], as demonstrated by in vivo experiments [18,19], and the absence of hydrogen release during the biodegradation process. However, Fe-based alloys exhibit a low biodegradation rate. In order to increase the electrochemical activity, iron is alloyed with manganese (Mn), whose electrode potential is much lower (−1.18 V) than that of pure Fe (−0.44 V) [20]. The further increase in the biodegradation rate and mechanical properties is achieved by adding silicon (Si) to Fe-Mn binary alloys, which form non-metallic inclusions of SiO_2 acting as cathode structural components, increasing the biodegradation rate. The enhancement of biodegradation rate can also be increased by adding nitrogen (N) to the alloys of the Fe-Mn system. The authors found [21] that N addition (0.3 . . . 0.6 wt.%) leads to a significant increase in the corrosion rate of the Fe-30Mn alloy in both Hank's solution and simulated gastric fluid.

The Fe-Mn-Si alloys exhibit a shape memory effect due to the reversible martensitic transformation FCC γ-austenite $\leftrightarrow$ HCP ε-martensite. In these alloys, which are intended for use as potential materials for bone implants, it is advisable to consider only the temperature of the start of forward $\gamma \rightarrow \varepsilon$ martensitic transformation (M_s) since near the temperature M_s pre-transition crystal lattice softening effect takes place, which leads to a decrease in Young's modulus, thus increasing the biomechanical compatibility. According to [22], Fe-Mn-Si alloy containing from 28 to 33 wt.% Mn and from 4 to 6 wt.% Si exhibits a nearly perfect shape memory effect. In [23], Fe-Mn-Si alloys with different Mn content were studied. It was shown that Fe-30Mn-5Si (hereinafter in wt.%) exhibits the most appropriate combination of functional properties: biodegradation rate (0.8 mm/year); Young's modulus (118 GPa); and M_s (50 °C) temperature close to human body temperature.

The subsequent studies were focused on improving the mechanical properties and biomechanical compatibility of the Fe-30Mn-5Si alloy by means of thermomechanical treatment (TMT). The best combination of functional properties was achieved by hot longitudinal rolling at 800 °C with a true strain of $e = 0.3$, resulting in a dynamically recrystallized structure with an average grain size of about 100 μm [24,25]. This TMT mode resulted in the M_s temperature lowering to the human body temperature (38 °C), achieving high mechanical properties (ultimate tensile strength 725 MPa, apparent yield stress 260 MPa, Young's modulus about 120 GPa, elongation to failure about 15%), suitable biodegradation rate of 0.5-mm/year, and high functional fatigue characteristics both in normal conditions in an air atmosphere and Hanks' solution. However, all these experiments were carried out in laboratory conditions involving small-size specimens. Moreover, the possibilities of TMT to form fine-grained, particularly ultra-fine-grained (UFG) structures in the long semi-products have not been investigated yet.

The conventional methods to obtain long semi-products with fine-grained and UFG structures include radial–shear rolling, multiaxial forging, and longitudinal rolling. These methods show excellent combinations of functional properties for Ti-, Al-, Mg-, and Fe-based alloys [26–37]. However, to determine the favorable temperature-rate conditions for implementing the above-mentioned TMT processes, it is necessary to carry out the rheological tests in wide temperature and strain rate ranges. For example, Zambrano et al. [38] studied the deformation behavior of Fe-Mn-Si-C alloys in the temperature range from 800 °C to 1050 °C with strain rates of 0.01, 0.1, and 1 s^{-1}. They found that in all cases, a dynamic recrystallization process was observed, accompanied by a decrease in the stress peak. Additionally, it was shown that at the same deformation temperature, such as 900 °C, an increase in the strain rate resulted in a higher amount of recrystallized grains. It is important to note that the process of dynamic recrystallization is accompanied by a non-uniform distribution of the applied external load, which can lead to the formation of various defects in the semi-products. In this regard, our previous work aimed at the study of the deformation behavior, structure formation, and mechanical properties of Fe-30Mn-5Si alloy subjected to the compression test in a wide deformation temperature range from 250 to 900 °C, and strain rates of 0.1, 1, and 10 s^{-1} [39]. Based on the relationship between the structure formation and the shape of obtained stress–strain diagrams, favorable regimes for subsequent radial–shear rolling, multiaxial forging, and longitudinal rolling were formulated: deformation temperature of 900 °C and a strain rate of 1 s^{-1} [39].

The biodegradation rate is one of the most important functional properties of materials designed for biodegradable bone implants. Therefore, the study of the Fe-30Mn-5Si alloy electrochemical behavior in relation to the structure formed after the rheological test in a wide deformation temperature range and strain rate of 1 s^{-1} deserves its own study. At this moment, there are no studies on the effect of the structure formed as a result of the rheological tests (compression test) on the electrochemical behavior of biodegradable

Fe-Mn-Si alloys. Thereon, the objective of this work was to study the relationship between the structure formation and the electrochemical behavior of biodegradable Fe-30Mn-5Si shape memory alloy subjected to compression tests in a wide range of temperatures.

2. Materials and Methods

The 12 kg ingot of 80 mm diam. and approximately 250 mm in length ingots of experimental Fe-30Mn-5Si alloy were produced by vacuum induction melting. The chemical composition and element distribution of elements were controlled by atomic emission spectroscopy with inductively coupled plasma and X-ray spectroscopy using a *JEOL JSM-6480 LV* (Jeol, Eching b. München, Germany) scanning electron microscope in different parts of the ingots. After verifying the chemical composition, the studied Fe-30Mn-5Si alloy was subjected to reference heat treatment (RHT), including homogenization annealing at 900 °C for 60 min and subsequent water cooling to avoid an as-cast structure [23,24,40]. The electrical discharge technique was used to cut 15 mm long and 10 mm diam. specimens for subsequent rheological compression tests.

In the compression tests (plastometry tests), the specimens were deformed up to the true (logarithmic) strain of $e = 0.7$ at a strain rate of 1 s^{-1} and temperatures of 350, 500, 700, and 900 °C using high-precision simulation laboratory equipment—*Gleeble System 3800* (Dynamic Systems Inc., Austin, TX, USA) equipped with the *Hydrawedge-II* high-temperature compression test module with auxiliary devices for registration and control of temperature, force, displacement, and strain rate. The compression test was carried out with the vacuumed work chamber of at least 10^{-5} atm to avoid the specimen's oxidation. To reduce the contact friction forces between the end surfaces of the samples and the deforming working surfaces of the strikers made of tungsten carbide, grease based on nickel, graphite, boron nitride, or special graphite foil was used. To prevent "sticking" of the samples to the striker surface, tantalum foil was placed between their contact areas. The studied Fe-30Mn-5Si specimens were directly heated by the controlled electric current application. This method provides fast heating and uniform heat distribution over the entire samples' volume. Upon completion of the compression tests, the specimens were air-cooled at a rate of 20 °C/s.

After compression tests, the specimens were cut along the load direction for subsequent study of the structure by optical microscopy, X-ray analysis, and electrochemical studies. After cutting, the specimens were ground with *P 320* to *P 4000* emery paper and further mirror-polished with diamond suspension. To reveal the structure, a solution of nitric acid in ethanol with a ratio of 4:96 wt.%, respectively, was used as an etchant. Three zones were studied in each specimen: the zone of action of tangential tensile stresses (zone A); the peripheral zone (Zone B); and the central zone (Zone C) of localized compressive stresses [41].

The X-ray diffraction (XRD) analysis was carried out at room temperature using a *DRON-3* (Burevestnik Innovation Centre, St. Petersburg, Russia) diffractometer with $Cu_{K\alpha}$ radiation ($\lambda = 0.154178$ nm). The studied 2θ-angle range from 30 to 100 deg. was scanned with 0.1 2θ deg. step and 3 s for exposure. A graphite monochromator was used for noise level lowering. In order to remove the stress-induced martensite formed by specimens' preparation, chemical etching was used. The thickness of the removed layer was 0.5 mm.

An electronic *IPC-Pro* potentiostat with a three-electrode 200 mL glass cell was used for electrochemical study to measure open circuit potential (OCP) and potentiodynamic polarization curves. During the electrochemical tests, a constant temperature of 37 °C was maintained using a *TW-2 Elmi* (ELMI Ltd., Riga, Latvia) thermostat. Hanks' solution, simulating the inorganic composition of bone tissue medium, was used as the electrolyte [24,25]. To ensure stable electrical contact, a copper wire was soldered to the studied specimens, with the subsequent embedment in epoxy resin. Then, samples were further ground with P320 to P1200 emery paper. The working area for the electrochemical experiment was 25 mm^2. Each electrochemical experiment was carried out at 37 °C and repeated at least three times to ensure reproducibility of the results. The exposure time of the samples to obtain steady-state OCP values was 7000 s. The polarization diagrams were obtained from the cathodic region, ranging from 250 mV negative to the OCP value of the studied alloy. The top potential values were approximately 100 mV more positive compared to the OCP values. The potential sweep rate for the polarization test was 0.2 mV/s. The saturated $Ag/AgCl_2$ electrode was used as the reference electrode, and the platinum one served as an auxiliary electrode. For the corrosion rate (C_r) estimation, the corrosion current density (i_{corr}) was extrapolated from the polarization curves as follows [42]:

$$C_r = \frac{i_{corr} 10^4 A_{Me}}{n F d_{Me}} 8.76 \text{ mm/year} \quad (1)$$

where i_{corr} is the corrosion current density, A/cm^2; n—is ionic charge (n = 2 for Fe^{2+}); F = 26.8 A·h/gram-equivalent is the Faraday constant; A_{Me} is the atomic weight of metal, g; d_{Me} is the metal density, g/cm.

The choice to use the method of corrosion rate estimation by extrapolation of the potentiodynamic polarization curves using Equation (1) is justified by the following factors. Firstly, Equation (1) is widely used to calculate the biodegradation rate of metals and alloys corroding in active dissolution mode. Secondly, previous studies conducted by the authors of this manuscript have shown that long-term corrosion testing, as described in ASTM G31-21 [43], results in the accumulation of corrosion products on the sample surface and in the electrolyte volume. However, in real conditions during the process of biodegradation and bone tissue regeneration, there is a constant release of low-valence ions, such as Fe^{2+} and Mn^{2+} (typical for the corrosion of these metals in neutral media), which prevents the formation of solid precipitates as corrosion products on the implant surface. The authors of this manuscript have previously conducted corrosion tests using this standard method. However, the corrosion rate calculated based on the mass loss of the samples was much lower than the corrosion rate of the Fe-30Mn-5Si alloy obtained by extrapolating polarization curves. Analysis of the samples with the X-ray photoelectron spectroscopy showed that after ASTM G31-21 testing, corrosion products in the form of iron (Fe_2O_3, Fe_3O_4, FeOOH) and manganese (MnO) compounds were present on the surface of the samples, with the most abundant Fe^{2+} peaks being most common in the Fe_3O_4 multiplet. It is known that the corrosion of iron in neutral media proceeds through oxygen depolarization. The anodic and cathodic processes are shown below, respectively [42,44,45]:

$$Fe - 2e \rightarrow Fe^{2+}\text{—oxidation;}$$

$$2H_2O + O_2 + 4e \rightarrow 4OH^-\text{—reduction;}$$

The following reactions take place on the metal surface in the test at full immersion in the electrolyte:

$$Fe^{2+} + 2OH^- \rightarrow Fe(OH)_2$$

$$4Fe(OH)_2 + O_2 + 2H_2O \rightarrow 4Fe(OH)_3$$

3. Results and Discussion

3.1. Structure and Phase Analysis

Figure 1 shows optical microscopy images of the Fe-30Mn-5Si alloy after the reference heat treatment (RHT), which provides a homogeneous state of chemical composition and structure [23–25]. As can be seen, the Fe-30Mn-5Si alloy consists of γ-austenite grains with an average grain size of about 200–300 µm with chaotic shape and ε-martensite plates inside some of the grains. In some areas, smaller equiaxed grains are observed.

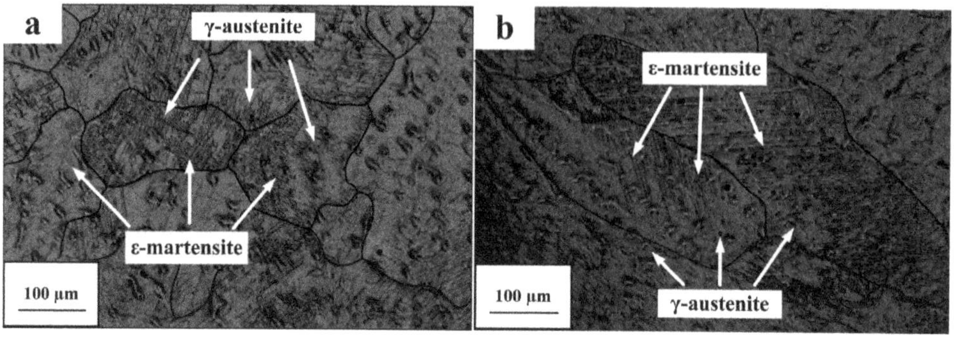

Figure 1. Typical optical microstructure images of the Fe-30Mn-5Si alloy after RHT were taken from various areas of specimens: (**a**) grains with average size of 200 μm; (**b**) grains with average size of 300 μm.

Figure 2 shows the optical microscopy images of the Fe-30Mn-5Si alloy after the compression tests at 350, 500, 700, and 900 °C with the strain rate of 1 s^{-1}. It can be observed that, regardless of the studied zone and deformation temperature, the grains' shape changes, and their size becomes smaller along the deformation axis and larger in the perpendicular direction, which could indicate the absence of the dynamic recrystallization process and, consequently, the presence of the dynamic polygonization process. A similar state is observed up to 900 °C. It is noteworthy to note that there are no differences in structure formation between the three studied zones. This indicates a uniform strain distribution during compression testing. One of the main characteristics that can be observed from Figure 2 is the absence of equiaxed recrystallized grains in all studied zones and deformation temperatures up to 900 °C, which probably indicates high resistance of the Fe-30Mn-5Si alloy to the development of dynamic recrystallization processes up to high temperatures. However, starting at the deformation temperature of 700 °C zone A and zone B (Figure 2g,h), the grains with smaller size (about 100–200 μm) can be observed. At first, it might be assumed that it is the result of partial recrystallization. However, after the deformation at 900 °C, only non-recrystallized grains are observed, which indicates the impossibility of the process of dynamic recrystallization at lower deformation temperature (700 °C). Thus, the smaller non-equiaxed grains in Figure 2g,h at temperature 700 °C are the deformed smaller grains that remained after the RHT regime (Figure 1). It is important to note that the dynamic recrystallization process during thermomechanical treatment can cause a non-uniform structure formation. This may result in uneven load application and lead to the destruction of test specimens.

Figure 3 shows the X-ray diffractograms of the Fe-30Mn-5Si alloy after the RHT and compression test in the deformation temperature range from 350 to 900 °C with a strain rate of 1 s^{-1}. In the diffraction patterns of the Fe-30Mn-5Si alloy after the RHT regime, the presence of two phases, FCC γ-austenite and HCP ε-martensite, is observed. Note that after compression tests in the deformation temperature range from 350 to 700 °C, only a single FCC γ-austenite phase is present. This may be related to the fact that after the deformation with true logarithmic strain e = 0.7 in the above-mentioned deformation temperature range, the temperature of the start of forward martensitic transformation $\gamma \rightarrow \varepsilon$ (M_s) falls below room temperature. It is noteworthy that the M_s temperature after the RHT is about 21 °C [39]. In this regard, the true logarithmic strain of e = 0.7 should be enough to decrease the M_s below room temperature due to the formation of a highly dislocated substructure, and correspondingly, the ε-martensite phase is absent in the X-ray diffractograms obtained at room temperature. It is also necessary to note the absence of iron carbide ($Fe_{23}C_6$ or Fe_3C) lines in the X-ray diffractograms; this will be important for the discussion of the subsequent electrochemical experiments' results. However, weak ε-martensite is observed in the X-ray diffractograms taken after the deformation at 900 °C (Figure 4), which indicates that at this deformation temperature, the softening processes (dynamical recovery and poligonization) prevail over the hardening process. Therefore, the M_S temperature remains above room temperature.

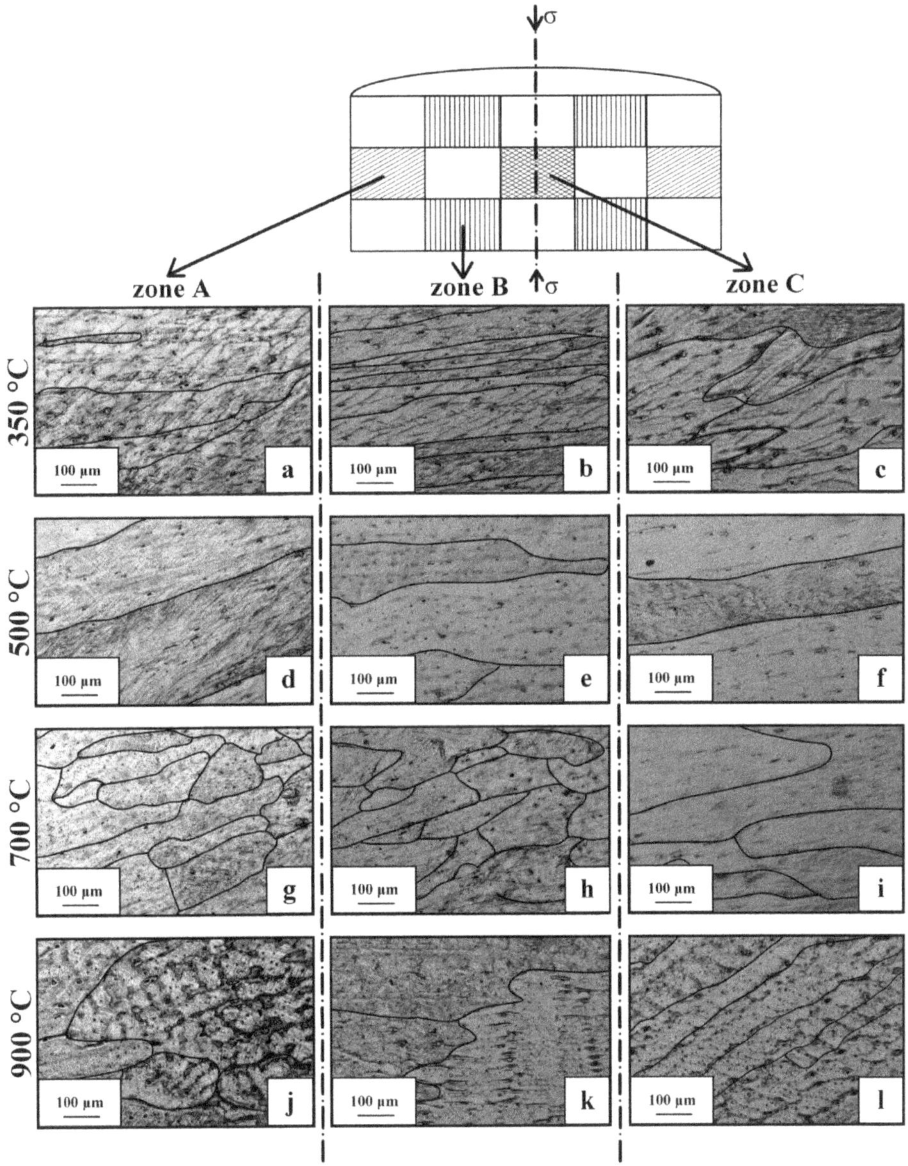

Figure 2. Schematic representation of different zones (the arrow indicates the load axis) and optical microscopy images of the Fe-30Mn-5Si alloy microstructure after compression test at strain rate 1 s^{-1} and deformation temperatures of 350 °C (**a–c**), 500 °C (**d–f**), 700 °C (**g–i**), and 900 °C (**j–l**) in zone A (zone of action of tangential tensile stresses), zone B (peripheral zone of localized compressive stresses), and zone C (central zone of localized compressive stresses).

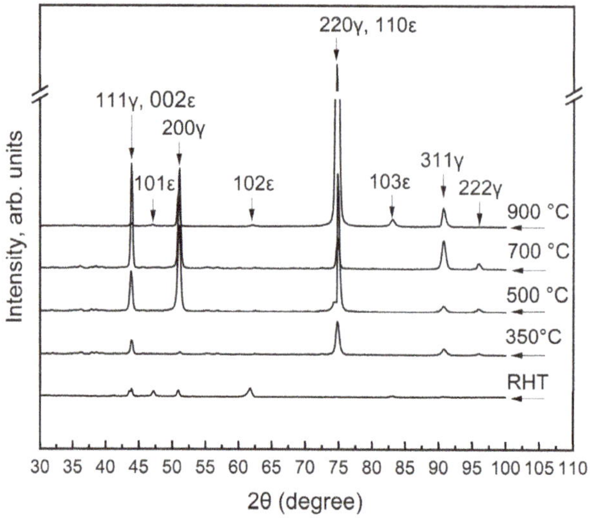

Figure 3. X-ray diffractograms of the Fe-30Mn-5Si alloy after RHT and compression tests in temperature range from 350 to 900 °C.

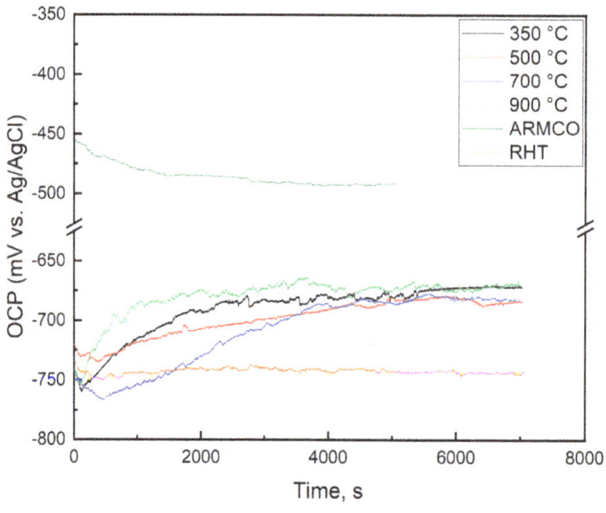

Figure 4. OCP curves of the Fe-30Mn-5Si alloy after compression test obtained at temperatures of deformation of 350, 500, 700, and 900 °C and strain rate 1 s^{-1} immersed in Hanks' solution at 37 °C.

Table 1 shows the half-height width of {200}γ ($B200_\gamma$) X-ray diffraction line after the RHT and compression tests. According to Table 1, the RHT produces a less broadening of the {200}γ X-ray line than alloy after compression tests at deformation temperatures from 350 to 700 °C, similar to deformation temperature of 900 °C. This confirms the presence of a more developed dislocation substructure and, consequently, the absence of the ε-martensite phase due to the above-described mechanism. The decrease in the γ-austenite X-ray line width $B200_\gamma$ is associated with the reduction in crystal lattice defect concentration via softening processes, such as dynamic recovery and dynamic polygonization in the dislocation substructure. With the increase in deformation temperature, the formation of subgrains with low-angle (<15°) sub-boundaries occurs. The higher the deformation temperature, the lower the density of free dislocations in the subgrain bodies, and the larger the subgrain size, which leads to a decrease in the X-ray line broadening [39].

Table 1. X-ray line half-height width $B200_\gamma$ of Fe-30Mn-5Si alloy after RHT and compression tests at deformation temperatures 350–900 °C with strain rate 1 s^{-1}.

RHT and TMT Regimes	$B_{200\gamma}$, 2θ deg.
RHT	0.39 ± 0.02
350 °C	0.51 ± 0.03
500 °C	0.50 ± 0.03
700 °C	0.43 ± 0.03
900 °C	0.39 ± 0.02

3.2. Electrochemical Behavior

The electrochemical behavior of the Fe-30Mn-5Si alloy was studied in Hanks' solution at 37 °C after the compression tests at deformation temperatures of 350 °C, 500 °C, 700 °C, and 900 °C and strain rate 1 s^{-1}. In order to clarify the features of the alloy's electrochemical behavior after compression tests, the RHT specimens, as well as pure ARMCO Fe, were also tested for comparison.

Figure 4 shows the OCP curves of the Fe-30Mn-5Si alloy after compression tests, RHT, and pure ARMCO Fe. As can be seen, during the exposure in the solution, the OCP curves of pure Armco Fe monotonically shift down toward the negative value until the stationary value is established (−480 mV). Moreover, with the general tendency of its change on polarization curves, small fluctuations of the potential within ±3 mV are observed, which reflect the processes that occur on the surface of the specimens. Thus, potential shifts to the negative direction (cathodic polarization) can be associated with the filling of primary pores in the film of corrosion products with electrolyte, resulting in an increase in the area of the metal in contact with the solution. The further anodic polarization is associated with the formation of corrosion products in the pores, which shield the metal surface from the electrolyte. The subsequent potential shifts in the negative direction can be associated with the destruction of the corrosion product film and the formation of secondary pores. In the steady-state stage, OCP fluctuates around a certain value, which reflects the stationary state of the corrosion process.

The observed OCP kinetics is characteristic of metals that corrode in active mode. Pure ARMCO Fe contains a small amount of cementite ($Fe_{23}C_6$ or Fe_3C [40]), which acts as an effective cathodic structural component. During the exposure, corrosion products (iron hydroxides) form on the surface of the studied specimens [46], which impedes the access of the depolarizer (dissolved oxygen) to the cathode areas of the surface and leads to the inhibition of the cathodic depolarization process due to the diffusion overvoltage generation and, as a result, to a negative potential shift and the decrease in the corrosion rate.

The steady-state OCP values of the Fe-30Mn-5Si alloy, after all studied TMT modes, exhibit much more negative values (−675 ± 7 mV), which is obviously due to the presence of strongly electronegative Mn (the standard electrode potential of the $Mn \leftrightarrow Mn^{2+} + 2e$ reaction is −1.18 V) in the alloy composition. The corrosion of the Fe-30Mn-5Si alloy also proceeds by the active dissolution mode; however, the nature of the OCP changing during the exposure, in this case, indicates some specific corrosion process features. In all cases, in the initial period of exposure (100–200 s), there is a slight but quite noticeable OCP shift toward the negative values, which indicates a similar nature of the effect of primary corrosion products.

In this case, the OCP values of the RHT specimens after the initial cathodic polarization are established at a value of −748 ± 2 mV and are stable during further exposure. At the same time, alloy specimens after the compression test experience anodic polarization of more than 100 mV. Such behavior of the alloy is related to its fully austenitic (FCC γ-phase) structure after all modes of compression testing. The exception is the specimens after the compression test at 900 °C, after which (according to the results of X-ray phase analysis, Section 3.1) a small amount of the HCP ε-martensite phase is formed, which acts as the anodic structural component [24,47]. However, this does not alter the overall OCP change trend during the exposure. The absence of the ε-martensite in the alloy (or its low content) suggests that active grain boundaries, subgrains, and subgrains themselves will act as the main anode structural component, which should lead to their selective dissolution. This should lead to a relative increase in the cathode γ-austenite area fraction and, consequently, the accelerated cathodic process and a corresponding corrosion rate increase with the observed potential shift towards the positive direction.

After undergoing the RHT, the Fe-30Mn-5Si alloy exhibits a two-phase structure consisting of γ-austenite and ε-martensite, with a higher volume fraction of the ε-martensite phase [40,44,48,49]; therefore, the "cathodc efficiency" of γ-austenite is less manifested, and the corrosion process occurs mainly by ε-martensite phase dissolution. It should be noted that both for the Fe-30Mn-5Si alloy and for pure Fe, the process of establishing the steady-state corrosion stage, the potential fluctuations are observed, but in the case of Fe-30Mn-Si alloy, the oscillation magnitude reaches 20 mV. Larger OCP fluctuations may indicate the formation of more friable corrosion products with low adhesion and, accordingly, poor protection ability.

It also seems appropriate to consider another option to justify the spontaneous anodic polarization observed during the exposure of the alloy specimens after all modes of the compression test. In theory, the formation of sustainable corrosion products exhibiting even minor protective ability can lead to inhibition of the anodic process and, accordingly, to the corrosion rate decrease. The structure of the alloy after compression testing is represented only by the cathodic γ-austenite, and the active grain and subgrain boundaries [39], which occupy an insignificant part of the surface area compared to the area of the grains themselves, act as the anodic structural component. Since the anodic process turns out to be highly localized, it is impossible to cover the predominant part of the specimen surface with corrosion products. But, due to the metal atom ionization becoming largely inhibited, we should expect the anodic process to slow down and, therefore, spontaneous anodic polarization to take place. However, this explanation contradicts the data obtained for the RHT mode. Since the anodic structural component (ε-martensite) dominates in the two-phase structure of the alloy after RHT, the corrosion products should occupy a significant part of the specimen surface during the dissolution process. This should lead to a significant inhibition of the anodic process and cause a large OCP shift toward the positive direction. However, according to Figure 4, the OCP value of the RHT is almost constant during the exposure and is about −750 mV, which is 50–70 mV lower than the steady-state OCP values of the alloy after compression tests. The inconsistency of such an explanation for the observed regularities of OCP changes during the exposure is supported by the results of the corrosion current density measurements presented below, according to which the highest corrosion rate is achieved for the alloy right after RHT.

Figure 5 shows the potentiodynamic polarization diagrams of pure ARMCO Fe and Fe-30Mn-5Si alloy after compression tests and RHT mode. It can be seen that regardless of the test mode, there is a similar pattern of OCP change upon the potential application. The anodic and cathodic branches of the polarization curves represent, respectively, the reaction of the dissolution of metal (alloy) atoms and the reaction involving oxygen in the depolarization process. A significant slope of the cathodic (lower) branches of the polarization curves indicates that the corrosion process, in all cases, proceeds with cathodic control. The cathodic process in neutral media is described by the following reaction:

$$O_2 + 2H_2O + 4e^- = 4OH^-, \quad (2)$$

which is the sum of sequential elementary reactions, four of which are electrochemical with the participation of one electron in each [42,44,45]. It is believed that the rate-limiting stage of the process is the formation of a molecular oxygen ion by an electrochemical reaction

$$O_2 + e^- \rightarrow O_2^-, \quad (3)$$

whose slowness causes the cathodic process overvoltage. As follows from the polarization diagrams plotted in the Tafel representation (Figure 5), the cathodic branches are straight lines in a wide potential range, with the same slope b of the Tafel equation, accounting for 0.2 ± 0.01, but with different cathode current densities under the same applied polarization potentials. Taking into account that the mechanism of the cathode process is the same, one can indicate the difference in kinetic parameters.

$$b = \frac{RT}{anF} \cdot 2.303; \quad (4)$$

where $R = 8.314 \frac{J}{mol \cdot K}$; T is the temperature, K; a is the charge transfer coefficient; n is the number of electrons involved in the reaction; $F = 96,500 \cdot \frac{c}{g-equivalent}$, Faraday constant.

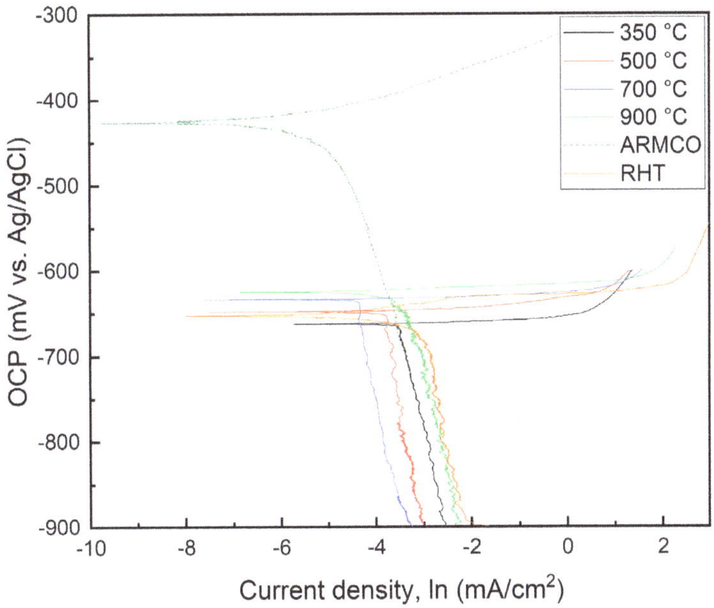

Figure 5. Polarization diagrams of the Fe-30Mn-5Si alloy immersed in Hanks' solution at 37 °C after compression tests at deformation temperatures 350, 500, 700, and 900 °C and strain rate of 1 s^{-1}.

Considering that $n = 1$ and $T = 310$ K, the charge transfer coefficient for the cathodic reaction is $\alpha = 0.3 \pm 0.01$, as follows from Equation (4). The α coefficient characterizes the degree of impact of the electrode electric field on the activation energy of the electrochemical stage, and the obtained value indicates the asymmetry of the activation barrier of the electrochemical reaction.

The electrochemical parameters of Fe-30Mn-5Si alloy specimens after compression tests and the RHT mode, compared to pure ARMCO Fe, obtained by extrapolation of the Tafel parts of the polarization curves in Hanks' solution at 37 °C, are listed in Table 2.

Table 2. The electrochemical parameters of Fe-30Mn-5Si alloy specimens after compression tests and the RHT mode, compared to pure ARMCO Fe, obtained by extrapolation of the Tafel parts of the polarization curves in Hanks' solution at 37 °C.

Treatment	Corrosion Potential E_{corr}, mV	$i_{corr} \cdot 10^5$, A/cm^2	Corrosion Rate *, *Cr*, mm/year
compression test temperature 350 °C	−663 ± 5	2.75 ± 0.16	0.33
compression test temperature 500 °C	−648 ± 4	2.11 ± 0.1	0.25
compression test temperature 700 °C	−633 ± 6	1.24 ± 0.09	0.14
compression test temperature 900 °C	−625 ± 4	3.58 ± 0.14	0.42
RHT	−653 ± 7	5.11 ± 0.26	0.6
Pure ARMCO Fe	−427 ± 3	0.85 ± 0.05	0.10

* Calculated values according to the Equation (1).

As can be seen from Table 2, the following regularity takes place: for Fe-30Mn-5Si alloy after compression tests, with an increase in deformation temperature from 350 to 900 °C, the corrosion potential shifts toward the positive values, which indicates a decrease in the electrochemical activity of the alloy; at the same time, the expected decrease in the corrosion current density is observed only with an increase in the deformation temperature to 700 °C, while deformation at 900 °C, on the contrary, leads to a significant increase in the corrosion current density despite the apparent higher thermodynamic (electrochemical) stability of the studied alloy.

The observed trends in the electrochemical parameters are related to the structure of the specimens. As shown above, the structure of Fe-30Mn-Si alloy after compression tests in the temperature range from 350 to 700 °C is characterized by the single-phase state (FCC γ-austenite). However, the observed changes in E_{corr} and i_{corr} indicate the necessity to consider the structural features that affect the electrochemical characteristics of the alloy. The compression test at the temperature of 350 °C is accompanied by the formation of a dynamically polygonized dislocation substructure in the deformed grains of the γ-austenite matrix, characterized by an increased density of dislocations in the subgrain body. The strongest broadening of the X-ray lines also indicates the most strengthened state of the structure (Table 1). With an increase in the temperature of compression tests to 500 and 700 °C, dynamic recovery is observed, leading to a decrease in the concentration of point defects and a decrease in dislocation density not associated with the formation and movement of grain boundaries and dynamic polygonization—redistribution of dislocations by slip and diffusion, which is accompanied by their partial annihilation and the formation of subgrains inside crystallites free of dislocations [39,50,51]. The observed changes in the state of the structure cause an "elevation" of the potential and a decrease in the corrosion rate.

The electrochemical parameters of specimens after compression test at 900 °C are influenced by two factors: (1) the ongoing process of dynamic recovery and dynamic recrystallization, accompanied by the formation of fine equiaxed grains, which should cause further "elevation" of the corrosion potential; (2) the appearance of an additional anodic phase—ε-martensite in the structure of the specimen, which should lead to an increased corrosion rate. However, judging by the significant increase in the corrosion current density, in this case, the determining factor in the electrochemical behavior of the specimens is the appearance of the ε-martensite. This is indicated by the electrochemical parameters of the RHT mode, which is in a two-phase (γ-austenite and ε-martensite) state of the structure, in which the ε-phase predominates. The present data correlate well with studies of the Fe-29Mn-5Si alloy, which also contains only FCC γ-austenite phase and consequently has a corrosion rate in the range of 0.21–0.25 mm/year [52].

4. Conclusions

1. The theoretical justification for the atypical changes in the electrode potential during the exposure of the Fe-30Mn-5S alloy, which corrodes via active dissolution mode in a biological solution, was provided. It is shown that the cause of the spontaneous anodic polarization of the Fe-30Mn-5Si alloy after compression tests at temperatures of 350, 500, 700, and 900 °C is the absence (or low content) of the anodic structural component (HCP ε-martensite) after the compression test at 900 °C. Wherein the active grains and subgrains boundaries, and subgrains themselves act as the anodic structural components, with their selective dissolution increasing the relative amount of cathodic structural components on the surface of the studied specimens that leads to an accelerated cathodic process and corresponding increase in biodegradation rate with an observed shift toward the positives values of the electrode potential;
2. The biodegradation rates of the Fe-30Mn-5Si alloy were calculated using the extrapolation method of polarization curves after compression tests at deformation temperatures of 350, 500, 700, and 900 °C and a strain rate of 1 s^{-1}. The analysis of polarization curves allowed for the establishment of the causes and patterns of changes in the electrochemical parameters of the studied alloy, depending on the thermomechanical treatment regimes. The observed elevation of the corrosion potential and the decrease in the corrosion current density of the Fe-30Mn-5Si alloy after compression tests at the deformation temperatures of 350, 500, and 700 °C, which consist only of the FCC γ-austenite phase, are due to dynamic recovery, which leads to a decrease in the concentration of point defects, and dynamic polygonization;
3. Based on the results of the rheological compression tests, taking into account the biodegradation rate of the Fe-30Mn-5Si alloy, the radial–shear rolling, multiaxial forging, and longitudinal rolling should be carried out at a deformation temperature of 900 °C and strain rate of 1 s^{-1} due to the following reason: these chosen deformation conditions result in a two-phase state, comprising FCC γ-austenite + HCP ε-martensite, which provides the highest biodegradation rate. It is worth noting that the results of this study on electrochemical behavior established that thermomechanical treatment at a specific deformation temperature from 350 to 900 °C and strain rate 1 s^{-1} can regulate the biodegradation rate of the Fe-30Mn-5Si alloy, which ranges from 0.14 to 0.42 mm/year, depending on the desired recovery rate for damaged bone tissue.

Author Contributions: Conceptualization, S.P. and Y.P.; methodology, Y.P., V.S. and P.K.; validation, S.P., Y.P. and P.K.; formal analysis, P.K. and Y.Z.; investigation, P.K., M.K., A.K. and A.B.; resources, S.P. and Y.Z.; data curation, Y.P. and V.S.; writing—original draft preparation, P.K. and Y.P.; writing—review and editing, Y.P. and S.P.; supervision, S.P. and Y.P.; project administration, P.K.; funding acquisition, Y.Z. All authors have read and agreed to the published version of the manuscript.

Funding: The present work has been carried out with the financial support of the Strategic Academic Leadership Program "Priority 2030" at NUST "MISIS" (project K7-2023-013).

Data Availability Statement: Data sharing not applicable.

Conflicts of Interest: The authors declare no conflict of interest. The funders had no role in the design of this study, in the collection, analyses, or interpretation of data, in the writing of this manuscript, or in the decision to publish the results.

References

1. Venezuela, J.; Dargusch, M.S. Addressing the slow corrosion rate of biodegradable Fe-Mn: Current approaches and future trends. *Curr. Opin. Solid State Mater. Sci.* **2020**, *24*, 100822. [CrossRef]
2. Hermawan, H. *Biodegradable Metals—From Concept to Applications*; Springer: Berlin/Heidelberg, Germany, 2012. [CrossRef]
3. Park, J.; Lakes, R.S. *Biomaterials—An Introduction*; Springer: New York, NY, USA, 2007. [CrossRef]

4. Narushima, T. *Metals for Biomedical Devices*, 2nd ed.; Niinomi, M., Ed.; Elsevier: Amsterdam, The Netherlands, 2019; pp. 495–521. [CrossRef]
5. Witte, F. The history of biodegradable magnesium implants: A review. *Acta Biomater.* **2010**, *6*, 1680–1692. [CrossRef]
6. Seitz, J.-M.; Durisin, M.; Goldman, J.; Drelich, J.W. Recent advances in biodegradable metals for medical sutures: A critical review. *Adv. Healthc. Mater.* **2015**, *4*, 1915–1936. [CrossRef]
7. Li, C.; Guo, C.; Fitzpatrick, V.; Ibrahim, A.; Zwierstra, M.J.; Hanna, P.; Lechtig, A.; Nazarian, A.; Lin, S.J.; Kaplan, D.L. Design of biodegradable, implantable devices towards clinical translation. *Nat. Rev. Mater.* **2020**, *5*, 61–81. [CrossRef]
8. Heiden, M.; Walker, E.; Stanciu, L. Magnesium, Iron and Zinc Alloys, the Trifecta of Bioresorbable Orthopaedic and Vascular Implantation—A Review. *J. Biotechnol. Biomater.* **2015**, *5*, 178. [CrossRef]
9. Li, H.; Zheng, Y.; Qin, L. Progress of biodegradable metals. *Prog. Nat. Sci. Mater. Int.* **2014**, *24*, 414–422. [CrossRef]
10. Hyuk, S.; Jung, Y.; Kim, S.H. Current status and future direction of biodegradable metallic and polymeric vascular scaffolds for next-generation stent. *Acta Biomater.* **2017**, *60*, 3–22. [CrossRef]
11. Witte, F.; Hort, N.; Vogt, C.; Cohen, S.; Kainer, K.U.; Willumeit, R.; Feyerabend, F. Degradable biomaterials based on magnesium corrosion. *Curr. Opin. Solid State Mater. Sci.* **2008**, *12*, 63–72. [CrossRef]
12. Sezer, N.; Evis, Z.; Kayhan, S.M.; Tahmasebifar, A.; Koç, M. Review of magnesiumbased biomaterials and their applications. *J. Magnes. Alloys* **2018**, *6*, 23–43. [CrossRef]
13. Yang, H.; Jia, B.; Zhang, Z.; Qu, X.; Li, G.; Lin, W.; Zhu, D.; Dai, K.; Zheng, Y. Alloying design of biodegradable zinc as promising bone implants for load-bearing applications. *Nat. Commun.* **2020**, *11*, 401. [CrossRef]
14. Mostaed, E.; Sikora-Jasinska, M.; Mostaed, A.; Loffredo, S.; Demir, A.G.; Previtali, B.; Mantovani, D.; Beanland, R.; Vedani, M. Novel Zn-based alloys for biodegradable stent applications: Design, development and in vitro degradation. *J. Mech. Behav. Biomed. Mater.* **2016**, *60*, 581–602. [CrossRef] [PubMed]
15. Kraus, T.; Moszner, F.; Fischerauer, S.; Fiedler, M.; Martinelli, E.; Eichler, J.; Witte, F.; Willbold, E.; Schinhammer, M.; Meischel, M.; et al. Biodegradable Fe-based alloys for use in osteosynthesis: Outcome of an in vivo study after 52 weeks. *Acta Biomater.* **2014**, *10*, 3346–3353. [CrossRef] [PubMed]
16. Francis, A.; Yang, Y.; Virtanen, S.; Boccaccini, A.R. Iron and iron-based alloys for temporary cardiovascular applications. *J. Mater. Sci. Mater. Med.* **2015**, *26*, 138. [CrossRef] [PubMed]
17. He, J.; He, F.; Li, D.; Liu, Y.-L.; Liu, Y.Y.; Yea, Y.-J.; Yin, D.C. Advances in Fe-based biodegradable metallic materials. *RSC Adv.* **2016**, *6*, 112819–112838. [CrossRef]
18. Fantanariu, M.; Trinca, L.C.; Solcan, C.; Trofin, A.; Strungaru, S.; Şindilar, E.V.; Plavan, G.; Stanciu, S. A new Fe–Mn–Si alloplastic biomaterial as bone grafting material: In vivo study. *Appl. Surf. Sci.* **2015**, *352*, 129–139. [CrossRef]
19. Trinca, L.C.; Burtan, L.; Mareci, D.; Fernández-Perez, B.M.; Stoleriu, I.; Stanciu, T.; Stanciu, S.; Solcan, C.; Izquierdo, J.; Souto, R.M. Evaluation of in vitro corrosion resistance and in vivo osseointegration properties of a Fe-Mn-Si-Ca alloy as potential degradable implant biomaterial. *Mater. Sci. Eng. C* **2021**, *118*, 111436. [CrossRef]
20. Revie, W.R.; Uhlig, H.H. *Corrosion and Corrosion Control: An Introduction to Corrosion Science and Engineering*, 4th ed.; John Wiley and Sons Inc.: Hoboken, NJ, USA, 2008.
21. Lu, S.; Wang, Q.; Zhang, Y.; Li, H.; Feng, H.; Tan, L.; Yang, K. A novel biodegradable high nitrogen iron alloy with simultaneous enhancement of corrosion rate and local corrosion resistance. *J. Mater. Sci. Technol.* **2023**, *152*, 94–99. [CrossRef]
22. Otsuka, K.; Wayman, C.M. *Shape Memory Materials*; Cambridge University Press: Cambridge, UK, 1998; p. 284.
23. Drevet, R.; Zhukova, Y.; Malikova, P.; Dubinskiy, S.; Korotitskiy, A.; Pustov, Y.; Prokoshkin, S. Martensitic Transformations and Mechanical and Corrosion Properties of Fe-Mn-Si Alloys for Biodegradable Medical Implants. *Metall. Mater. Trans. A* **2018**, *49*, 1006–1013. [CrossRef]
24. Prokoshkin, S.; Pustov, Y.; Zhukova, Y.; Kadirov, P.; Dubinskiy, S.; Sheremetyev, V.; Karavaeva, M. Effect of thermomechanical treatment on functional properties of biodegradable Fe-30Mn-5Si shape memory alloy. *Metall. Mater. Trans. A* **2021**, *52*, 2024–2032. [CrossRef]
25. Prokoshkin, S.; Pustov, Y.; Zhukova, Y.; Kadirov, P.; Karavaeva, M.; Prosviryakov, A.; Dubinskiy, S. Effect of thermomechanical treatment on structure and functional fatigue characteristics of biodegradable Fe-30Mn-5Si 3 (wt.%) shape memory alloy. *Materials* **2021**, *14*, 3327. [CrossRef]
26. Sheremetyev, V.; Kudryashova, A.; Cheverikin, V.; Korotitskiy, A.; Galkin, S.; Prokoshkin, S.; Brailovski, V. Hot radial shear rolling and rotary forging of metastable beta Ti-18Zr-14Nb (at. %) alloy for bone implants: Microstructure, texture and functional properties. *J. Alloys Compd.* **2019**, *800*, 320–326. [CrossRef]
27. Kudryashova, A.; Sheremetyev, V.; Lukashevich, K.; Cheverikin, V.; Inaekyan, K.; Galkin, S.; Prokoshkin, S.; Brailovski, V. Effect of a combined thermomechanical treatment on the microstructure, texture and superelastic properties of Ti-18Zr-14Nb alloy for orthopedic implants. *J. Alloys Compd.* **2020**, *843*, 156066. [CrossRef]
28. Naizabekov, A.B.; Lezhnev, S.N.; Arbuz, A.S.; Panin, E.A. Obtaining of long-length rods with ultrafine-grained structure by the radial-shear rolling. *IOP Conf. Ser. Mater. Sci. Eng.* **2019**, *461*, 012065. [CrossRef]
29. Dobatkin, S.; Galkin, S.; Estrin, Y.; Serebryany, V.; Diez, M.; Martynenko, N.; Lukyanova, E.; Perezhogin, V. Grain refinement, texture, and mechanical properties of a magnesium alloy after radial-shear rolling. *J. Alloys Compd.* **2019**, *774*, 969–979. [CrossRef]
30. Stefanik, A.; Szota, P.; Mroz, S.; Bajor, T.; Dyja, H. Properties of the AZ31 magnesium alloy round bars obtained in different rolling pricesses. *Arch. Metall. Mater.* **2015**, *60*, 3001–3005. [CrossRef]

31. Gryc, A.; Bajor, T.; Dyja, H. The analysis of influence the parameters of rolling process in three high skew rolling mill of AZ31 magnesium alloy bars on temperature distribution. *Metalurgija* **2016**, *55*, 772–774.
32. Akopyan, T.; Aleshchenko, A.S.; Belov, N.A.; Galkin, S.P. Effect of Radial-Shear Rolling on the Formation of Structure and Mechanical Properties of Al-Ni and Al-Ca Aluminum-Matrix Composite Alloys of Eutectic Type. *Phys. Met. Metallogr.* **2018**, *119*, 241–250. [CrossRef]
33. Karpov, B.V.; Patrin, P.V.; Galkin, S.P.; Kharitonov, E.A.; Karpov, I.B. Radial-Shear Rolling of Titanium Alloy VT-8 Bars with Controlled Structure for Small Diameter Ingots (≤200 mm). *Metallurgist* **2018**, *61*, 884–890. [CrossRef]
34. Negodin, D.A.; Galkin, S.P.; Kharitonov, E.A.; Karpov, B.V.; Kharkovsky, D.N.; Dubovitskaya, I.A.; Patrin, P.V. Testing of the Technology of Radial-Shear Rolling and Predesigning Selection of Rolling Minimills for the Adaptable Production of Titanium Rods with Small Cross Sections Under the Conditions of the "CHMP" JSC. *Metallurgist* **2019**, *62*, 1143. [CrossRef]
35. Lopatin, N.V. Effect of hot rolling by screw mill on microstructure of a Ti-6Al-4V titanium alloy. *Int. J. Mater. Form.* **2013**, *6*, 459–465. [CrossRef]
36. Galkin, S.P. Theory and Technology of Stationary Helical Rolling of Billets and Bars of Low-Plasticity Steels and Alloys; Dissertation Abstract for the Degree of Doctor of Technical Sciences. Ph.D. Thesis, Moscow Institute, Moscow, Russia, 1998.
37. Manjunath, G.A.; Shivakumar, S.; Fernandez, R.; Nikhil, R.; Sharath, P.C. A review on effect of multi-directional forging/multi-axial forging on mechanical and microstructural properties of aluminum alloy. *Mater. Today Proc.* **2021**, *47*, 2565–2569. [CrossRef]
38. Zambrano, O.A.; Logé, R.E. Dynamic recrystallization study of a Fe-Mn-Si based shape memory alloy in constant and variable thermomechanical conditions. *Mater. Charact.* **2019**, *152*, 151–161. [CrossRef]
39. Kadirov, P.; Zhukova, Y.; Pustov, Y.; Karavaeva, M.; Sheremetyev, V.; Korotitskiy, A.; Shcherbakova, E.; Baranova, A.; Komarov, V.; Prokoshkin, S. Effect of plastic deformation in various temperature-rate conditions on structure and mechanical properties of biodegradable Fe-30Mn-5Si alloy. *Metall. Mater. Trans. A*, 2023; *submitted*.
40. Pustov, Y.; Zhukova, Y.; Malikova, P.; Prokoshkin, S.; Dubinskii, S. Structure and Corrosion-Electrochemical Behavior of Bioresorbable Alloys Based on the Fe–Mn System. *Prot. Met. Phys. Chem. Surf.* **2018**, *58*, 469–476. [CrossRef]
41. Sheremetev, V.A.; Akhmadkulov, O.B.; Komarov, V.S.; Korotitskii, A.V.; Lukashevich, K.E.; Galkin, S.P.; Andreev, V.A.; Prokoshkin, S.D. Thermomechanical Behavior and Structure Formation of Shape Memory Ti–Zr–Nb Alloy for Medical Applications. *Met. Sci. Heat Treat.* **2021**, *63*, 403–413. [CrossRef]
42. Winston, R.R. (Ed.) *Uhlig's Corrosion Handbook*, 3rd ed.; John Wiley & Sons, Inc.: Hoboken, NJ, USA, 2011. [CrossRef]
43. *ASTM G31-21*; Standard Guide for Laboratory Immersion Corrosion Testing of Metals. ASTM International: West Conshohocken, PA, USA, 2021.
44. Isaev, N.I. *Theory of Corrosion Processes*; Metallurgiya: Moscow, Russia, 1997; p. 360. (In Russian)
45. Zhuk, N.P. *A Course in Corrosion Theory and Protection of Metals*; Metallurgiya: Moscow, Russia, 1976; p. 472. (In Russian)
46. Jiang, X.; Zhang, L.; Yin, L.; Yang, G.; Xie, J.; Zhang, L.; Lu, H.; Liang, D.; Deng, L. Corrosion behavior of fluorinated carbonyl iron-hydrophobic composites in neutral salt spray environment. *Corros. Sci.* **2023**, *210*, 110823. [CrossRef]
47. Sokol, I.Y.; Ulyanin, E.A.; Feldgandler, E.G. *Structure and Corrosion of Metals and Alloys, Atlas: A Handbook*; Metallurgy: Moscow, Russia, 1989; p. 400. (In Russian)
48. Hermawan, H.; Dube, D.; Mantovani, D. Degradable metallic biomaterials: Design and development of Fe–Mn alloys for stents. *J. Biomed. Mater. Res. Part A* **2010**, *93A*, 1–11. [CrossRef]
49. Xuemei, Z.; Yansheng, Z. Effect of ε-martensite on the electrochemical corrosion behavior of an Fe-Mn-Si shape memory alloy in aqueous solutions. *J. Mater. Sci. Lett.* **1997**, *16*, 1516–1517. [CrossRef]
50. Li, Y.; Jahr, H.; Pavanram, P.; Bobbert, F.S.L.; Puggi, U.; Zhang, X.-Y.; Pouran, B.; Leeflang, M.A.; Weinans, H.; Zhou, J.; et al. Additively manufactured functionally graded biodegradable porous iron. *Acta Biomater.* **2019**, *96*, 646–661. [CrossRef]
51. Obayi, C.S.; Tolouei, R.; Mostavan, A.; Paternoster, C.; Turgeon, S.; Okorie, B.A.; Obikwelu, D.O.; Mantovani, D. Effect of grain sizes on mechanical properties and biodegradation behavior of pure iron for cardiovascular stent application. *Biomatter* **2016**, *6*, e959874. [CrossRef]
52. Rybalchenko, O.; Anisimova, N.; Martynenko, N.; Rybalchenko, G.; Belyakov, A.; Shchetinin, I.; Lukyanova, E.; Chernogorova, O.; Raab, A.; Pashintseva, N.; et al. Biocompatibility and Degradation of Fe-Mn-5Si Alloy after Equal-Channel Angular Pressing: In Vitro and In Vivo Study. *Appl. Sci.* **2023**, *13*, 9628. [CrossRef]

metals

MDPI

Article

Enhancing Mechanical and Biocorrosion Response of a MgZnCa Bulk Metallic Glass through Variation in Spark Plasma Sintering Time

Bin Shi Jie Bryan [1], Kai Soon Fong [2], Chua Beng Wah [2], Sravya Tekumalla [3], Min Kyung Kwak [4], Eun Soo Park [4] and Manoj Gupta [1,*]

1 Department of Mechanical Engineering, National University of Singapore, Singapore 117576, Singapore
2 Singapore Institute of Manufacturing Technology (SIMTech), Agency for Science, Technology and Research (A*STAR), 5 CleanTech Loop #01-01, CleanTech Two Block B, Singapore 636732, Singapore
3 Department of Mechanical Engineering, University of Victoria, 3800 Finnerty Road, Victoria, BC V8W 2Y2, Canada
4 Department of Materials Science and Engineering, Research Institute of Advanced Materials & Institute of Engineering Research, Seoul National University, Seoul 08826, Republic of Korea
* Correspondence: mpegm@nus.edu.sg

Abstract: Development of metallic glasses is hindered by the difficulties in manufacturing bulk parts large enough for practical applications. Spark plasma sintering (SPS) has emerged as an effective consolidation technique in the formation of bulk metallic glasses (BMGs) from melt-spun ribbons. In this study, $Mg_{65}Zn_{30}Ca_5$ melt-spun ribbons were sintered at prolonged sintering times (15 min to 180 min) via SPS under a pressure of 90 MPa and at a temperature of 150 °C (which is below the crystallization temperature), to provide an insight into the influence of sintering time on the consolidation, structural, and biodegradation behavior of Mg-BMGs. Scanning Electron Microscopy was used to characterize the microstructure of the surface, while the presence of the amorphous phase was characterized using X-ray diffraction and Electron Backscatter Diffraction. Pellets 10 mm in diameter and height with near-net amorphous structure were synthesized at 150 °C with a sintering time of 90 min, resulting in densification as high as 98.2% with minimal crystallization. Sintering at extended durations above 90 min achieved higher densification and resulted in a significant amount of local and partial devitrification. Mechanical properties were characterized via compression and microhardness testing. Compression results show that increased sintering time led to better structural integrity and mechanical properties. Notably, SPS150_90 displayed ultimate compressive strength (220 MPa) that matches that of the cortical bone (205 MPa). Corrosion properties were characterized via potentiodynamic polarization with Phosphate Buffered Solution (PBS). The results suggest that the sintered samples have significantly better corrosion resistance compared to the crystalline form. Overall, SPS150_90 was observed to have a good balance between corrosion properties (10× better corrosion resistance to as-cast alloy) and mechanical properties.

Keywords: magnesium-based bulk metallic glass; spark plasma sintering; melt spinning; mechanical properties; biocorrosion properties

1. Introduction

Magnesium-based metallic glasses have been studied extensively in recent years for a wide range of biomedical applications [1,2]. These metallic glasses are different from traditional metallic alloys as they exist in the amorphous state and are shown to exhibit excellent corrosion resistance and yield strength [1,3]. Coupled with their good biocompatibility in vivo [4], magnesium-based bulk metallic glasses have great potential in the biomedical sector as an alternative class of absorbable implant biomaterials [5].

Despite the numerous advantages of bulk metallic glasses, very few have made it to the manufacturing stage. This is because the synthesis of bulk metallic glasses

(BMGs) is extremely challenging, owing to the need for extremely high cooling rates of ~10^4–10^5 °C/s [6]. Different densification methods have been proposed to obtain BMGs, including warm extrusion [7,8], equal channel angular extrusion [9], selective laser sintering [10], and spark plasma sintering [11]. Among them, spark plasma sintering (SPS) has garnered special interest as a post-consolidation method due to its high throughput power consolidation, which is assisted by elevated temperature and pressure [12].

Sintering is a process in which materials are compacted into solid-state objects. The bonding of sintered particles via atomic diffusion mechanics is governed by both heat and pressure [13]. Since Mg-based BMGs have a relatively low working temperature range of up to 150 °C in Mg-Zn-Ca metallic glass, SPS has shown to be extremely useful in the synthesis of Mg-based BMGs owing to its precise temperature control and high pressure [11]. Any temperature above 150 °C would result in the nucleation of crystals, destroying the amorphous nature of the BMG [11].

In our previous paper, we devised a two-step synthesis process involving melt spinning and spark plasma sintering to successfully fabricate fully amorphous Mg-based bulk metallic glass pellets with a diameter of 10 mm and high corrosion resistance [11]. However, their structural integrity and mechanical properties were not ideal because of the presence of significant porosity and the lack of inter-atomic bonding during the SPS process.

Accordingly, this research focuses on refining the SPS process through three objectives: (i) improving the structural integrity and mechanical properties of Mg-Zn-Ca BMG, while maintaining a complete amorphous nature by increasing sintering time; (ii) understanding the effect of sintering time on diffusion mechanics; and (iii) maintaining good biocorrosion resistance post SPS.

2. Materials and Methods

2.1. Materials

An alloy ingot with a composition of $Mg_{65}Zn_{30}Ca_5$ (as-cast) was synthesized via Disintegrated Melt Deposition (DMD), and the raw materials used are detailed in Table 1 below. The raw materials were melted at a target superheat temperature of 750 °C, followed by stirring of the melt at 420 rpm for 5 min before casting.

Table 1. Raw materials used in the synthesis of $Mg_{65}Zn_{30}Ca_5$.

Raw Material	Supplier
Magnesium (Mg) turnings, 99.9% purity	ACROS Organics, Waltham, MA, USA
Zinc (Zn) granules, 99.8% purity	Alfa Aesar, Waltham, MA, USA
Calcium (Ca) granules, 99.8% purity	Alfa Aesar, Waltham, MA, USA

2.2. Synthesis

The master alloy (5 g) was melt-spun in a quartz crucible under an argon atmosphere with a copper spinning wheel turning at 2000 rpm, with the argon injection pressure set at 450 mbar. The distance between the nozzle of the quartz crucible and the copper wheel was controlled at 0.5 mm. Amorphous ribbons of the alloy were produced with a thickness and width of ~0.1 mm and 3 mm, respectively. A schematic of the melt spinning setup is shown in Figure 1.

The amorphous ribbons obtained were mechanically milled into powder and compressed at room temperature using a 10 mm compaction die in a hydraulic press under 500 psi pressure for 1 min.

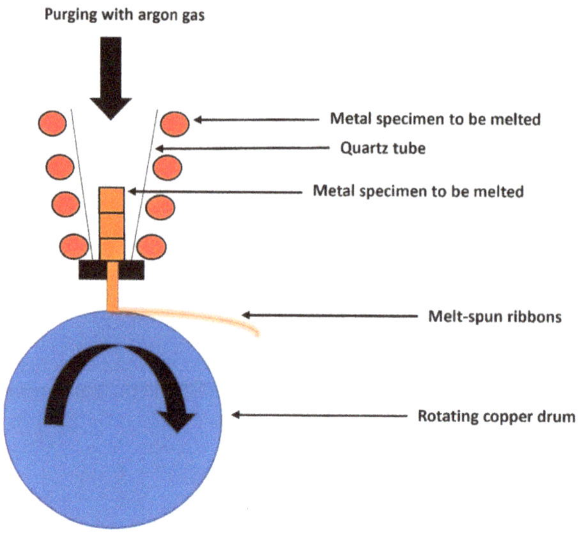

Figure 1. Schematic illustration of the melt spinning process.

Finally, the compacted powder was sintered into pellets, with a diameter of 10 mm and a height of 10 mm, via SPS. A DR. SINTER LAB Jr. SERIES 632LX machine from Fuji Electronics was utilized for this process, operating under vacuum conditions (<3 Pa) and featuring a maximum sintering DC pulse of 3000 A. A K-type thermocouple was employed to monitor and regulate the temperature. During the sintering process, the ribbons were placed in a high-strength carbon die at a pressure of 90 MPa and at 150 °C. The temperature of 150 °C was established as the optimal operating temperature, beyond which crystallization occurs [11]. A schematic illustration of the SPS technique is shown in Figure 2.

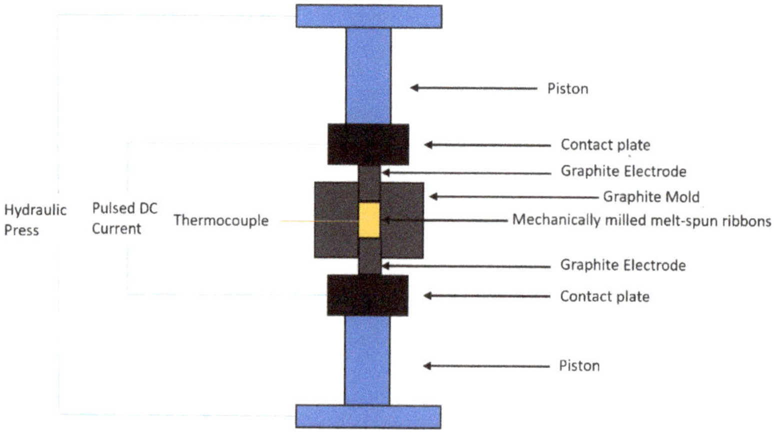

Figure 2. Schematic illustration of spark plasma sintering.

The initial heating rate was held at 15 K/min but was gradually reduced to 7 K/min near the end of the heating process to avoid overheating. Table 2 presents the specific processing parameters for each sample, with designations such as SPS150_15 representing a sample sintered at 150 °C for 15 min.

Table 2. Spark plasma sintering operating parameters and experimental procedure.

	Starting Temp (°C)	Heating Rate (°C/min)	Time (min)	Heating Rate (°C/min)	Time (min)	Final Temperature (°C)	Sintering Time (min)	Force (kN)	Pressure (MPa)
SPS150_15	30	15	4	7	8.6	150	15	7	90
SPS150_30	30	15	4	7	8.6	150	30	7	90
SPS150_60	30	15	4	7	8.6	150	60	7	90
SPS150_90	30	15	4	7	8.6	150	90	7	90
SPS150_180	30	15	4	7	8.6	150	180	7	90

2.3. *Characterization*

2.3.1. Surface Morphology

The spark plasma-sintered samples were ground with 1200 grit SiC paper for XRD analysis and polished up to a 0.05 μm finish using a diamond suspension for SEM/EBSD studies. During each polishing step, the samples were cleaned ultrasonically for 5 min with ethanol. The microstructure of the spark plasma-sintered samples was analyzed using both an OLYMPUS metallographic optical microscope and a JEOL JSM-6010 Scanning Electron Microscope (SEM). The spark plasma-sintered samples were also analyzed using a Shimadzu LAB-XRD 6000 X-ray diffractometer for phase identification. The diffractometer had a Cu K_α radiation source (λ = 0.1541 nm), and a scan speed of 2°/min was used for phase identification. To confirm the amorphous nature of the samples, electron backscattered diffraction (EBSD) was performed using a Gemini SEM 450 (Carl Zeiss Microscope GmbH, Germany) high-resolution scanning electron microscope equipped with an EDAX EBSD system with a velocity camera. Prior to EBSD, the samples were polished using a water-free oxide polishing suspension and were cleaned using lab grade ethanol. Additionally, the samples also underwent ion polishing for 2 h using the PECS II system by Gatan to remove any residual surface deformation/oxide layers.

2.3.2. Density and Porosity

Density measurements were conducted following Archimedes' principle. The samples were initially weighed in air and then in distilled water using an A&D ER-182A electronic balance (Bradford, Massachusetts) with an accuracy of ±0.0001 g. Table 3 below shows the composition and density of the alloy synthesized in this work.

The experimental density of the alloy was determined via Equation (1) below:

$$p_E = \frac{w_{air} \times p_{water} - w_{water} \times p_{air}}{w_{air} - w_{water}} \quad (1)$$

where p_E represents the experimental density of the alloy in g/cm^3, w is the weight of specimen, and p_{air} and p_{water} refer to the density of air and water, respectively, in g/cm^3.

The difference between experimental densities of the as-cast alloy compared to the experimental density of the sintered samples can be used to calculate the porosity levels within the sintered sample, as governed by Equation (2) below:

$$\text{Porosity}(\%) = \frac{p_1 - p_2}{p_2} \times 100 \quad (2)$$

where p_1 and p_2 represent the experimental density of the as-cast master alloy and sintered samples, respectively, in g/cm^3.

Table 3. Composition and densities of the $Mg_{65}Zn_{30}Ca_5$ synthesized in this work.

Material Composition	Raw Materials Composition by Weight %	Theoretical Density (g/cm^3)	Actual Density (g/cm^3)
$Mg_{65}Zn_{30}Ca_5$	Mg 42.2% Zn 52.42% Ca 5.36%	2.86	2.85

2.3.3. Mechanical Properties

Vickers hardness measurements were made using a Shimadzu-HMW automatic digital microhardness tester. A load of 245.2 mN and a 15s dwell time were utilized for each indentation across the sample surface. This set of parameters is in line with ASTM standard E384-08. At least 20 indentations were obtained for each sample.

Following this, compression testing was conducted using an MTS 810 fully automated servo hydraulic mechanical testing machine. The strain rate was set at 8.33×10^{-5} s^{-1}, with height and diameter set to 10 mm (height-to-diameter ratio of 1).

2.3.4. Biocorrosion Tests

The biocorrosion tests were performed via an electrochemical cell setup. The spark plasma-sintered sample was used as the working electrode, the counter electrode was a graphite rod, and a Saturated Calomel Electrode (SCE) was used as the reference electrode. Lastly, the electrolyte used in this study was Phosphate Buffered Solution (PBS) from Life Technologies, Singapore, with a pH of 7.4.

The system was left as an open circuit for 60 min before being polarized at a voltage of −200 mV vs. their open circuit potential (OCP) to +200 mV vs. their OCP, with a scan rate of 1 mV s^{-1}. To simulate body conditions, the electrolyte was maintained at 37 °C via a water bath. A Gamry Reference 600 potentiostat was used for data collection, while a Solatron Analytical SI 1287 was used for data analysis.

3. Results

3.1. Spark Plasma Sintering of Mg-Zn-Ca BMGs

There are a few parameters that can be optimized with SPS, namely pressure, sintering temperature, and sintering time. In the present study, pressure was kept constant at 90 MPa, which is the maximum pressure that can be safely exerted on the high-strength carbon die. Sintering temperature was maintained at 150 °C, which is the optimal working temperature of $Mg_{65}Zn_{30}Ca_5$, above which crystallization occurs that destroys the amorphous nature [11]. Therefore, this study focused on the effect of elevated sintering time on the amorphous nature, mechanical properties, and biocorrosion properties of MgZnCa BMGs.

The BMGs were successfully sintered by SPS at 150 °C with a sintering time of up to 180 min. The effect of sintering time on the amorphous characteristics of MgZnCa BMGs can be observed by the XRD patterns in Figure 3. The amorphous nature of the BMG was largely maintained for up to 90 min of sintering time, as shown by the diffused broad peak at ~2θ = 38°. The XRD results also show very minute peaks across the diffused region at ~2θ = 37° in all sintered samples. Here, diffraction peaks superimposed onto the amorphous halo for all sintered samples could be observed, indicating that some form of crystallinity exists and that the sintered samples are predominantly but not completely amorphous. At an extended sintering time of above 90 min, significant crystalline peaks start to emerge such as that in SPS150_180. These peaks correspond to crystalline Mg-Zn and α-Mg phases which indicate that crystallization has occurred, disrupting the homogenous amorphous matrix.

To account and check for the amorphous nature of the samples post sintering, a differential scanning calorimeter was utilized. Here, selected DSC scans of SPS150_90 and SPS150_180 are shown in Figure 4a,b. The two samples were used for comparison because there was a sharp change in the XRD patterns from SPS150_90 to SPS150_180, as shown in Figure 3. Contrast in the DSC trace could be observed. SPS150_90 showed an onset of glass transition (T_g) and subsequent crystallizations, with the first onset of crystallization denoted as T_x at ~160–170 °C in Figure 4b. This glass transition and the first crystallization temperatures are in line with studies reported on similar Mg-Zn-Ca metallic glass systems [14–16]. On the other hand, SPS150_180 is lacking the onset of glass transition and crystallization, as shown in Figure 4b. Therefore, the DSC scans prove that SPS150_90 is predominantly in the amorphous state with a very small amount of crystallinity, while SPS150_180 recrystallized to a certain degree.

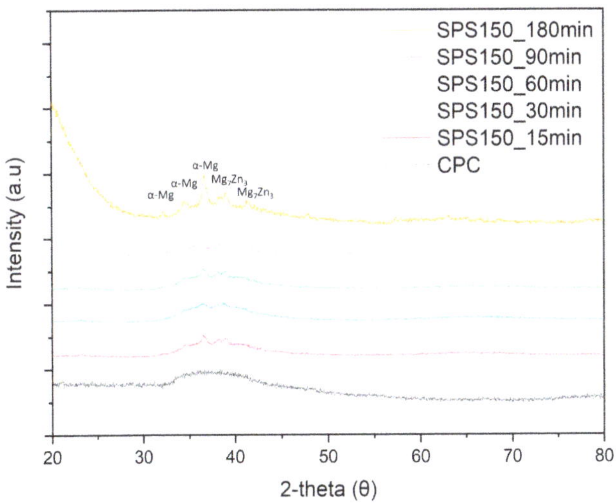

Figure 3. XRD patterns of pre-SPS cold compacted (CPC) and spark plasma-sintered Mg-Zn-Ca ribbons at varying sintering times from 15 to 180 min.

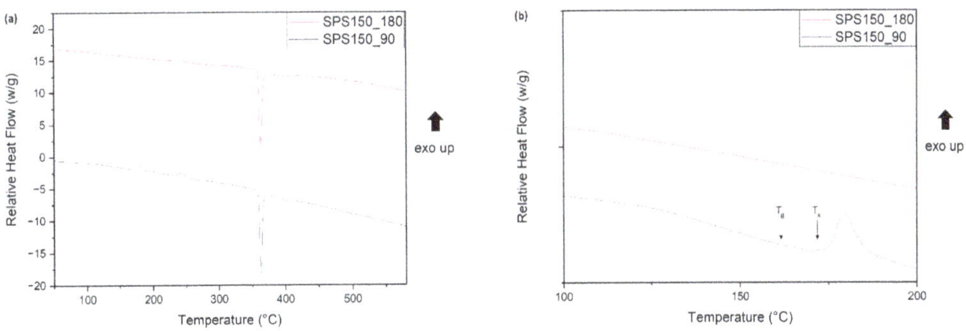

Figure 4. Comparison of DSC traces between SPS150_90 and SPS150_180 (**a**) across the full range of temperatures (50 °C to 600 °C) and (**b**) a magnified view of 100 °C to 200 °C.

EBSD was also performed to analyze microstructural features in SPS150_180 to confirm the presence of grains, which showed the presence of crystallinity. From the EBSD Inverse Pole Figure (IPF) map in Figure 5a, grains can clearly be indexed in SPS150_180; α-Mg crystalline phase was identified with a final grain size of 2.70 ± 1.9 μm and a significant amount of crystallinity, as shown in Figure 5b. On the other hand, no grains could be indexed with SPS150_90, which suggests that the presence of crystallinity, while present, is minimal. These results are consistent with our data collected from XRD, suggesting that significant dynamic crystallization occurred with prolonged sintering times above 90 min.

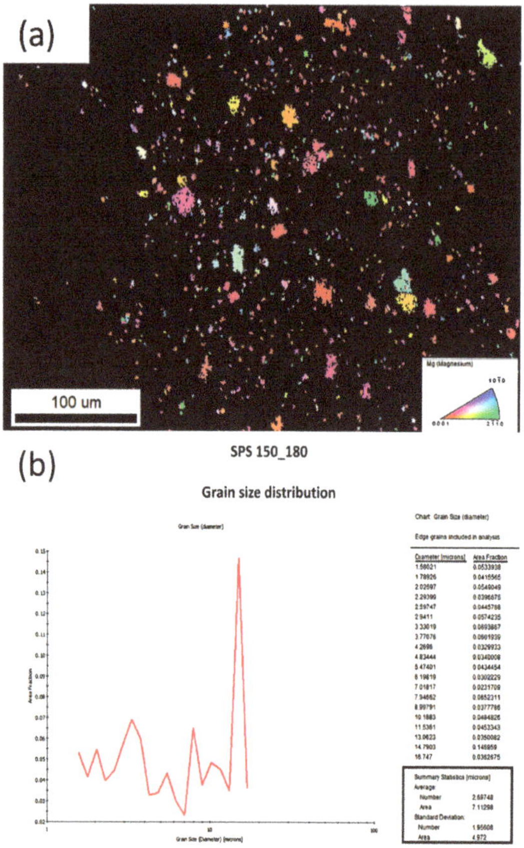

Figure 5. Electron backscattered diffraction patterns of MgZnCa: (**a**) IQ + IPF map and (**b**) grain size distribution.

The surface morphologies of the pre-SPS powder and the spark plasma-sintered BMGs at different sintering times (15 to 180 min) compared to the as-cast alloy are shown in Figure 6. Here, we can look at the effect of sintering time on surface morphology. At lower sintering times (15–30 min), it can be observed that the powders are not bonded together, as shown by the large gaps between the powders in Figure 6b,c. In addition, the contrasting layers seen in Figure 6b,c show that any form of bonding between layers is lacking, with the presence of a significant amount of porosity on the surface also shown.

At higher sintering times (60–180 min), the interparticle distances are significantly reduced and there is minimal porosity present, as shown in Figure 6d–f. Unlike the flaky appearance due to the presence of contrasting layers seen at lower sintering times, a smooth and homogenous texture was observed at higher sintering times. Denoted by the circles in Figure 6d–f, these are areas in which the interfaces among powder particles start to fade away, forming a more homogeneous matrix. These could suggest that, at elevated sintering times, short-order diffusion of atoms may have occurred between adjacent powder particles, resulting in some form of chemical and mechanical interlocking which is likely to promote better structural integrity.

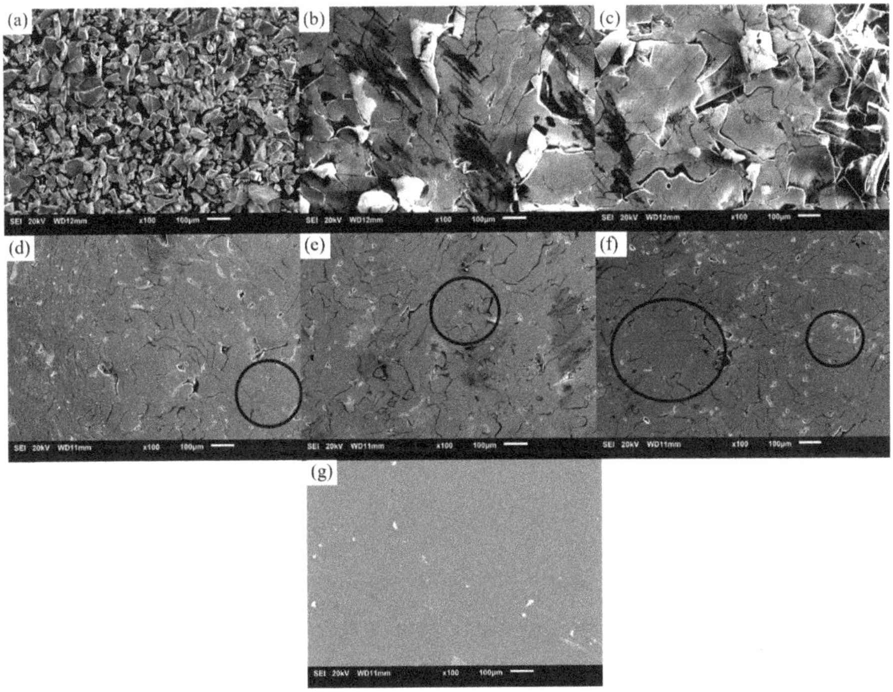

Figure 6. SEM micrographs of (**a**) pre-sintered Mg-Zn-Ca powder, spark plasma-sintered Mg-Zn-Ca BMGs with (**b**) 15 min, (**c**) 30 min, (**d**) 60 min, (**e**) 90 min, and (**f**) 180 min sintering times, and (**g**) as-cast alloy. Here, the dark circles show regions in which the interfaces between ribbons could have been diffused.

Comparing Figure 6d–f, we can observe that the further increase in sintering time from 60 min to 180 min had minimal impact on porosity (1.88 vol. % in SPS150_60 to 1.22 vol. % in SPS150_180). This may be attributed to a lack of long-range diffusion in atoms that typically require higher temperature [17].

3.2. Density and Porosity

The measured densities and derived porosities of Mg-Zn-Ca spark plasma-sintered BMGs compared to as-cast Mg-Zn-Ca alloy are shown in Figure 7. The densities of SPS150_15, SPS150_30, SPS150_60, SPS150_90, and SPS150_180 are 2.74 g/cm^3, 2.76 g/cm^3, 2.81 g/cm^3, 2.82 g/cm^3, and 2.83 g/cm^3, respectively, while their respective porosities are 4.29%, 3.70%, 1.88%, 1.53%, and 1.22%. The porosity values are derived by comparing density values to the density of the as-cast Mg-Zn-Ca alloy. Here, increased sintering time has shown to be successful in reducing porosity by ~70% while maintaining the amorphous nature of the BMGs. With an extended sintering time above 60 min, the porosities of spark plasma-sintered samples remained low and represent those of near-net-shaped products (~2%).

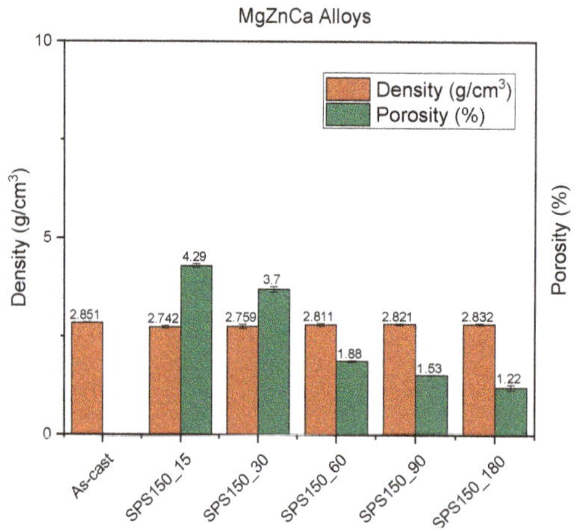

Figure 7. Density and porosity of Mg-Zn-Ca BMGs at different sintering times.

3.3. Mechanical Properties

Figure 8 shows the Vickers hardness results of spark plasma-sintered BMGs under different sintering times. The results have shown that the hardness value increases with increasing sintering time, with a maximum value of ~430 HV exhibited by SPS150_180. Generally, Vickers hardness highly depends on the volume fraction of pores [18]. From SPS150 to SPS150_180, the porosity reduced significantly, as discussed earlier (4.29% compared to 1.22%). Therefore, we can observe that the hardness of the material increased from ~350 HV to ~430 HV. While there was a slight reduction in hardness from ~361 HV in SPS150_15 to ~353 HV in SPS150_30, the difference is insignificant in the overall trend.

However, the hardness of the material did not change significantly from SPS150_90 to SPS150_180 despite doubling the sintering time and further reducing porosity (1.53% to 1.22%). This could be due to the presence of grains that could allow for easier dislocation motion in amorphous matrices within SPS150_180, resulting in a drop in hardness.

The as-cast sample is expected to have a significantly lower hardness of 75 HV as it is in a complete crystalline state. This hardness value for the as-cast alloy is in line with other studies conducted on crystalline Mg-based alloys [18–20].

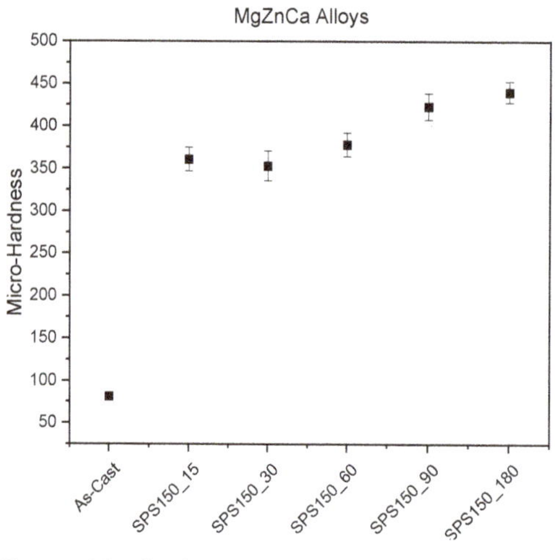

Figure 8. Microhardness of Mg-Zn-Ca BMGs at different sintering times.

The mechanical properties of sintered samples were also investigated by compression tests. Engineering stress–strain curves of all samples are shown in Figure 9. Calculated values of ultimate compression strength (UCS), fracture strain, and energy absorbed are shown in Table 4. The results show a notable improvement in UCS and fracture strain with increasing sintering time. The increase in sintering time from 15 min to 180 min resulted in an increase in UCS from 108 MPa to 254 MPa.

Comparing both the hardness and compressive strength of the materials, we can observe a discrepancy in the results: the hardness value of the as-cast alloy is the lowest, while compressive strength is the highest. This could be because the hardness measurement was more of a surface indentation, which is relatively independent of the bulk sample and does not provide a full representation of the structural integrity of the bulk sample. The nature of the sintering process (which is significantly below the melting temperature of the alloy) would mean that voids within the sintered samples are present. Additionally, the lack of melting would mean that the interlocking between particles would not be as strong (think of two particles being melting together compared to two particles compacted together). This phenomenon would not be captured by the surface indenter but would be captured by the compression tests as it involves the entire bulk sample.

Table 4. Compressive response of Mg-Zn-Ca BMGs sintered for various lengths of time.

Sample Name	UCS (MPa)	Reference
As-cast	324 ± 10	-
Cortical bone	205 ± 17	[21]
SPS150_15	108 ± 10.5	-
SPS150_30	143 ± 7.4	-
SPS150_60	187 ± 8.9	-
SPS150_90	220 ± 5.3	-
SPS150_180	254 ± 7.2	-

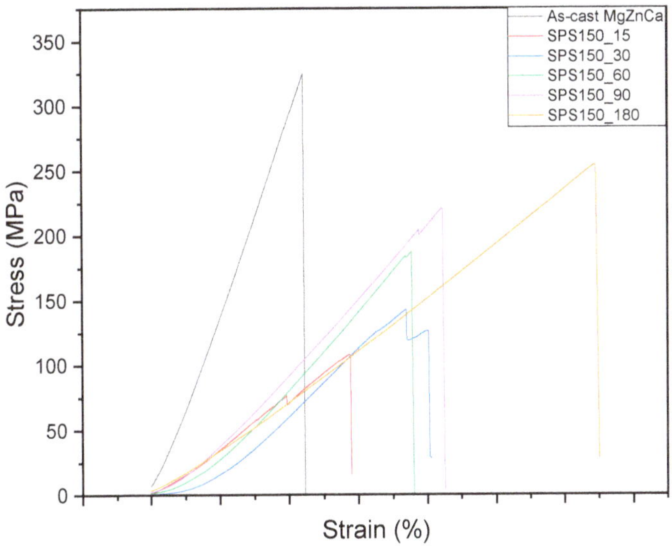

Figure 9. Compressive stress–strain curves of Mg-Zn-Ca BMGs with various sintering times.

In our earlier discussion, the SEM images of the sintered samples, particularly those with more than 60 min of sintering time, had ribbon interfaces that started fading away, forming a more homogenous matrix. This, coupled with a reduction in porosity, could possibly explain why we see a significant improvement in UCS with prolonged sintering time. The mechanical properties of the sintered samples, particularly SPS150_90, exhibited compressive strength that is similar to cortical bone (220 MPa and 205 MPa, respectively). The similarity in compressive strength can reduce the effect of stress shielding [22], which commonly occurs due to a difference in mechanical properties between the bone and the implant material [23]. Expectedly, SPS150_180 has better mechanical properties among all sintered samples due to its significantly longer sintering time compared to the other samples. However, the presence of crystallization, as shown in Figure 3, will have a detrimental effect on its corrosion resistance, which will be discussed in the next section. Therefore, a careful optimization between mechanical properties and corrosion resistance can be tailored for different applications.

3.4. Biocorrosion Properties

The changes in open circuit potential (OCP) as a function of time after 1 h of immersion in Phosphate Buffer Solution (PBS) at 37 °C are shown in Figure 10. Here, OCP refers to the potential difference between the working electrode (metal surface to be studied) and the reference electrode [24]. The E_{ocp} values for as-cast crystalline MgZnCa and SPS150_15, SPS150_30, SPS150_60, SPS150_90, and SPS150_180 MgZnCa BMGs are −1.47 V, −1.32 V, −1.27 V, −1.25 V, −1.35 V, and −1.4 V, respectively. The less negative the corrosion potential, the lower the tendency for corrosion in PBS [25]. Minor changes in E_{ocp} were noticed from SPS150_15 to SPS150_90, with SPS150_60 having the most positive potential among the sintered samples. SPS150_180 and crystalline as-cast MgZnCa, on the other hand, have significantly more negative potential values, signaling a significant reduction in corrosion resistance as compared to the other sintered samples.

From the OCP data, it can be observed that the OCP was fairly smooth and stable from SPS150_15 to SPS150_90. This suggests the formation of a passive film on the surface. SPS150_180 and the crystalline as-cast alloy, on the other hand, showed significant fluctuations in OCP across the entire duration. This would suggest that the passive film formed on the surface is unstable, resulting in the continuous re-passivation and de-passivation of the oxide film [26].

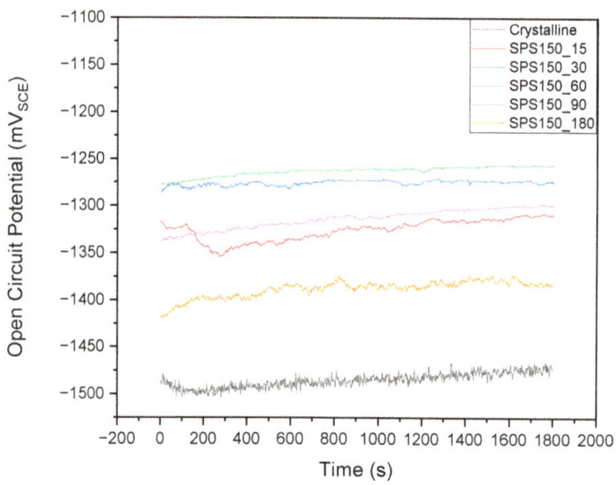

Figure 10. Time dependence of open circuit potential (OCP) across different sintering times.

The potentiodynamic polarization (PDP) curves for the spark plasma-sintered samples as compared to the as-cast crystalline MgZnCa are shown in Figure 11. For a more accurate depiction of results, three scans were performed for each sample. The median trace with the most representative corrosion current (i_{corr}) and corrosion potential (E_{corr}) was selected and plotted.

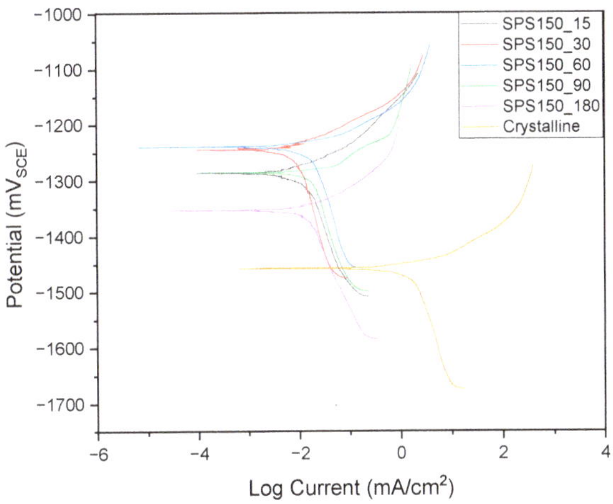

Figure 11. Potentiodynamic plot of Mg-Zn-Ca across various sintering times.

From the PDP scans, it was found that the BMGs have more noble corrosion potentials (−1200 to −1350 mV$_{SCE}$) compared to the crystalline MgZnCa (−1470 mV$_{SCE}$), indicating lower corrosion activity. This is likely due to the lack of crystalline phases and because the BMGs are generally free of microstructure defects such as grain boundaries and dislocations, reducing galvanic couples and preventing intergranular and localized pitting corrosion [27]. In addition, the absence of structural defects suppresses ion diffusion, thereby improving corrosion resistance [28].

A closer inspection of the PDP plots reveals that the cathodic reaction kinetics are comparable from SPS150_15 to SPS150_90, indicating similar cathodic activity (hydrogen evolution). When compared to SPS150_180, significantly higher cathodic activity could be observed for the latter. Cathodic activity is extremely important in the application of a biomaterial as the rapid hydrogen evolution can lead to the formation of gas cavities and swelling at the implant site [29].

Comparing the anodic reaction kinetics, the anodic curves showed similar current density values from SPS150_15 to SPS150_90. Corrosion current density is a parameter that closely relates to the degradation resistance of the material. Therefore, a lower corrosion current density would represent better corrosion resistance. However, in SPS150_180, the presence of crystallinity results in a shift to higher current density values and enhanced anodic dissolution.

Table 5 illustrates the Tafel fitting results of the potentiodynamic polarization curves. The corrosion current density (I_{corr}) is a parameter that strongly correlates to degradation resistance, while the corrosion rate in mm/year is derived from the following general formula (Equation (3)) [30]:

$$\text{Corrosion Rate in mm/year} = 3.27 \times I_{corr} \times \frac{EW}{p} \quad (3)$$

where I_{corr} represents the corrosion current density (mA/cm^2), EW represents the equivalent weight of the material, and p represents the density of the material in g/cm^3.

Table 5. Corrosion response of Mg-Zn-Ca at various sintering temperatures.

Sample Name	I_{corr} (mA/cm^2)	E_{corr} (mV$_{SCE}$)	Corrosion Rate (mm/Year)
CRYSTALLINE	19.2×10^{-2}	−1469	4.086
SPS150_15	1.25×10^{-2}	−1276	0.265
SPS150_30	1.01×10^{-2}	−1240	0.214
SPS150_60	1.95×10^{-2}	−1200	0.413
SPS150_90	1.86×10^{-2}	−1290	0.394
SPS150_180	3.91×10^{-2}	−1351	0.829

The corrosion current density and the corrosion rate are directly related, meaning that a higher corrosion current density indicates a higher rate of metal dissolution, which results in a higher corrosion rate [31]. The corrosion rates in Table 5 suggest that corrosion resistance is the most optimal at sintering times of up to 30 min, with a ~20× improvement in corrosion resistance compared to the as-cast crystalline alloy (0.214 mm/year in SPS150_30 compared to 4.086 mm/year in the crystalline MgZnCa alloy).

With prolonged sintering time, however, the corrosion rate increased significantly. The corrosion rate of SPS150_180 was 0.829 mm/year, suggesting a comparable corrosion rate to as-cast crystalline MgZnCa (4.086 mm/year). The results suggest that the presence of a significant amount of crystallinity in SPS150_180, as identified by the EBSD in Figure 5, accelerated the degradation that is otherwise very limited in samples with a 90 min sintering time. It is interesting to note that by increasing sintering time from 30 min to 90 min, the corrosion rate increased ~2× from 0.214 mm/year in SPS150_30 to ~0.413 mm/year in SPS150_90. This increase in corrosion activity may be attributed to the presence of more crystallinity at elevated sintering times.

Nonetheless, all sintered samples in this work showed better corrosion resistance than the crystalline as-cast samples, indicating that a more stable passive film formed on the surfaces. However, the presence of crystalline phases in SPS150_180, as confirmed with the XRD and EBSD studies, had a detrimental effect on corrosion resistance. Therefore, by optimizing sintering time, SPS150_90 has been shown to produce a good balance between mechanical properties and corrosion resistance.

4. Discussion

Influence of Sintering Time on the Densification of Melt-Spun Ribbons

It is widely known that metallic glasses are prone to structural relaxation and crystallization. Diffusion plays a major role in these processes, and even more so towards the understanding behind the consolidation mechanism during SPS. While the knowledge of diffusion in metallic glasses has been rather limited, temperature-dependent Arrhenius plots on diffusion have been well discussed [32–34]. Often, people have interpreted this as a sign of a diffusion process that resembles that in crystals. In crystals, defects that behave like vacancies at thermal equilibrium are the carriers for diffusion. However, the resulting diffusivity values have sometimes differed greatly from those that are typical of a vacancy mechanism [35].

Spark plasma sintering of BMGs is a relatively new concept. In recent years, some studies have been performed on spark plasma-sintered Zr-based [36] and Fe-based [37] alloys. However, unlike many other materials sintered via SPS, Mg-based BMGs are extremely challenging to obtain. This is due to the temperature constraint during sintering, which is brought forth by the inherently low crystallization temperature of ~160 °C. As such, any sintering temperature used should be below the crystallization temperature of the alloy to prevent the onset of crystallization.

To improve the diffusion rate in amorphous matrices, we can increase the pressure during sintering to force atoms to move through particulate barriers. Increasing pressure can also cause brittle fractures in metallic glass ribbons or particles with little ductility, resulting in increased surface area and density during sintering. High pressure can overcome limitations set by temperature and has been studied in detail for densification during SPS [38]. The current study was limited to a maximum pressure of 90 MPa due to the constraints imposed by the equipment and compaction die used. However, it is worth noting that industrial grade SPS equipment has been reported to exert much higher pressures of 400–500 MPa. For instance, a paper published by Perriere et al. reported the use of 500 MPa to achieve full densification of amorphous powders [39].

Time dependence on the diffusion kinetics of metallic glasses is less studied. For instance, in a work on diffusion coefficients in metallic glasses by Tyagi et al., it was suggested that enhanced diffusion over the relaxed state of the amorphous solid can be attributed to the diffusion of atoms via excess vacancies, which could be generated due to the mismatch of atomic size during the glass forming process [40]. As a consequence of this enhanced diffusion, which is time- and energy-dependent, an irreversible increase in volume density could be observed [40].

Our results indicate that nearly full densification could be obtained with prolonged sintering time, as shown in the density values in Figure 6. Densification via SPS can be achieved through two steps. Firstly, the initial stages of sintering occur at narrow necks of particle interfaces and would generally result in a localized increase in temperature, where relaxation occurs rapidly due to the high thermal diffusivity of the material. This results in some form of cohesion between particles [11,41]. As the sintering progresses, complete densification could then happen during the second step. This occurs when the necks (regions of contact between neighboring particles) are wide enough, thus allowing for homogeneous heating and Newtonian flow (enhanced diffusion) of the particles within the BMGs, therefore leading to complete densification [37,42].

While our studies have shown that increased sintering time results in better densification, what is notably interesting is that prolonged sintering time can have an effect on the amorphous behavior of the spark plasma-sintered samples. When sintering time was increased from 90 min to 180 min, a significant amount of crystallization was observed by both XRD and EBSD studies, as shown in Figure 3 and Figure 4, respectively. Generally, crystallization requires the nucleation of grains, which is energy-dependent [43]. However, the fact that SPS150_90 shows little signs of crystallization shows that a temperature of 150 °C provides insufficient energy for the nucleation of grains with 90 min of sintering time. While this phenomenon is yet to be studied, a possible explanation could be the enhanced diffusion attributed to prolonged sintering time at elevated pressure (90 MPa) and temperature (150 °C) coupled with the presence of free excess volume in metallic glasses [44].

5. Summary

The influence of sintering time on the production of spark plasma-sintered Mg-Zn-Ca BMGs was studied. The experimental results reveal the following:

(1) Increasing the sintering time during the SPS process is effective in improving the density and structural integrity of Mg-Zn-Ca BMGs at 150 °C and 90 MPa.

(2) A predominantly amorphous structure was obtained post SPS with a sintering time of up to 90 min (SPS150_90), with density close to that of the master alloy (2.82 g/cm^3 vs. 2.85 g/cm^3, ~98.2% densification). Densification increases with increasing sintering time.

(3) SPS150_90 achieved UCS of 220 MPa, which is similar to that of cortical bone, thereby eliminating the issue of stress shielding.

(4) Sintered samples with a sintering time of 90 min or less have similar corrosion resistance. There was a slight reduction in corrosion resistance (~2×) from SPS150_30 to SPS150_90 (0.214 mm/year vs. 0.394 mm/year). Nonetheless, SPS150_90 exhibited ~10× better corrosion resistance than as-cast MgZnCa (0.394 mm/year vs. 4.086 mm/year).

Author Contributions: Conceptualization, K.S.F. and M.G.; methodology, K.S.F. and C.B.W.; validation, S.T.; formal analysis, B.S.J.B., K.S.F., S.T. and E.S.P.; investigation, B.S.J.B., M.K.K. and E.S.P.; resources, S.T. and M.K.K.; writing—original draft, B.S.J.B.; writing—review and editing, M.G.; supervision, C.B.W. and M.G. All authors have read and agreed to the published version of the manuscript.

Funding: M.K.K. and E.S.P. were supported by the Creative Materials Discovery Program through the National Research Foundation of Korea (NRF) funded by the Korean Government (MSIT) (no. NRF-2019M3D1A1079215).

Data Availability Statement: Not applicable.

Conflicts of Interest: The authors declare no conflict of interest.

References

1. González, S. Improved mechanical performance and delayed corrosion phenomena in biodegradable Mg–Zn–Ca alloys through Pd-alloying. *J. Mech. Behav. Biomed. Mater.* **2012**, *6*, 53–62. [CrossRef] [PubMed]
2. Kraus, T.; Fischerauer, S.F.; Hänzi, A.C.; Uggowitzer, P.J.; Löffler, J.F.; Weinberg, A.M. Magnesium alloys for temporary implants in osteosynthesis: In vivo studies of their degradation and interaction with bone. *Acta Biomater.* **2012**, *8*, 1230–1238. [CrossRef]
3. Lu, W.; He, M.; Yu, D.; Xie, X.; Wang, H.; Wang, S.; Yuan, C.; Chen, A. Ductile behavior and excellent corrosion resistance of Mg-Zn-Yb-Ag metallic glasses. *Mater. Des.* **2021**, *210*, 110027. [CrossRef]
4. Babaremu, K.O.; John, M.E.; Mfoh, U.; Akinlabi, E.T.; Okokpujie, I.P. Behavioral Characteristics of Magnesium as a Biomaterial for Surface Engineering Application. *J. Bio Tribo Corros.* **2021**, *7*, 142. [CrossRef]
5. Denkena, B.; Lucas, A. Biocompatible Magnesium Alloys as Absorbable Implant Materials—Adjusted Surface and Subsurface Properties by Machining Processes. *CIRP Ann.* **2007**, *56*, 113–116. [CrossRef]
6. Bin, S.J.B.; Fong, K.S.; Chua, B.W.; Gupta, M. Mg-based bulk metallic glasses: A review of recent developments. *J. Magnes. Alloys* **2021**, *10*, 899–914. [CrossRef]
7. Cai, A.H.; Xiong, X.; Yong, L.; An, W.K.; Zhou, G.J.; Yun, L.; Tie-Lin, L.; Li, X.S. Effect of consolidation parameters on mechanical properties of Cu-based bulk amorphous alloy consolidated by hot pressing. *Trans. Nonferrous Met. Soc. China* **2012**, *22*, 2032–2040. [CrossRef]
8. Lee, P.Y.; Kao, M.C.; Lin, C.K.; Huang, J.C. Mg–Y–Cu bulk metallic glass prepared by mechanical alloying and vacuum hot-pressing. *Intermetallics* **2006**, *14*, 994–999. [CrossRef]
9. Robertson, J.; Im, J.T.; Karaman, I.; Hartwig, K.T.; Anderson, I.E. Consolidation of amorphous copper based powder by equal channel angular extrusion. *J. Non Cryst. Solids* **2003**, *317*, 144–151. [CrossRef]
10. Pauly, S. Processing metallic glasses by selective laser melting. *Mater. Today* **2013**, *16*, 37–41. [CrossRef]
11. Bin, S.J.B.; Fong, K.S.; Chua, B.W.; Gupta, M. Development of Biocompatible Bulk MgZnCa Metallic Glass with Very High Corrosion Resistance in Simulated Body Fluid. *Materials* **2022**, *15*, 8989. [CrossRef] [PubMed]
12. Anselmi-Tamburini, U. Spark Plasma Sintering. In *Encyclopedia of Materials: Technical Ceramics and Glasses*; Pomeroy, M., Ed.; Elsevier: Oxford, UK, 2021; pp. 294–310. [CrossRef]
13. Cardinal, S.; Pelletier, J.M.; Qiao, J.C.; Bonnefont, G.; Xie, G. Influence of spark plasma sintering parameters on the mechanical properties of Cu50Zr45Al5 bulk metallic glass obtained using metallic glass powder. *Mater. Sci. Eng. A* **2016**, *677*, 116–124. [CrossRef]
14. Zhang, Y.N.; Rocher, G.J.; Briccoli, B.; Kevorkov, D.; Liu, X.B.; Altounian, Z.; Medraj, M. Crystallization characteristics of the Mg-rich metallic glasses in the Ca–Mg–Zn system. *J. Alloys Compd.* **2013**, *552*, 88–97. [CrossRef]
15. Jin, C.; Liu, Z.; Yu, W.; Qin, C.; Yu, H.; Wang, Z. Biodegradable Mg–Zn–Ca-Based Metallic Glasses. *Materials* **2022**, *15*, 2172. [CrossRef]
16. Shamlaye, K.F.; Löffler, J.F. Synthesis and characterization of Mg-based bulk metallic glasses in the Mg–Ag–Y–(Cu) system. *J. Alloys Compd.* **2021**, *859*, 157803. [CrossRef]
17. Yang, D.; Zhang, Y.; Song, X.; Chen, Y.; Shen, Z.; Yang, C. Effects of sintering temperature and holding time on porosity and shrinkage of glass tubes. *Ceram. Int.* **2016**, *42*, 5906–5910. [CrossRef]
18. Muhammad, W.N.A.W.; Sajuri, Z.; Mutoh, Y.; Miyashita, Y. Microstructure and mechanical properties of magnesium composites prepared by spark plasma sintering technology. *J. Alloys Compd.* **2011**, *509*, 6021–6029. [CrossRef]
19. Putra, A.G.; Manaf, A.; Anawati, A. Enhancing the Hardness of Mg-9Al-1Zn Cast Alloy by Solution Treatment. *IOP Conf. Ser. Mater. Sci. Eng.* **2019**, *515*, 012088. [CrossRef]
20. Minarik, P.; Stráský, J.; Veselý, J.; Lukáč, F.; Hadzima, B.; Kral, R. AE42 magnesium alloy prepared by spark plasma sintering. *J. Alloys Compd.* **2018**, *742*, 172–179. [CrossRef]
21. Morgan, E.F.; Unnikrisnan, G.U.; Hussein, A.I. Bone Mechanical Properties in Healthy and Diseased States. *Annu. Rev. Biomed. Eng.* **2018**, *20*, 119–143. [CrossRef]
22. Be'ery-Lipperman, M.; Gefen, A. A method of quantification of stress shielding in the proximal femur using hierarchical computational modeling. *Comput. Methods Biomech. Biomed. Engin.* **2006**, *9*, 35–44. [CrossRef]

23. Raffa, M.L.; Nguyen, V.H.; Hernigou, P.; Flouzat-Lachaniette, C.H.; Haiat, G. Stress shielding at the bone-implant interface: Influence of surface roughness and of the bone-implant contact ratio. *J. Orthop. Res.* **2021**, *39*, 1174–1183. [CrossRef] [PubMed]
24. Leslie, N.; Mauzeroll, J. Spatially resolved electrochemical measurements. In *Reference Module in Chemistry, Molecular Sciences and Chemical Engineering*; Elsevier: Amsterdam, The Netherlands, 2023. [CrossRef]
25. Chen, J.; Zhu, X.; Etim, I.P.; Siddiqui, M.A.; Su, X. Comparative study of the effects of MAO coating and Ca-P coating on the biodegradation and biocompatibility of Mg 69 Zn 27 Ca 4 metal glass. *Mater. Technol.* **2020**, *37*, 21–27. [CrossRef]
26. Zhang, Y.; Yan, C.; Wang, F.; Li, W. Electrochemical behavior of anodized Mg alloy AZ91D in chloride containing aqueous solution. *Corros. Sci.* **2005**, *47*, 2816–2831. [CrossRef]
27. Gu, X.; Zheng, Y.; Zhong, S.; Xi, T.; Wang, J.; Wang, W. Corrosion of, and cellular responses to Mg–Zn–Ca bulk metallic glasses. *Biomaterials* **2010**, *31*, 1093–1103. [CrossRef]
28. Ford, D.C.; Hicks, D.; Oses, C.; Toher, C.; Curtarolo, S. Metallic glasses for biodegradable implants. *Acta Mater.* **2019**, *176*, 297–305. [CrossRef]
29. Noviana, D.; Paramitha, D.; Ulum, M.F.; Hermawan, H. The effect of hydrogen gas evolution of magnesium implant on the postimplantation mortality of rats. *J. Orthop. Transl.* **2016**, *5*, 9–15. [CrossRef] [PubMed]
30. Rybalka, K.V.; Beketaeva, L.A.; Davydov, A.D. Determination of corrosion current density by the rate of cathodic depolarizer consumption. *Russ. J. Electrochem.* **2016**, *52*, 268–272. [CrossRef]
31. Balani, K.; Verma, V.; Agarwal, A.; Narayan, R. Corrosion Behavior of Metals. In *Biosurfaces*; John Wiley & Sons, Ltd.: New York, NY, USA, 2014; pp. 345–352. [CrossRef]
32. Schmidt, H.; Gruber, W.; Gutberlet, T.; Ay, M.; Stahn, J.; Geckle, U.; Bruns, M. Structural relaxation and self-diffusion in covalent amorphous solids: Silicon nitride as a model system. *J. Appl. Phys.* **2007**, *102*, 043516. [CrossRef]
33. Lee, D.; Vlassak, J.J. Diffusion kinetics in binary CuZr and NiZr alloys in the super-cooled liquid and glass states studied by nanocalorimetry. *Scr. Mater.* **2019**, *165*, 73–77. [CrossRef]
34. Syutkin, V.M.; Grebenkin, S. Diffusion in bulk metallic glasses. *Appl. Phys. Lett.* **2020**, *117*, 134104. [CrossRef]
35. Lazarus, D. Diffusion in Crystalline and Amorphous Solids. *MRS Online Proc. Libr. (OPL)* **1985**, *57*, 297. [CrossRef]
36. Suárez, M.; Fernández-González, D.; Díaz, L.A.; Diologent, F.; Verdeja, L.F.; Fernández, A. Consolidation and mechanical properties of ZrCu39.85Y2.37Al1.8 bulk metallic glass obtained from gas-atomized powders by spark plasma sintering. *Intermetallics* **2021**, *139*, 107366. [CrossRef]
37. Paul, T.; Singh, A.; Littrell, K.C.; Ilavsky, J.; Harimkar, S.P. Crystallization Mechanism in Spark Plasma Sintered Bulk Metallic Glass Analyzed using Small Angle Neutron Scattering. *Sci. Rep.* **2020**, *10*, 2033. [CrossRef]
38. Zheng, B.; Ashford, D.; Zhou, Y.; Mathaudhu, S.N.; Delplanque, J.P.; Lavernia, E.J. Influence of mechanically milled powder and high pressure on spark plasma sintering of Mg–Cu–Gd metallic glasses. *Acta Mater.* **2013**, *61*, 4414–4428. [CrossRef]
39. Perrière, L.; Champion, Y.; Bernard, F. Spark Plasma Sintering of Metallic Glasses. In *Spark Plasma Sintering of Materials: Advances in Processing and Applications*; Cavaliere, P., Ed.; Springer International Publishing: Cham, Switzerland, 2019; pp. 291–335. [CrossRef]
40. Tyagi, A.K.; Macht, M.P.; Naundorf, V. Diffusion coefficients of 63Ni in Fe40Ni40B20 metallic glass. *Acta Metall. Mater.* **1991**, *39*, 609–617. [CrossRef]
41. Perrière, L.; Thaï, M.T.; Tusseau-Nenez, S.; Ochin, P.; Blétry, M.; Champion, Y. Spark plasma sintering for metallic glasses processing. *Rev. Métallurgie* **2012**, *109*, 5–10. [CrossRef]
42. Ajenifuja, E.; Odetola, P.; Popoola, A.P.; Popoola, O. Spark plasma sintering and structural analysis of nickel-titanium/coconut shell powder metal matrix composites. *Int. J. Adv. Manuf. Technol.* **2020**, *108*, 3465–3473. [CrossRef]
43. Olson, G.L.; Roth, J.A. Kinetics of solid phase crystallization in amorphous silicon. *Mater. Sci. Rep.* **1988**, *3*, 1–77. [CrossRef]
44. Yavari, A.R.; Le Moulec, A.; Inoue, A.; Nishiyama, N.; Lupu, N.; Matsubara, E.; José Botta, W.; Vaughan, G.; Di Michiel, M.; Kvick, Å. Excess free volume in metallic glasses measured by X-ray diffraction. *Acta Mater.* **2005**, *53*, 1611–1619. [CrossRef]

metals

MDPI

Article

Mechanical and Corrosion Properties of Mg–Gd–Cu–Zr Alloy for Degradable Fracturing Ball Applications

Jiahao Jiang [1], Xue Geng [1] and Xiaobo Zhang [1,2,*]

[1] School of Materials Science and Engineering, Nanjing Institute of Technology, Nanjing 211167, China
[2] Jiangsu Key Laboratory of Advanced Structural Materials and Application Technology, Nanjing 211167, China
* Correspondence: xbzhang@njit.edu.cn

Abstract: Generally, excellent mechanical properties of Mg alloys are desired, but their rapid degradation properties are seldom utilized. Petroleum fracturing techniques are required to take full advantage of this rapid degradation. Therefore, we have prepared an as-extruded Mg–6.0Gd–1.2Cu–1.2Zr (wt.%) alloy and treated it with peak aging to analyze its potential as a degradable fracture ball. The results show that the as-extruded alloy mainly consists of an α-Mg matrix, second phase, and large elongated α-Mg grains (LEGs). After aging, the LEGs undergo static recrystallization, which improves the mechanical properties of the alloy, and a lamellar long period stacking ordered (LPSO) phase is observed. Under simulated underground temperature conditions (93 °C), the ultimate tensile strength and elongation of both as-extruded and as-aged alloys are over Ȧ MPa and 11.1%, respectively, and the ultimate compressive strength and elongation of both alloys are over 336 MPa and 16.9%, respectively. The corrosion rate of the as-extruded alloy in 3 wt.% KCl solution at 93 °C reaches 1660.8 mm/y by mass loss test, and that of the as-aged alloy increases to 1955.1 mm/y. The atomic force microscope analysis result confirms that the second phase shows the highest corrosion potential, followed by the lamellar LPSO phase and α-Mg matrix. The as-extruded and as-aged Mg–6.0Gd–1.2Cu–1.2Zr alloy with good mechanical properties and a high corrosion rate in this work shows promising potential for degradable fracturing ball applications.

Keywords: Mg–Gd–Cu–Zr alloy; aging; extrusion; mechanical properties; corrosion; atomic force microscope; degradable fracturing ball

1. Introduction

In the industrial field, hydraulic fracturing technology is mainly used in unconventional oil and gas exploitation [1,2]. The use of hydraulic fracturing technology can effectively increase oil and gas production and recoverable reserves. Therefore, the related technologies and tools of hydraulic fracturing have become the focus of oil and gas development research. The fracturing ball is the key factor that determines the success of fracturing technology [3]. In general, during the process of fracturing, the fracturing ball can temporarily block the lower layer, separate the construction oil layer from the production oil layer, and ensure the smooth development of the fracturing operation. Meanwhile, when the fracturing ends in the oil and gas production stage, the fracturing ball needs to accelerate its dissolution rate and achieve complete degradation as soon as possible [4]. Finally, due to the high working environment temperature, the fracturing ball is needed to support high-temperature operation. Most of the materials currently used for fracturing balls are stainless steel, aluminum alloy, and other non-degradable materials [5]. They should be removed at last by traditional milling or drilling, which is complex and difficult to control and also increases production costs. There are also polymer materials used to prepare fracturing tools, but they are prone to serious plastic deformation, and the degradation products of these polymer materials may have a serious impact on the environment. In addition, it is difficult to control their degradation behavior, hence seriously affecting production safety [6,7]. The preparation of fracturing balls using degradable materials is considered to be the most effective solution. Therefore, it is urgent to develop environmentally friendly fracturing ball materials with sufficient strength and a fast degradation rate that can effectively reduce construction costs and risks, shorten the construction period, and improve construction efficiency.

Mg alloys have low density, high specific strength and stiffness, good damping, and electromagnetic shielding performance, which makes them candidate materials for many practical applications such as automobiles and aerospace [8–10]. However, the standard electrode potential of Mg alloys is low, and their corrosion resistance is poor in the chloride ion environment, which restricts their broad engineering applications [11,12]. As for degradable fracturing ball materials, controllable fast degradation and sufficient mechanical properties at high working temperatures are necessary [13]. Mg alloys have good feasibility as fracturing tools, nevertheless, compared with traditional fracturing materials, the mechanical properties of Mg alloys need to be further improved, and the degradation of Mg alloys should be significantly increased to meet the practical application requirements. Recently, researchers have focused on how to improve the mechanical properties and accelerate the degradation of Mg alloys for fracturing ball application.

Generally, alloying can effectively improve the mechanical properties of Mg alloys by solid solution strengthening, second phase strengthening, and grain refinement strengthening [14,15]. Zr and Cu are often used to strengthen Mg alloys by refining grains and promoting the recrystallization of Mg alloys [16–18]. The addition of Zr can further increase the peak aging hardness of Mg–RE alloy. In particular, the introduction of Cu, Zn, Ni, and other transition elements in rare earth (such as Gd, Y, Ho, Er, Dy, etc.) containing Mg alloys can also form long-period stacking ordered (LPSO) structures, which are beneficial to purify the alloy, refine the microstructure, and improve the mechanical properties [19,20]. In addition, extrusion and heat treatment are also effective methods to improve the strength of Mg alloys due to the combined effects of grain refinement, solution strengthening, and precipitation strengthening [21–23]. At present, only a few researchers choose extruded alloys as temporary plugging materials for fracturing [7,16,24].

It is well known that when the content of Fe, Ni, or Cu in Mg alloys exceeds the tolerance limit, the corrosion of Mg alloys will be greatly accelerated [4,7,24,25]. These elements have low solubility in Mg alloys, and as the cathode active position of galvanic corrosion, they accelerate corrosion of the substrate. Xiao et al. [17] have found that the addition of Cu in Mg–Al–Zn alloys affects the quantity, morphology, and distribution of the $Al_7Mg_8Cu_3Zn_1$ phase, and the corrosion rate is accelerated due to the intensification of micro-galvanic corrosion. Zhang et al. [7] found that the corrosion rate of Mg–3Zn–1Y alloy increased by 38 times after adding Cu. Tan et al. [14] also found that the corrosion rate Mg–2.5Cu alloy in 3 wt.% KCl solution reached 1113 mm/y. Zhang et al. [4] used the micro-galvanic effect of Fe to accelerate the corrosion, which aggravated the corrosion degree of Mg–6Al–1Zn alloy, and the mechanical properties of the alloy were more stable after annealing. In addition, the composition, distribution, quantity, and morphology of the LPSO phase have beneficial or adverse effects on the corrosion properties of Mg alloys: on the one hand, it can promote degradation of Mg alloys by the formation of a galvanic couple; on the other hand, it may restrict corrosion as a barrier [26,27]. In general, the LPSO structure exists in the Mg–RE–Zn/Ni/Cu/Co system (RE = Y, La, Ce, Pr, Sm, Nd, Dy, Ho, Er, Gd, and Tm) [28]. In addition, the solubility of gadolinium (Gd) in Mg is 23.49 wt.% at the eutectic temperature, which plays a key role in the solid solution strengthening of Mg alloys [29]. Ma et al. [16] changed the distribution of the LPSO structure in a Mg–7.2Y–2.8Ni alloy using annealing treatment and improved the degradation rate and tensile strength of the alloy (280 MPa). This can make it become one of the candidate materials for a fracturing ball. Therefore, the introduction of the LPSO phase is beneficial to regulate the degradation rate of Mg alloys, which can provide new insights for the new type of fracturing temporary plugging Mg alloys. At present, there are few works in the literature on Mg–Gd–Cu alloys with high degradation rates. In order to meet the self-degradation of the fracturing ball and high mechanical properties in a high-temperature environment, in this work, Mg–6Gd–1.2Cu–1.2Zr alloy was prepared using gravity casting and then hot extruded, followed by aging treatment. By simulating the underground working environment, we studied whether both the as-extruded and as-aged alloys have stable mechanical properties at high temperatures and fast degradation rates and evaluated whether it has reliable performance for degradable fracturing ball applications.

2. Experimental Methods

The Mg–6.0Gd–1.2Cu–1.2Zr (wt.%) alloy was prepared with gravity casting using pure Mg (99.99%), Mg–30Gd (wt.%), Mg–30Cu (wt.%), and Mg–30Zr (wt.%) master alloys [18,30]. The ingots (Φ 60 mm × 50 mm) were hot extruded at 400 °C into rods (Φ 20 mm) with an extrusion ratio of 9:1. Before hot extrusion, the as-cast alloy was homogenized at 400 °C for 24 h. The as-extruded alloy was optimized using a reasonable aging treatment. Finally, the alloy with the highest microhardness was obtained with aging at 170 °C for 8 h.

The samples were cut along perpendicular and parallel extrusion directions, ground with SiC paper, polished with 2.5 μm diamond paste, and then etched in a picric acid solution for microstructure observation. The microstructure was characterized using an optical microscope (OM, OLYMPLUS GX51) and a scanning electron microscope (SEM, Zeiss-Merlin Compact) coupled with an energy dispersion spectrometer (EDS, Oxford Atec X-Max 50). The SEM observation was conducted at 7.0 kV under the secondary electron mode. X-ray diffraction (XRD, Rigaku Ultima-IV) was used to analyze the phases of the alloy with a working voltage of 40 kV, scanning angle of 20°~80°, and scanning speed of 6°/min.

The tensile properties of the alloy at ambient temperature and the tensile and compressive properties at simulated underground temperature (93 °C) were evaluated using an electronic material testing machine (Zwick/Roell, Z030TH) equipped with a heating chamber. Each sample with a gauge length of 15 mm, width of 3.5 mm, and thickness of 2 mm was held for 30 min to ensure uniform heating prior to the tensile test at 93 °C. The samples with a dimension of Φ 8 mm × 12 mm were also heated to 93 °C and held for 30 min for the compression test. The tensile and compression speeds were 1 mm/min. The fracture morphologies of the specimens were observed using SEM.

Commonly, 3 wt.% KCl solution is used as a base fluid to stabilize the clay and prevent its expansion in the oil industry, which is helpful to maintain the chemical environment of the clay particles. Moreover, the fracturing ball works underground, and the ambient temperature is higher than room temperature [14]. To evaluate the corrosion rate of the alloy under simulated underground conditions, three parallel samples with a dimension of Φ 13 mm × 3.5 mm were polished and then immersed in 3 wt.% KCl solution at 93 °C for hydrogen evolution and mass loss tests.

The hydrogen was collected using a burette which was filled with KCl solution and placed on top of a funnel with a specimen in the center of the funnel. The volume of hydrogen evolution was recorded every 10 min, and then the burette was filled with KCl solution immediately again to collect hydrogen. After 60 min of immersion, the samples were placed in a boiling chromic acid solution (200 g/L) to remove the corrosion products. The corrosion rate of the alloy was calculated using the following formula [8,30]:

$$v = 95.05\frac{v_H}{\rho AT} \quad (1)$$

$$v = 87.6\frac{\Delta w}{\rho AT} \quad (2)$$

where v is the corrosion rate (mm/y), v_H is the amount of hydrogen evolution (mL), Δw is the mass loss (mg), ρ is the density of the alloy (g/cm^3), A is the surface area of the specimen (cm^2), and T is the immersion time (h).

The potentiodynamic polarization test was conducted in 3 wt.% KCl with an electrochemical workstation (Parstat 2273). A platinum rod was used as the counter electrode, the saturated calomel electrode as the reference electrode, and an alloy sample as the working electrode. The scanning speed was 1 mV/s. An atomic force microscope (AFM, Bruker Multimode 8) was used to measure the corrosion potential of the different phases, and Nanoscope Analysis 1.7 software was used to analyze the data.

3. Results and Discussion

3.1. Microstructure and Phase

Figure 1 shows optical micrographs of the as-extruded and as-aged Mg–6.0Gd–1.2Cu–1.2Zr alloy from both perpendicular and parallel extrusion directions. It can be seen from Figure 1a,b that the microstructure of the as-extruded alloy is mainly composed of dynamic recrystallized (DRXed) equiaxed α-Mg grains, a discontinuously distributed second phase, and some long-elongated α-Mg grains (LEGs). After the aging treatment, lamellar structures appear near the black second phase that can be inferred to be LPSO according to our previous studies [30], the LEGs cannot be observed, and much finer grains are formed, indicating the occurrence of static recrystallization. Moreover, the second phase seems to be reduced and more discontinuously distributed along the matrix grain boundaries. Compared with the as-cast alloy [30], the as-extruded alloy and as-aged alloy have obvious grain refinement, which is mainly due to the recrystallization during the extrusion and aging treatment.

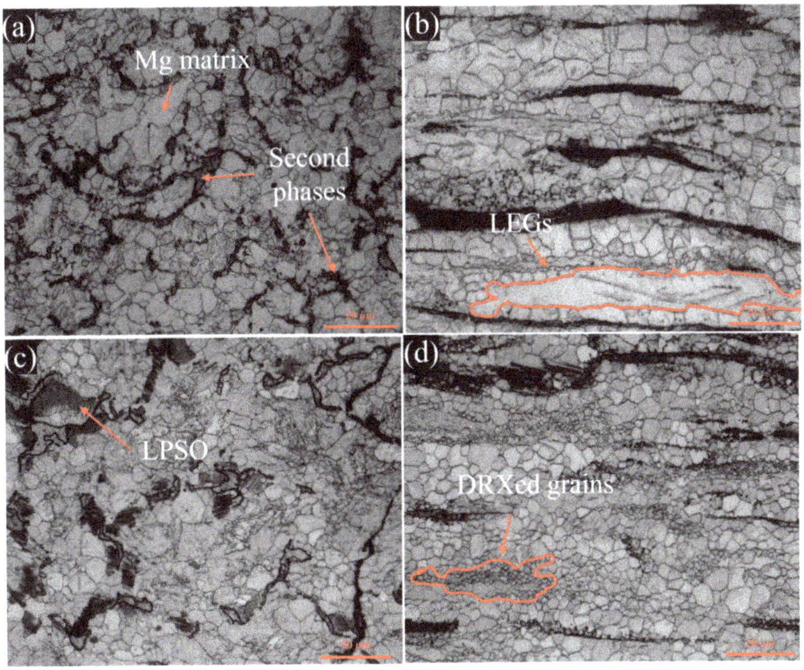

Figure 1. Optical micrographs of the alloy perpendicular (**a**,**c**) and parallel (**b**,**d**) to the extrusion direction for the (**a**,**b**) as-extruded alloy and (**c**,**d**) as-aged alloy.

SEM images with higher magnification are presented in Figure 2 to better show the second and LPSO phases. It is clear that the second phase in the as-extruded alloy distributes more dispersive as compared to that in the as-aged alloy. Importantly, LPSO structures located near the second phase are formed (marked as C1 in Figure 2c) after the aging treatment. EDS analysis was performed to compare the chemical compositions of different regions marked in Figure 2a,c, and the results are listed in Table 1. The second phase is rich in Gd and Cu in both the as-extruded and as-aged alloys, and the content of Gd is slightly higher than that of Cu. After the aging treatment, the content of Gd and Cu slightly increases in the matrix but decreases in the second phase. The LPSO structure marked as C1 contains 15.06 wt.% Gd and 8.17 wt.% Cu, which is lower than the second phase but much higher than the α-Mg matrix. In addition, Zr distributes in all the phases.

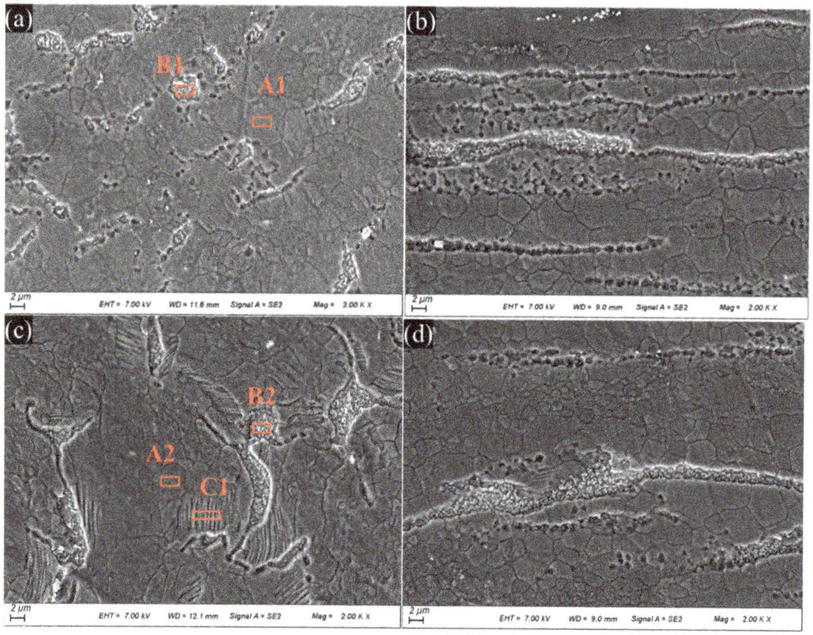

Figure 2. SEM micrographs of the alloy perpendicular (**a**,**c**) and parallel (**b**,**d**) to the extrusion direction for the (**a**,**b**) as-extruded alloy and (**c**,**d**) as-aged alloy.

Table 1. Chemical compositions (wt.%) of the marked areas in Figure 2.

Areas	Gd	Cu	Zr	Mg
A1	3.95	0.86	0.83	Balance
B1	25.88	24.65	0.18	Balance
A2	4.29	1.58	0.41	Balance
B2	24.85	22.31	0.32	Balance
C1	15.06	8.17	0.48	Balance

The XRD patterns of the as-extruded and as-aged alloys are shown in Figure 3. The Mg_2(Gd, Cu) phase was detected in addition to the α-Mg matrix, which is consistent with the phases under as-cast and solution-treated conditions [30]. Mg_2(Gd, Cu) corresponds to the second phase observed using OM and SEM. The volume fraction of the LPSO structure is low so that its diffraction peaks cannot be observed clearly.

The alloy has obvious partial dynamic recrystallization during extrusion. In general, under the action of extrusion stress, a large number of dislocations in Mg alloys encounter obstacles to reorganize and merge, forming sub-grain structures along grain boundaries. With the increase in sub-grain orientation difference, large angle grain boundaries are formed, and new recrystallized grains are formed [31]. It can be seen from Figure 1b that some LEGs did not undergo dynamic recrystallization but show coarse and elongated morphology along the extrusion direction. This may be due to a relatively low extrusion ratio or extrusion temperature, which results in insufficient deformation energy storage to trigger recrystallization in some local regions [32]. After extrusion, the dislocation density around the LEGs is high, which belongs to the sub-structure grains. The aging treatment promotes thermodynamic recrystallization, transforming these substructural grains into new finer equiaxed grains known as static recrystallization [33]. According to the EDS results in Table 1, Gd and Cu atoms diffuse from the second phase with higher content to the α-Mg matrix, resulting in an increase in Gd and Cu in the matrix and a decrease in them in the second phase during the aging treatment. Consequently, the outer edges of some matrix grains near the second phase are rich in Gd and Cu elements and form the LPSO structure that contains moderate Gd and Cu among the α-Mg, the second phase, and the LPSO phase. Moreover, the second phase can provide the area for LPSO growth and reduce the free energy barrier required for LPSO formation [34,35]. Wu et al. [19] also found that lamellar LPSO structures were formed in the matrix of the Mg–10Gd–1Zn–0.5Zr alloy after aging treatment.

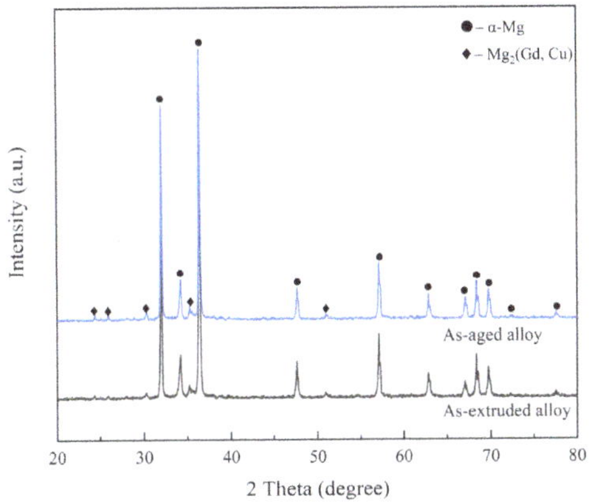

Figure 3. XRD patterns of the Mg–6.0Gd–1.2Cu–1.2Zr alloy.

3.2. Mechanical Properties

The tensile yield strength (TYS), ultimate tensile strength (UTS), compressive yield strength (CYS), ultimate compressive strength (UCS), and elongation of the as-extruded alloy and as-aged alloy under different test conditions are shown in Table 2. The TYS (231 MPa) and UTS (254 MPa) of the as-aged alloy are increased by 11.0% and 11.6%, respectively, compared to those of the as-extruded alloy at room temperature. The elongation in the as-aged alloy (10.4%) is much higher than that in the as-extruded alloy (5.6%), which shows an 85.7% increase. Under the 93 °C test condition, both TYS and UTS of both alloys decrease, the TYS (203.8 MPa) and UTS (234.3 MPa) of the as-aged alloy are still slightly higher than those of the as-extruded alloy (TYS = 191.4 MPa, UTS = 226.0 MPa), but the elongation in the as-aged alloy is slightly lower than that in the as-extruded alloy. Moreover, the TYS and UTS of both alloys are a little lower, but the elongation is higher at 93 °C than those at 25 °C. Compared with the tensile mechanical properties, the CYS of both alloys shows little difference, but the UCS and elongation increase significantly. It is clear that the UCS and elongation of the alloy decrease after the aging treatment.

After the aging treatment, the TYS increases by 11.0%, which can be explained by the following comprehensive reasons: solid solution strengthening, fine grain strengthening caused by static recrystallization, LPSO strengthening, and the decrease in the second phase which weaken the precipitation strengthening [36–40]. It is well known that Gd has a high solid solubility in Mg alloys, which can be used as an effective solute enhancer to strengthen Mg alloys [36]. Combined with the EDS results from Table 1, more Gd and Cu are dissolved into the matrix of the as-aged alloy, which can cause lattice distortion and hinder dislocation slip, thus playing a role in solid solution strengthening. Moreover, there are more fine static recrystallization grains in the as-aged alloy that can contribute to strength improvement through grain refinement strengthening. In addition, when the LPSO structure is parallel to the extrusion direction, the short fiber reinforcement mechanism can effectively hinder the dislocation movement and even accumulate dislocations, thereby improving the TYS of Mg alloys [40]. However, after the aging treatment, the second phase shows aggregated blocks, and the amount is reduced. Therefore, the strengthening effect of second-phase precipitation is weakened. Based on the above strengthening mechanism, the TYS of the as-aged alloy at room temperature is better than that of the as-extruded alloy.

Table 2. Mechanical properties of the Mg–6.0Gd–1.2Cu–1.2Zr alloy at room temperature and 93 °C.

Tests	Alloys	T (°C)	TYS/CYS (MPa)	UTS/UCS (MPa)	Elongation (%)
Tensile	As-extruded	25	208.2 ± 2.1	228.0 ± 2.2	5.6 ± 1.3
	As-extruded	93	191.4 ± 1.5	226.0 ± 2.3	12.9 ± 1.2
	As-aged	25	231.0 ± 2.4	254.4 ± 2.2	10.4 ± 1.4
	As-aged	93	203.8 ± 2.2	234.3 ± 2.1	11.1 ± 1.1
Compression	As-extruded	93	194.3 ± 1.8	374.5 ± 13.8	19.7 ± 1.3
	As-aged	93	193.3 ± 2.9	336.1 ± 16.8	16.9 ± 0.9

At 93 °C, the TYS and UTS of both alloys decrease, and the TYS and UTS of the as-aged alloy decrease by 11.8% and 7.9%, respectively, compared with those at ambient temperature. This is because the second phase of Mg alloys will soften at high temperatures, and the dislocation movement of the alloy is more easily activated at 93 °C, resulting in a slight decrease in strength. Moreover, the decrease in UTS is also related to dislocation activation. The increase in temperature is beneficial to activate the dislocation motion, thus reducing TYS [41].

As for elongation at ambient temperature, the tensile elongation in the as-aged alloy (10.4%) is almost twice that in the as-extruded alloy (5.9%). It is well known that the elongation in Mg alloys is related to microstructure [42]. The distribution of the second phase in the as-extruded alloy is relatively uniform, but due to the presence of LEGs, they are difficult to coordinate deformation. When dislocations slip to these grain boundaries, dislocations are hindered and accumulate severely, which is prone to cause stress concentration and eventually result in fracture. Although, these LEGs transform into equiaxed fine grains due to static recrystallization during the aging process that are conducive to coordination and transmission of deformation, thus leading to much better ductility. The fracture morphologies of both alloys under ambient temperature and 93 °C are shown in Figure 4. The cleavage planes for the LEGs and secondary microcracks can be observed in Figure 4a, which play a negative role in ductility, but the fracture morphology of the as-aged alloy at ambient temperature is full of tiny dimples, indicating good ductility. At 93 °C, the tensile elongation in the as-aged alloy increases slightly (only about 7%), but that in the as-extruded alloy increases by 130%, which is mainly due to the occurrence of static recrystallization at this tested temperature since no coarse cleavage planes can be observed in Figure 4c. Furthermore, the residual stress release during the test at 93 °C may also play a positive role in the elongation improvement of the as-extruded alloy.

At 93 °C, The CYS/TYS of the as-extruded and as-aged alloys are 1.01 and 0.94, respectively, showing good tension-compression symmetry. Due to the symmetric nature of sliding, prismatic and basal sliding should behave identically under tension and compression, resulting in good tension-compression yield symmetry [43,44] Grain refinement is also beneficial to weaken the tension and compression asymmetry of Mg alloys. Moreover, the UCS and compressive elongation in both alloys are much higher than the UTS and tensile elongation. This is largely attributed to the different tensile and compressive deformation mechanisms of Mg alloys: Tensile deformation is mainly matrix slip, hindered by dislocations and grain boundaries, but according to the Mohr–Coulomb criterion, the specimen will be cut off at an angle of 45° during compression [45], and the friction related to compressive stress will be generated inside the alloy, which will hinder the movement in the section and make it difficult for the crack tip to expand. Compared with the tensile tests, this friction is more effective than the hindrance of dislocations and grain boundaries.

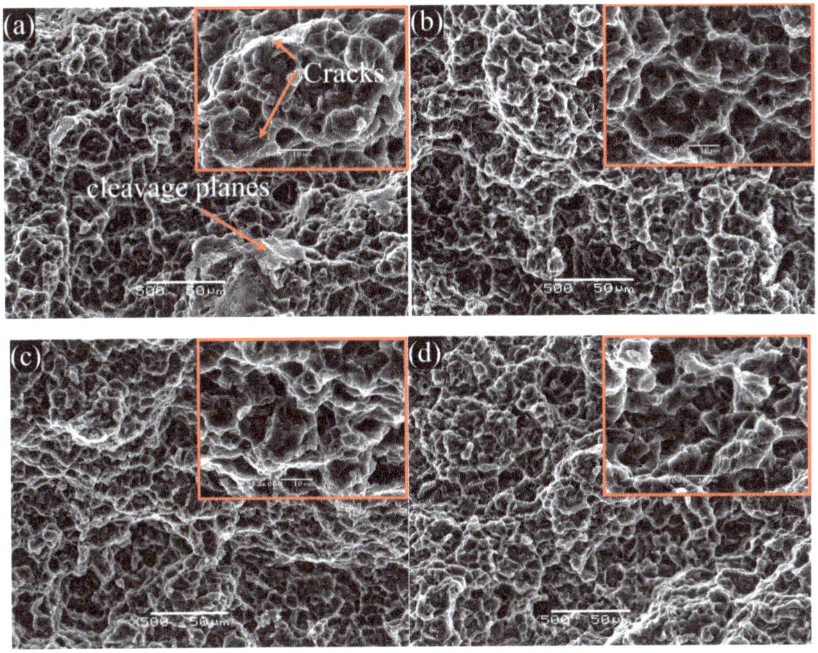

Figure 4. Tensile fracture morphologies of the as-extruded (**a**,**c**) and the as-aged (**b**,**d**) alloys at 25 °C (**a**,**b**) and 93 °C (**c**,**d**).

3.3. Corrosion Behavior

The hydrogen evolution curves of both alloys are shown in Figure 5a. It shows that the hydrogen evolution volume of the as-aged alloy is larger than that of the as-extruded alloy during the 60 min immersion test, indicating a higher corrosion rate. Three corrosion stages can be inferred from the hydrogen evolution curves: (1) in the first 10 min stage, the hydrogen evolution curves of both alloys are relatively flat and stable; (2) during the 10–40 min stage, the slope of hydrogen evolution curve of both alloys increases with prolonging immersion time, indicating an increased corrosion rate at this stage; and (3) during the 40–60 min stage, the hydrogen evolution curves present nearly linear, demonstrating a stable corrosion rate at this stage. This is probably related to the corrosion product on the surface of the alloys [45,46]. At the beginning of immersion, $Mg(OH)_2$ film is formed on the surface of the samples which can protect against further corrosion to some extent. As the immersion time prolongs, the insoluble $Mg(OH)_2$ film reacts with Cl^- and generates soluble $MgCl_2$, and the samples undergo rapid corrosion accompanied by thick corrosion products on the surfaces. Then, the continuous penetration of the corrosion solution into the uncorroded surface and the increased corrosion products keep balance, resulting in a stable corrosion rate at the last immersion stage.

Figure 5b shows the corrosion rates of the alloys calculated based on the mass loss test and hydrogen evolution test. The mass loss corrosion rate of the as-extruded alloy (1660.8 mm/y) and the as-aged alloy (1955.1 mm/y) are a little higher than their corrosion rates calculated from hydrogen evolution (1518.6 mm/y and 1814.6 mm/y, respectively). This may be due to the fact that the corrosion reaction is still ongoing when the gas-collecting burette is replaced during the hydrogen evolution experiment, and this part of the hydrogen has not been collected. In addition, some incompletely corroded second phase falls off after pickling, which increases the corrosion rate of the alloy using the mass loss test compared with the hydrogen evolution test.

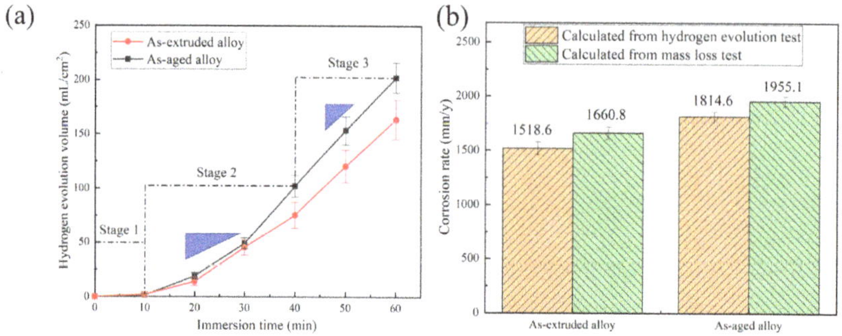

Figure 5. Hydrogen evolution curves (**a**) and corrosion rates (**b**) of the Mg–6.0Gd–1.2Cu–1.2Zr alloy immersed in 3 wt.% KCl for 60 min.

Figure 6 shows the corrosion morphologies of both alloys immersed in 3 wt.% KCl solution at 93 °C for 60 min. The corrosion products on the surfaces can be clearly observed, and there are some bright acicular corrosion products on the top layer. According to the analysis of EDS results in Figure 6b,d, the corrosion products mainly contain Mg, O, and Cl, as well as some K and Gd elements. Since EDS cannot detect the presence of H, combined with other corrosion studies, the surface of the corrosion layer mainly consisted of $Mg(OH)_2$, $MgCl_2{\cdot}6H_2O$, and MgO, and the acicular corrosion products are a mixture of MgO and $Mg(OH)_2$ [3,47].

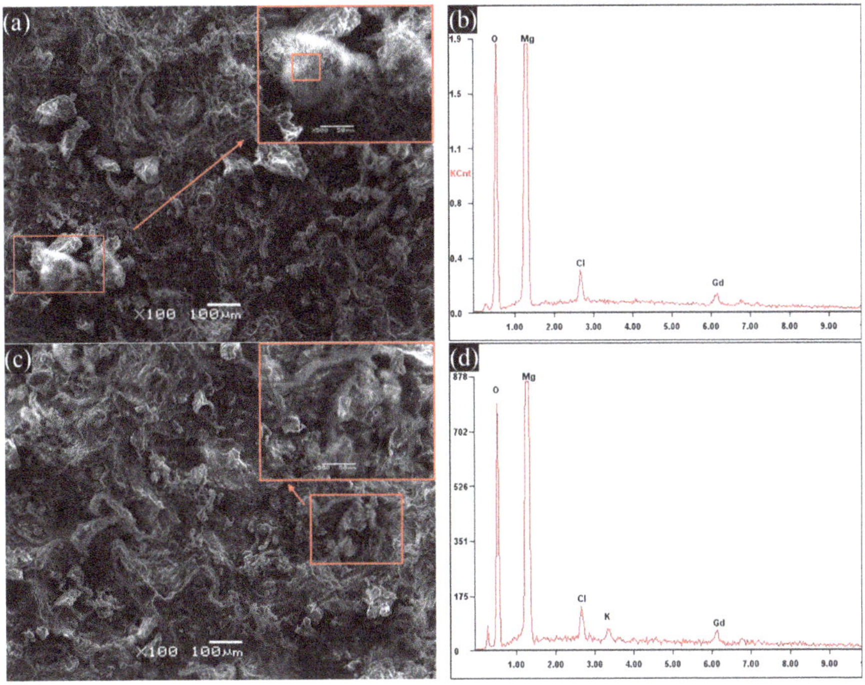

Figure 6. Surface morphologies of the as-extruded alloy (**a**) and the as-aged alloy (**c**) alloy immersed in 3 wt.% KCl for 60 min and selected area EDS spectra (**b**,**d**).

In order to further determine the corrosion behavior of both alloys, the electrochemical polarization curves are shown in Figure 7, and the corresponding fitting results are listed in Table 3. The corrosion potential of the as-aged alloy (−1.50 V) is lower than that of the as-extruded alloy (−1.45 V), which indicates a faster corrosion tendency from thermodynamics, and the corrosion current density of the as-aged alloy is higher as compared to that of the as-extruded alloy, which means a higher corrosion rate from corrosion dynamics [48]. The electrochemical results are consistent with those obtained using the hydrogen evolution and mass loss tests.

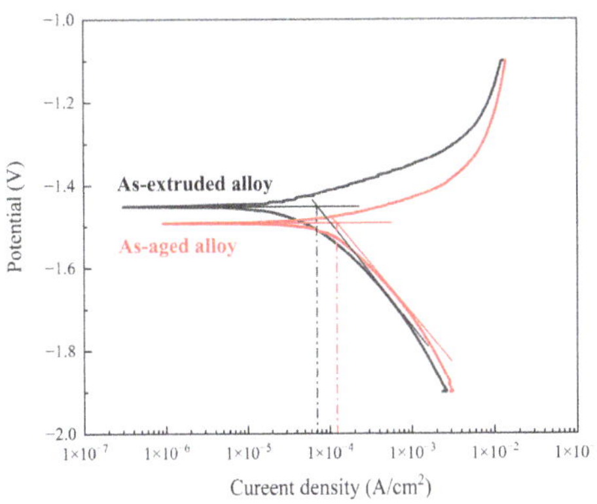

Figure 7. Polarization curves of the Mg–6.0Gd–1.2Cu–1.2Zr alloy in 3 wt.% KCl solution.

Table 3. Electrochemical parameters of the Mg–6.0Gd–1.2Cu–1.2Zr alloy obtained using Tafel extrapolation from Figure 7.

Alloy	E_{corr} (V)	I_{corr} (A/cm^2)
As-extruded	−1.45	6.3×10^{-5}
As-aged	−1.50	1.2×10^{-4}

The corrosion rates of the as-extruded and as-aged alloys in 3 wt.% KCl solution at 93 °C are 1660.8 and 1955.1 mm/y, respectively, which is desired for degradable fracturing ball applications. Generally, grain refinement can decrease the corrosion rate of Mg alloys because Mg alloys with fine grains more easily form a dense corrosion product film on a corroded surface, thereby protecting the substrate from corrosion [49]. Moreover, fine grains may also reduce the corrosion rate of Mg alloys if they have obvious passivation behavior [50]. However, in the present work, both alloys have no obvious passivation stage in Figure 7, so fine grains do not reduce the corrosion rate of the alloy. In addition, studies have shown that refined grains mean more grain boundaries. The grain boundary is a typical surface defect with high energy that has priority corrosion and can be used as a corrosion channel to accelerate the corrosion rate [51]. The as-extruded alloy obtained many fine grains by dynamic recrystallization, so they have many grain boundaries, which provides a favorable channel for corrosion propagation and accelerates the corrosion of the alloy [52]. Due to the static recrystallization during aging, the number of fine grains is obviously more than that of the as-extruded alloy, which owns more driving force for corrosion and thus leads to faster corrosion.

It is well known that the corrosion barrier and micro-galvanic effect are two different aspects of the second phase in corrosion. A uniformly distributed tiny second phase can reduce the corrosion rate, while a discontinuously distributed bulky second phase will accelerate the corrosion rate of the Mg alloy [53]. During corrosion, the second phase usually acts as a cathode and accelerates the corrosion of the anode substrate. After the aging treatment, the second phase is more aggregated, forming a "large cathode and small anode" galvanic corrosion, which accelerates the galvanic corrosion process.

The effect of the LPSO phase on the corrosion of a Mg alloy is also dual: the LPSO phase could act as a micro-cathode to accelerate corrosion if it distributes discontinuously; however, if it distributes continuously along the grain boundaries or at the outer edge of matrix grains, it could act as a bridge and restrict corrosion caused by galvanic corrosion between the matrix and eutectic phase [30]. Furthermore, it has been found that the lamellar LPSO phase can also be used as a cathode phase to provide a driving force for galvanic corrosion of the α-Mg matrix [54]. The discontinuous lamellar LPSO phase in the as-aged alloy makes it unable to protect the matrix and accelerates corrosion.

Figure 8 shows the local potential distribution between the α-Mg matrix, second phase, and LPSO phase observed in the parallel extrusion direction. The line in Figure 8a intersects different phases, the three-dimensional volt potential of the tested regions is shown in Figure 8b, and the corresponding volt potential curve of the analyzed line is shown in Figure 8c. The volt potential of the three phases shows the following order: second phase > LPSO phase > α-Mg matrix, which indicates that both the second phase and the LPSO act as cathodes in the microcell to accelerate the corrosion of the matrix.

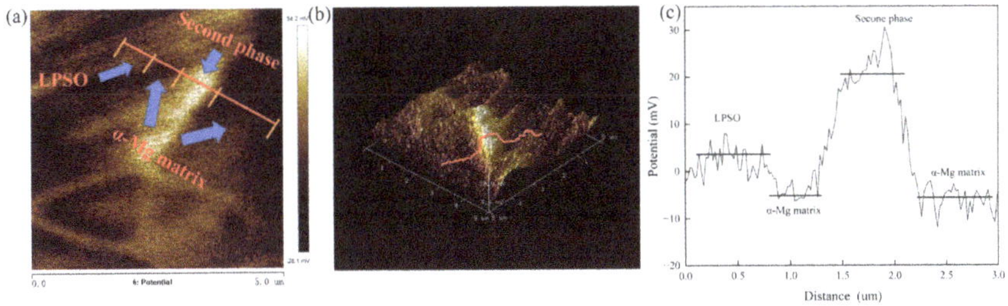

Figure 8. Volt potential distribution of as-aged alloy: (**a**) two-dimensional volt potential distribution, (**b**) three-dimensional volt potential distribution, (**c**) volt potential distribution of the selected line.

Figure 9 shows the comparison of the ultimate compressive strength and corrosion rate of Mg–6.0Gd–1.2Cu–1.2Zr alloy in this work and other Mg alloys [14,15,25,37,46,55–57]. It is clear that the Mg–6.0Gd–1.2Cu–1.2Zr alloy with moderate strength and a high corrosion rate exhibits excellent comprehensive properties that can be effectively applied to the field of temporary plugging. A much higher corrosion rate of the studied alloy in this work may be attributed to the addition of Cu and particular microstructures.

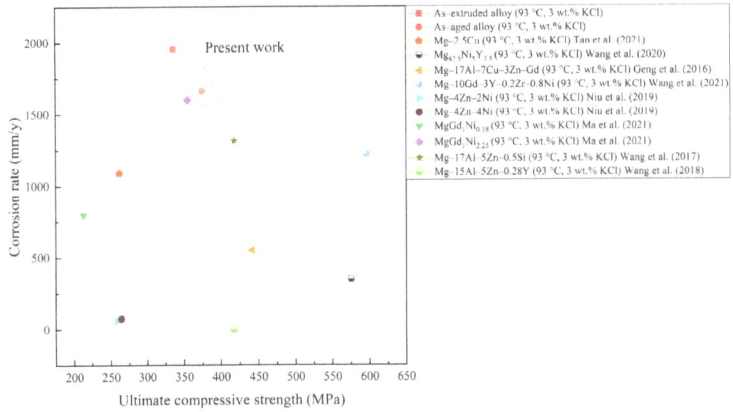

Figure 9. Comparison of the ultimate compressive strength and corrosion rate among the present Mg–Gd–Cu–Zr alloy and other Mg alloys for degradable fracturing ball applications [14,15,25,37,46,55–57].

4. Conclusions

(1) The microstructure of the as-extruded Mg–6.0Gd–1.2Cu–1.2Zr alloy consists of an α-Mg matrix, a second phase, and un-recrystallized large, elongated grains. After the aging treatment at 170 °C for 8 h, the large, elongated grains disappear due to static recrystallization and a lamellar LPSO structure is formed because of the diffusion of the Gd and Cu atoms.

(2) The tensile yield strength of the as-extruded Mg–6.0Gd–1.2Cu–1.2Zr alloy is slightly improved after the aging treatment due to the compromise between solution strengthening, grain refinement strengthening, LPSO strengthening, and second phase strengthening. The significant improvement in elongation after aging or testing at 93 °C is mainly attributed to static recrystallization. The ultimate compressive strength of the as-extruded and as-aged alloys at 93 °C is 374.5 MPa and 336.1 MPa, respectively, which may meet the requirements of underground temperature work.

(3) The corrosion rate of the as-extruded and as-aged Mg–6.0Gd–1.2Cu–1.2Zr alloy in 3 wt.% KCl solution at 93 °C is 1660.8 mm/y and 1955.1 mm/y, respectively. Galvanic

corrosion plays a leading role in the whole corrosion process. The aggregated second phase, the formation of the LPSO phase, and the refined grains by static recrystallization are responsible for the more rapid corrosion after the aging treatment.

(4) The comprehensive mechanical properties and corrosion rate of the Mg–6.0Gd–1.2Cu–1.2Zr alloy under the as-extruded and as-aged conditions exceeds most reported Mg alloys, and the alloy shows promising potential for degradable fracturing ball applications.

Author Contributions: Conceptualization, J.J. and X.Z.; methodology, J.J.; validation, X.G. and J.J.; formal analysis, J.J.; investigation, J.J. and X.G.; resources, X.Z.; data curation, J.J. and X.G.; writing—original draft preparation, J.J.; writing—review and editing, X.Z.; visualization, J.J. and X.G.; supervision, X.Z.; project administration, X.Z.; funding acquisition, X.Z. All authors have read and agreed to the published version of the manuscript.

Funding: This project was supported by the National Natural Science Foundation of China (52071175), the Key Research & Development Plan (Social Development) of Jiangsu Province (BE2020702), and the Innovative Foundation Project for Students of Nanjing Institute of Technology (TB202217002).

Data Availability Statement: The research data are available upon reasonable request to the authors.

Conflicts of Interest: The authors declare no conflict of interest.

References

1. Jia, A. Progress and prospects of natural gas development technologies in China. *Nat. Gas Ind. B* **2018**, *5*, 547–557. [CrossRef]
2. Liu, B.; Yang, Y.; Zhang, Y.; Du, H.; Hou, L.; Wei, Y. Investigation of rapidly decomposable AZ91–RE–xCu (x = 0, 1, 2, 3, 4) alloys for petroleum fracturing balls. *J. Phys. Chem. Solids* **2020**, *144*, 109499. [CrossRef]
3. Kang, L.; Shi, Y.; Luo, X.; Liu, B. [Retracted] Influence of Al Content on Degradation Behavior of Cu-Doped Mg–Al Alloys for Drill-Free Plugging Applications. *Adv. Mater. Sci. Eng.* **2020**, *2020*, 8898669. [CrossRef]
4. Zhang, C.; Wu, L.; Huang, G.; Chen, L.; Xia, D.; Jiang, B.; Atrens, A.; Pan, F. Effects of Fe concentration on microstructure and corrosion of Mg–6Al–1Zn–xFe alloys for fracturing balls applications. *J. Mater. Sci. Technol.* **2019**, *35*, 2086–2098. [CrossRef]
5. Zhao, X.; Rui, Z.; Liao, X.; Zhang, R. The qualitative and quantitative fracture evaluation methodology in shale gas reservoir. *J. Nat. Gas Sci. Eng.* **2015**, *27*, 486–495. [CrossRef]
6. Li, X.; Zhou, L.; Shi, T.; Li, B.; Yu, W.; Zhang, C.; Yan, P.; Zhang, Q.; Tian, H.; Lv, Y. Corrosion decomposition and mechanical behaviors of As-cast Mg–*x*Zn–Zr alloys. *Mater. Corros.* **2020**, *71*, 1453–1461. [CrossRef]
7. Zhang, Y.; Wang, X.; Kuang, Y.; Liu, B.; Zhang, K.; Fang, D. Enhanced mechanical properties and degradation rate of Mg–3Zn–1Y based alloy by Cu addition for degradable fracturing ball applications. *Mater. Lett.* **2017**, *195*, 194–197. [CrossRef]
8. Hong, L.; Wang, R.; Zhang, X. The Role of Nd in Corrosion Properties of Mg–12Gd–2Zn–0.4Zr Alloys. *J. Mater. Eng. Perform.* **2021**, *30*, 6000–6008. [CrossRef]
9. Nie, Y.-J.; Dai, J.-W.; Zhang, X.-B. Effect of Ag Addition on Microstructure, Mechanical and Corrosion Properties of Mg–Nd–Zn–Zr Alloy for Orthopedic Application. *Acta Met. Sin.* **2023**, *36*, 295–309. [CrossRef]
10. Ben Tzion-Mottye, L.; Ron, T.; Eliezer, D.; Aghion, E. The Effect of Mn on the Mechanical Properties and In Vitro Behavior of Biodegradable Zn-2%Fe Alloy. *Metals* **2022**, *12*, 1291. [CrossRef]
11. Geng, X.; Jiang, J.; Zhang, X. Corrosion Behavior of Mg–*x*Gd–1Zn–0.4Zr Alloys with Different Gd Additions for Biomedical Application. *Metals* **2022**, *12*, 1763. [CrossRef]
12. Dai, J.; Dong, Q.; Nie, Y.; Jia, Y.; Chu, C.; Zhang, X. Insight into the role of Y addition in the microstructures, mechanical and corrosion properties of as-cast Mg-Gd-Y-Zn-Ca-Zr alloys. *Mater. Des.* **2022**, *221*, 110980. [CrossRef]
13. Pei, X.; Wei, S.; Shi, B.; Shen, Z.; Wang, X.; Tong, Z.; Fu, T. Disintegrating fracturing ball used in ball injection sliding sleeve for staged fracturing. *Pet. Explor. Dev.* **2014**, *41*, 805–809. [CrossRef]
14. Tan, W.; Li, T.; Li, S.; Fang, D.; Ding, X.; Sun, J. High strength-ductility and rapid degradation rate of as-cast Mg–Cu–Al alloys for application in fracturing balls. *J. Mater. Sci. Technol.* **2021**, *94*, 22–31. [CrossRef]
15. Wang, J.; Gao, S.; Liu, X.; Peng, X.; Wang, K.; Liu, S.; Jiang, W.; Guo, S.; Pan, F. Enhanced mechanical properties and degradation rate of Mg–Ni–Y alloy by introducing LPSO phase for degradable fracturing ball applications. *J. Magnes. Alloy.* **2020**, *8*, 127–133. [CrossRef]
16. Ma, K.; Wang, J.; Ren, J.; Dai, C.; Liu, S.; Peng, Y.; Pan, Y. Enhanced degradation properties of Mg-Y-Ni alloys by tailoring the LPSO morphology for fracturing tools applications. *Mater. Charact.* **2021**, *181*, 111489. [CrossRef]
17. Xiao, D.H.; Geng, Z.W.; Chen, L.; Wu, Z.; Diao, H.Y.; Song, M.; Zhou, P.F. Effects of Alloying Elements on Microstructure and Properties of Magnesium Alloys for Tripling Ball. *Met. Mater. Trans. A* **2015**, *46*, 4793–4803. [CrossRef]
18. Dai, J.; Nie, Y.; Zhang, X. Ambient and elevated temperature mechanical behavior of Mg–Gd–Cu–Zr alloy. *Mater. Res. Express* **2018**, *5*, 116524. [CrossRef]

19. Wu, X.; Pan, F.; Cheng, R. Formation of long period stacking ordered phases in Mg–10Gd–1Zn–0.5Zr (wt.%) alloy. *Mater. Charact.* **2018**, *147*, 50–56. [CrossRef]
20. Yuan, X.; Du, Y.; Dong, D.; Liu, D.; Jiang, B. Corrosion Resistance Improvement of an Extruded Mg–Gd–Y–Zn–Zr–Ca via Aging Treatment. *J. Mater. Eng. Perform.* **2021**, *31*, 2909–2917. [CrossRef]
21. Niu, Y.; Hou, J.; Ning, F.; Chen, X.; Jia, Y.; Le, Q. Hot deformation behavior and processing map of Mg–2Zn–1Al–0.2RE alloy. *J. Rare Earths* **2020**, *38*, 665–675. [CrossRef]
22. Niu, H.-Y.; Deng, K.-K.; Nie, K.-B.; Wang, C.-J.; Liang, W.; Wu, Y.-C. Degradation behavior of Mg–4Zn–2Ni alloy with high strength and high degradation rate. *Mater. Chem. Phys.* **2020**, *249*, 123131. [CrossRef]
23. Zhang, Z.; Zhang, J.-H.; Wang, J.; Li, Z.-H.; Xie, J.-S.; Liu, S.-J.; Guan, K.; Wu, R.-Z. Toward the development of Mg alloys with simultaneously improved strength and ductility by refining grain size via the deformation process. *Int. J. Miner. Met. Mater.* **2020**, *28*, 30–45. [CrossRef]
24. Zhong, S.; Zhang, D.; Wang, Y.; Chai, S.; Feng, J.; Luo, Y.; Hua, J.; Dai, Q.; Hu, G.; Xu, J.; et al. Microstructures, mechanical properties and degradability of Mg-2Gd-0.5(Cu/Ni) alloys: A comparison study. *J. Mater. Sci. Technol.* **2022**, *128*, 44–58. [CrossRef]
25. Geng, Z.; Xiao, D.; Chen, L. Microstructure, mechanical properties, and corrosion behavior of degradable Mg–Al–Cu–Zn–Gd alloys. *J. Alloys Compd.* **2016**, *686*, 145–152. [CrossRef]
26. Xie, J.; Zhang, J.; Zhang, Z.; Yang, Q.; Guan, K.; He, Y.; Wang, R.; Zhang, H.; Qiu, X.; Wu, R. New insights on the different corrosion mechanisms of Mg alloys with solute-enriched stacking faults or long period stacking ordered phase. *Corros. Sci.* **2022**, *198*, 110163. [CrossRef]
27. Chen, X.; Wang, H.; Le, Q.; Jia, Y.; Zhou, X.; Yu, F.; Atrens, A. The role of long-period stacking ordered phase on the discharge and electrochemical behaviors of magnesium anode Mg-Zn-Y for the primary Mg-air battery. *Int. J. Energy Res.* **2020**, *44*, 8865–8876. [CrossRef]
28. Nie, Y.; Dai, J.; Li, X.; Zhang, X. Recent developments on corrosion behaviors of Mg alloys with stacking fault or long period stacking ordered structures. *J. Magnes. Alloy.* **2021**, *9*, 1123–1146. [CrossRef]
29. Istrate, B.; Munteanu, C.; Cimpoesu, R.; Cimpoesu, N.; Popescu, O.; Vlad, M. Microstructural, Electrochemical and In Vitro Analysis of Mg-0.5Ca-xGd Biodegradable Alloys. *Appl. Sci.* **2021**, *11*, 981. [CrossRef]
30. Dai, J.; Zhang, X.; Wang, Z. Microstructure and enhanced corrosion resistance of biodegradable Mg–Gd–Cu–Zr alloy by solution treatment. *Mater. Technol.* **2018**, *33*, 301–310. [CrossRef]
31. Huang, K.; Logé, R. A review of dynamic recrystallization phenomena in metallic materials. *Mater. Des.* **2016**, *111*, 548–574. [CrossRef]
32. Luo, S.; Wang, N.; Wang, Y.; Chen, J.; Qin, H.; Kong, S.; Bai, T.; Lu, W.; Xiao, L.; Ma, X.; et al. Texture, microstructure and mechanical properties of an extruded Mg–10Gd–1Zn–0.4Zr alloy: Role of microstructure prior to extrusion. *Mater. Sci. Eng. A* **2022**, *849*, 143476. [CrossRef]
33. Yang, Z.; Ma, A.; Liu, H.; Sun, J.; Song, D.; Wang, C.; Yuan, Y.; Jiang, J. Multimodal Microstructure and Mechanical Properties of AZ91 Mg Alloy Prepared by Equal Channel Angular Pressing plus Aging. *Metals* **2018**, *8*, 763. [CrossRef]
34. Egusa, D.; Abe, E. The structure of long period stacking/order Mg–Zn–RE phases with extended non-stoichiometry ranges. *Acta Mater.* **2012**, *60*, 166–178. [CrossRef]
35. Kim, J.-K.; Ko, W.-S.; Sandlöbes, S.; Heidelmann, M.; Grabowski, B.; Raabe, D. The role of metastable LPSO building block clusters in phase transformations of an Mg–Y–Zn alloy. *Acta Mater.* **2016**, *112*, 171–183. [CrossRef]
36. Xu, C.; Nakata, T.; Qiao, X.G.; Zheng, M.Y.; Wu, K.; Kamado, S. Ageing behavior of extruded Mg–8.2Gd–3.8Y–1.0Zn–0.4Zr (wt.%) alloy containing LPSO phase and γ' precipitates. *Sci. Rep.* **2017**, *7*, 43391. [CrossRef]
37. Wang, J.; Li, T.; Li, H.; Ma, Y.; Zhao, K.; Yang, C.; Zhang, J. Effect of trace Ni addition on microstructure, mechanical and corrosion properties of the extruded Mg–Gd–Y–Zr–Ni alloys for dissoluble fracturing tools. *J. Magnes. Alloy.* **2021**, *9*, 1632–1643. [CrossRef]
38. Chen, X.; Jia, Y.; Le, Q.; Ning, S.; Li, X.; Yu, F. The interaction between in situ grain refiner and ultrasonic treatment and its influence on the mechanical properties of Mg–Sm–Al magnesium alloy. *J. Mater. Res. Technol.* **2020**, *9*, 9262–9270. [CrossRef]
39. Hong, L.; Wang, R.; Zhang, X. Effects of Nd on microstructure and mechanical properties of as-cast Mg–12Gd–2Zn–xNd–0.4Zr alloys with stacking faults. *Int. J. Miner. Met. Mater.* **2022**, *29*, 1570–1577. [CrossRef]
40. Zhang, X.; Ba, Z.; Wang, Z.; Wu, Y.; Xue, Y. Effect of LPSO structure on mechanical properties and corrosion behavior of as-extruded GZ51K magnesium alloy. *Mater. Lett.* **2016**, *163*, 250–253. [CrossRef]
41. Homma, T.; Hirawatari, S.; Sunohara, H.; Kamado, S. Room and elevated temperature mechanical properties in the as-extruded Mg–Al–Ca–Mn alloys. *Mater. Sci. Eng. A* **2012**, *539*, 163–169. [CrossRef]
42. Trivedi, P.; Nune, K.; Misra, R.; Goel, S.; Jayganthan, R.; Srinivasan, A. Grain refinement to submicron regime in multiaxial forged Mg–2Zn–2Gd alloy and relationship to mechanical properties. *Mater. Sci. Eng. A* **2016**, *668*, 59–65. [CrossRef]
43. Garces, G.; Perez, P.; Cabeza, S.; Lin, H.; Kim, S.; Gan, W.; Adeva, P. Reverse tension/compression asymmetry of a Mg–Y–Zn alloys containing LPSO phases. *Mater. Sci. Eng. A* **2015**, *647*, 287–293. [CrossRef]
44. Chi, Y.; Zhou, X.; Qiao, X.; Brokmeier, H.; Zheng, M. Tension-compression asymmetry of extruded Mg–Gd–Y–Zr alloy with a bimodal microstructure studied by in-situ synchrotron diffraction. *Mater. Des.* **2019**, *170*, 107705. [CrossRef]
45. Papanastasiou, P.; Durban, D. Singular crack-tip plastic fields in Tresca and Mohr–Coulomb solids. *Int. J. Solids Struct.* **2018**, *136–137*, 250–258. [CrossRef]

46. Niu, H.-Y.; Deng, K.-K.; Nie, K.-B.; Cao, F.-F.; Zhang, X.-C.; Li, W.-G. Microstructure, mechanical properties and corrosion properties of Mg–4Zn–xNi alloys for degradable fracturing ball applications. *J. Alloys Compd.* **2019**, *787*, 1290–1300. [CrossRef]
47. Ding, J.; Liu, X.; Wang, Y.; Huang, W.; Wang, B.; Wei, S.; Xia, X.; Liang, Y.; Chen, X.; Pan, F.; et al. Effect of Sn Addition on Microstructure and Corrosion Behavior of As-Extruded Mg–5Zn–4Al Alloy. *Materials* **2019**, *12*, 2069. [CrossRef]
48. Gusieva, K.; Davies, C.H.J.; Scully, J.R.; Birbilis, N. Corrosion of magnesium alloys: The role of alloying. *Int. Mater. Rev.* **2014**, *60*, 169–194. [CrossRef]
49. Li, C.; Xu, D.; Chen, X.-B.; Wang, B.; Wu, R.; Han, E.; Birbilis, N. Composition and microstructure dependent corrosion behaviour of Mg-Li alloys. *Electrochim. Acta* **2018**, *260*, 55–64. [CrossRef]
50. Zhang, X.; Dai, J.; Zhang, R.; Ba, Z.; Birbilis, N. Corrosion behavior of Mg–3Gd–1Zn–0.4Zr alloy with and without stacking faults. *J. Magnes. Alloy.* **2019**, *7*, 240–248. [CrossRef]
51. Yin, S.; Duan, W.; Liu, W.; Wu, L.; Yu, J.; Zhao, Z.; Liu, M.; Wang, P.; Cui, J.; Zhang, Z. Influence of specific second phases on corrosion behaviors of Mg–Zn–Gd–Zr alloys. *Corros. Sci.* **2019**, *166*, 108419. [CrossRef]
52. Yang, H.; Wu, L.; Jiang, B.; Lei, B.; Liu, W.; Song, J.; Huang, G.; Zhang, D.; Pan, F. Effects of Grain Size on the Corrosion and Discharge Behaviors of Mg–Y Binary Alloys for Mg-Air Batteries. *J. Electrochem. Soc.* **2020**, *167*, 130515. [CrossRef]
53. Shen, Y.; Chen, J.; Wang, G.; Cao, D.; Tan, L.; Misra, R.; Yang, K. Effects of ECAP extrusion on the mechanical and biodegradable properties of an extruded Mg–1.5Zn–0.5Y–0.5Zr alloy. *Mater. Technol.* **2020**, *37*, 135–142. [CrossRef]
54. Liu, J.; Yang, L.; Zhang, C.; Zhang, B.; Zhang, T.; Li, Y.; Wu, K.; Wang, F. Role of the LPSO structure in the improvement of corrosion resistance of Mg–Gd–Zn–Zr alloys. *J. Alloys Compd.* **2018**, *782*, 648–658. [CrossRef]
55. Ma, K.; Liu, S.; Dai, C.; Liu, X.; Ren, J.; Pan, Y.; Peng, Y.; Su, C.; Wang, J.; Pan, F. Effect of Ni on the microstructure, mechanical properties and corrosion behavior of MgGd1Nix alloys for fracturing ball applications. *J. Mater. Sci. Technol.* **2021**, *91*, 121–133. [CrossRef]
56. Wang, M.; Xiao, D.; Liu, W. Effect of Si addition on microstructure and properties of magnesium alloys with high Al and Zn contents. *Vacuum* **2017**, *141*, 144–151. [CrossRef]
57. Wang, M.; Xiao, D.; Zhou, P.; Liu, W.; Ma, Y.; Sun, B. Effects of rare earth yttrium on microstructure and properties of Mg–Al–Zn alloy. *J. Alloys Compd.* **2018**, *742*, 232–239. [CrossRef]

metals

MDPI

Review

Corrosion of Fixed Orthodontic Appliances: Causes, Concerns, and Mitigation Strategies

António Fróis [1,2,3], Ana Cristina Santos [1,2,3] and Cristina Santos Louro [1,*]

1 Department of Mechanical Engineering, CEMMPRE, ARISE, University of Coimbra, Rua Luis Reis Santos, 3030-177 Coimbra, Portugal; antonio.frois@student.uc.pt (A.F.); acsantos@fmed.uc.pt (A.C.S.)
2 Faculty of Medicine, Biophysics Institute, Coimbra Institute for Clinical and Biomedical Research/Centre for Innovative Biomedicine and Biotechnology (iCBR/CIBB), University of Coimbra, 3000-548 Coimbra, Portugal
3 Area of Environment Genetics and Oncobiology (CIMAGO), University of Coimbra, 3000-548 Coimbra, Portugal
* Correspondence: cristina.louro@dem.uc.pt

Abstract: The orthodontic supply market is a prosperous billion-dollar industry, driven by an increasing demand for orthodontic appliances. The supremacy of metallic first-generation biomaterials is evident for manufacturing brackets, archwires, bands, and other components due to their well-recognized chemical inertness, spontaneous passivation, biocompatibility, and favorable mechanical properties combination. However, the oral cavity is the ultimate corrosion-promoting environment for any metallic material. In this work, the general picture of the intraoral degradation of fixed orthodontic appliances is first addressed, from the causes to the harmful effects and their oral clinical implications. Current mitigation strategies are also pointed out, including the alloys' bulk composition adjustment combined with new and advanced manufacturing processes and/or their surface treatment or coating deposition. The versatile use of thin films and coatings stands out with different deposition technologies: Many in vivo and in vitro efforts have been devoted to oral aging, from monolithic to composite architectures and micro- to nano-scale materials, to meet the best and safest oral practice demands. Unfortunately, literature data suggest that even the existing commercially available protective coatings have drawbacks and are fallible. Further multidisciplinary research is still required to effectively mitigate the corrosion behavior of fixed orthodontic appliances.

Keywords: bioalloys; biocompatibility; corrosion; intraoral aging; orthodontics; protective coatings

1. Introduction

Orthodontics may be defined as the

> *"branch of dentistry that is concerned with the supervision, guidance and correction of the growing and mature dentofacial structures. It includes the diagnosis, prevention, interception and treatment of all forms of malocclusion of the teeth and associated alterations in their surrounding structures".* [1]

Malocclusions—usually referred to as *"crooked"* or *"misaligned teeth"*—are a worldwide dental problem [2–5]. Technically, a malocclusion is not a disease, but rather aesthetical and/or functional misalignments between the dental arches or teeth irregularities (beyond what is considered a normal biological variation), as shown in Figure 2. Still, malocclusions can cause susceptibility to trauma and periodontal diseases [2,4,6–9]. Standard treatments for dental malocclusions involve removable or fixed orthodontic appliances.

Fixed appliances are, in general, more effective than removable ones—especially for more complex situations and/or for adult patients—and incorporate brackets, archwires, tubes, and/or bands, tightened by metallic or polymeric ligatures [10,11] (Figure 1). During treatment, a constant load is transferred from the brackets to the teeth, by using orthodontic archwires (attached to the brackets), obtaining tooth movement while adjacent bone and tissue are remodeled [12]. Typical loads for tooth movement using fixed appliances are summarized in Table 1, usually involving values lower than 1 N.

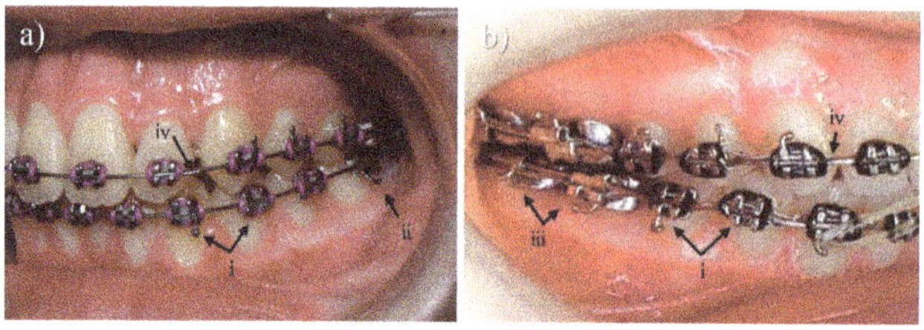

Figure 1. Typical Fixed Orthodontic Appliances: (**a**) by courtesy of Professor Sónia A. Pereira, Faculty of Medicine, University of Coimbra, Portugal; (**b**) used with permission of Dental Press Publishing from [22]. (i) Brackets with polymeric ligatures; (ii) tube; (iii) bands with attached tubes; (iv) archwires.

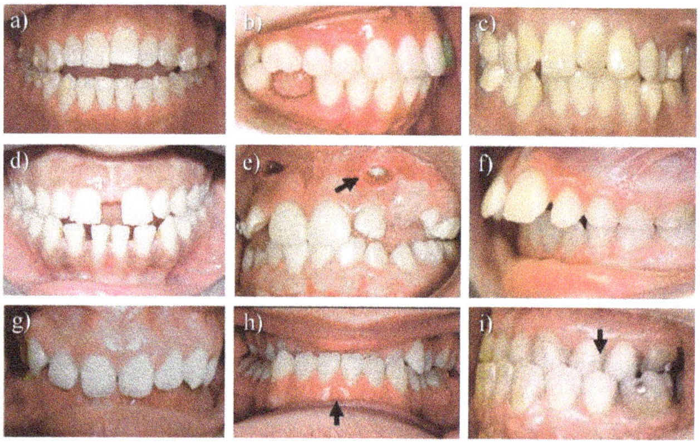

Figure 2. Some common malocclusion types: (**a**) open bite; (**b**) missing tooth; (**c**) crowding; (**d**) spacing; (**e**) impacted tooth; (**f**) overjet; (**g**) overbite; (**h**) underbite or anterior crossbite; (**i**) posterior crossbite. Adapted from [13–21]. Sources: (**a**) reproduced with permission of Oxford University Press on behalf of the European Orthodontic Society, from [13]; (**b**) reproduced with permission of Dental Press Publishing from [14]; (**c**) reproduced with permission from [15]; (**d**) reproduced with permission from [16], © Georg Thieme Verlag KG; (**e**) reproduced with permission of Dental Press Publishing from [17]; (**f**) reproduced with permission from [18]; (**g**) reprinted from [19], Copyright (2007), with permission from Elsevier; (**h**) reproduced with permission of Journal of Clinical Orthodontics from [20]; (**i**) reproduced with permission of Dental Press Publishing from [21].

A standard comprehensive orthodontic treatment may last approximately 2 years [23] and involves three sequential phases: (1st) leveling and aligning; (2nd) correction of molar relationship and space closure; and (3th) detailing and finishing [24].

In contemporary orthodontics, the market supply entails a worldwide billion-dollar industry that is expected to grow in the next few years [25]. Metallic materials are still the first choice for manufacturing fixed appliances due to their balanced set of mechanical, biological, and chemical properties [26]. Up to now, the most commonly used metallic alloys include stainless steel (SS), pure titanium (Ti) and its alloys—especially nickel–titanium (NiTi)—and cobalt–chromium (CoCr) alloys. Other metallic materials can also be found in fixed orthodontic appliances, but with a lower application range.

Table 1. Typical loads for orthodontic tooth movement (adapted with permission of Elsevier, from [27]; permission conveyed through Copyright Clearance Center, Inc.).

Movement	Description	Load (N)
Tipping	Predominant movement of the dental crown in the opposite direction.	0.34–0.59
Bodily movement (translation)	Movement that tilts the tooth until its root is in the vertical direction.	0.69–1.18
Root uprighting	Predominant movement of the root.	0.49–0.98
Rotation	Rotation of the tooth around its long axis.	0.34–0.59
Extrusion	Moving the tooth in the opposite direction to the supporting alveolar bone.	0.34–0.59
Intrusion	Moving the tooth into the supporting alveolar bone.	0.10–0.20

A clinical concern during orthodontic treatments is intraoral corrosion. Always associated with metallic ion release into the oral cavity, corrosion can be intensified by dental plaque accumulation and/or mechanical actions such as friction and fatigue stress. Several important consequences of this undesirable degradation may arise, namely enamel discoloration and demineralization, hypersensitivity, inflammatory reactions and local pain, and, in more severe cases, toxicity effects [28–32].

The need to modify the orthodontic alloys has been identified. Current research guidelines point in two main directions: (i) to adjust the alloys' bulk composition combined with new and advanced manufacturing processes; or (ii) to modify their surface, while taking advantage of the excellent mechanical properties of the bulk. The composition and microstructure of the surface can be altered by using chemical or physical methods, either by treatment or coating deposition.

The present overview is schematized in Figure 3 and partitioned into five sections. After this introductory section (Section 1), Section 2 focuses on metallic corrosion, the oral environment's aggressiveness, and its impact on orthodontic alloys. Section 3 comprises an overview of the harmful effects and clinical impact of intraoral corrosion on fixed appliances. A brief condensation concerning the mitigation strategies is presented in Section 4, describing current modifications of the orthodontic alloys. Finally, due to the utmost importance of the surface properties of any biomaterial, Section 5 provides a comprehensive overview of the protective coatings in orthodontics, from metals to ceramics and polymers, as well as their composite architectures with different reinforcement materials.

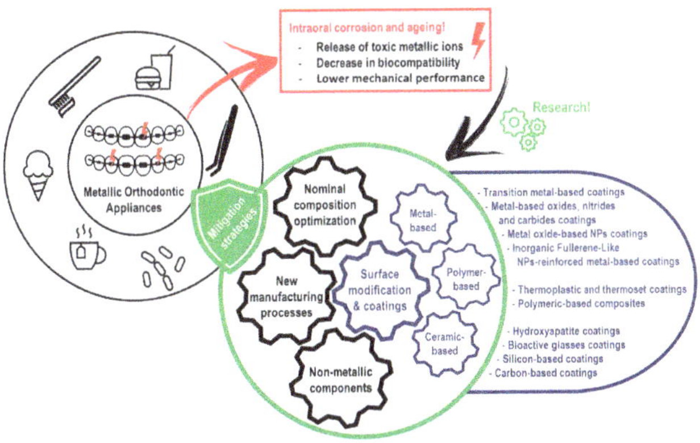

Figure 3. Schematic summary of the current review work, concerning corrosion of fixed orthodontic appliances.

2. Metallic Corrosion

This chapter focuses on the main alloys used for the manufacturing of orthodontic appliances, the characteristics of the oral environment, and their effects on the corrosion behavior of metallic alloys.

2.1. Orthodontic Alloys

Metals and alloys thrive in the medical field and are more employed as biomaterials than any other material type [26]. Today, the major metallic alloys used in orthodontic applications include stainless steel (SS), pure titanium (Ti) and its alloys—especially the nickel–titanium (NiTi)—and cobalt–chromium (CoCr) alloys. Some of the main characteristics of these bioalloys, in comparison to human molar tooth enamel, are summarized in Table 2.

Stainless steels are iron (Fe)-based alloys containing at least 12% chromium (Cr) and a maximum of 1.2% carbon (C), according to the European Standard EN 10088-1 [33] (Table 3). SS are outstanding materials for manufacturing brackets, bands, tubes, and ligatures [11,34,35], namely the austenitic 3xx series-AISI (American Iron and Steel Institute: 302, 303, 304L, and 316L), the precipitation hardening (PH) steels, as well as the duplex steels (SAF 2205) [10,36–40]. Together with Ti alloys, SS archwires are frequently used in an orthodontic treatment, especially during the 2nd and 3rd phases [41,42].

Table 2. Main characteristics of bioalloys used for manufacturing orthodontic components [37,43–46].

	Main Composition	Young's Modulus (GPa)	Yield Strength (MPa)
Human molar tooth enamel	Calcium phosphate hydroxyapatite	70–115	
Stainless steel (AISI 316L)	Fe–Cr–Ni	160–187	960–1500
Cobalt–chromium	Co–Cr–Fe–Ni	150–217	830–1200
α-Titanium	Ti (grade 4)	104	550
β-Titanium	Ti–Mo–Sn–Zr	60–68	620–690
Ti-6Al-4V	Ti–Al–V (grade 5)	100–110	830–1070
Nickel–titanium	Ni–Ti	32–36	200–500

Table 3. Composition of several orthodontic stainless-steel grades (reprinted from [10], with permission from Elsevier, after [47] (used with permission of Elsevier Science & Technology Journals; permission conveyed through Copyright Clearance Center, Inc.).

Designation	Composition (wt.%)								
AISI	Fe	Cr	Ni	Mn	Mo	C	P	Si	S	Other
303	Bal.	17–19	8–10	2	0.6	0.15	0.2	1.0	0.15	-
304L	Bal.	18–20	8–12	2	-	0.03	004	10	0.03	-
316L	Bal.	16–18	1014	2	2.5	0.03	0.04	1.0	0.03	-
630/17-4	Bal.	15–17	3–5	1	-	0.07	0.04	1.0	0.04	4 Cu/3 Nb
630/17-7	Bal.	16–18	6.5–7.5	1	-	0.09	0.04	1.0	0.04	0.08–1.5 Al
SAF 2205	Bal.	22	5.5	2	3	0.03	0.03	1.0	0.02	0.16 N
18–8 Plus	Bal.	8	0.16	18	1	0.15	0.045	1.0	0.03	0.5 N
431	Bal.	26	-	-	4	-	-	-	-	-
AI29	Bal.	29	0.3	0.5	4	0.02	0.035	0.35	0.01	0.5 Ti

The major advantages of these Fe–Cr alloys include their good corrosion resistance combined with their outstanding biomechanical behavior and affordable price. The key feature of the corrosion behavior is the Cr content, which is between 16 and 25 wt.% for austenitic (face-centered cubic structure, FCC) Fe–Ni–Cr alloys (Table 3). The Cr element in the solid solution phase of SS alloys allows the development of the typical external protective chromium oxide (Cr_2O_3) thin film. Other bulk alloying elements of SS include molybdenum (Mo) and nickel (Ni > 8%), which improve the corrosion resistance effectiveness: while Ni promotes the formation of the FCC structure, Mo stabilizes the Cr-based passive layer.

However, some concerns regarding oral corrosion resistance, despite the presence of a small molybdenum (Mo) content (Table 3), and the overall biocompatibility led to the emergence of alternatives [10,11,37,42]. The high Ni nominal content in SS alloys can cause contact dermatitis (see Section 3), which has been encouraging for the development of new Ni-"free" austenitic stainless steels (see Section 4).

CoCr-based alloys have been used in orthodontics since the 1960s for manufacturing brackets and archwires [37,42,48]. With higher Cr content (>20%), these alloys surpass the SS ones in corrosion resistance—mainly in chloride environments due to the Cr-rich oxide passive layer—and biocompatibility, with higher wear resistance [42]; yet, improved ductility and resilience may be achieved (Table 2). The foremost drawbacks reported in the literature include additional heat treatments to improve mechanical performance and a more complex soldering process [10]. Currently, CoCr-based wires are commercially available in four color-coded variations according to the heat treatment applied; the blue one ("soft") is the most used due to its low yield strength compared to stainless steels [37] (Table 2).

Ti and its alloys are among the most biocompatible materials and were introduced in orthodontics in the 1980s, gaining popularity for brackets, tubes, and archwires production [11,49]. This class of metallic materials presents outstanding mechanical properties, excellent corrosion resistance (better than SS), in addition to low density (4.5 against 7.8 g/cm^3 for SS), providing a very high strength-to-weight ratio and non-eliciting allergic responses. Commercially pure titanium (α-Ti, Grade 4) and/or Ti-6Al-4V (Grade 5) brackets and β-Ti (including titanium molybdenum alloy—TMA) archwires are examples of some Ni-free components with outstanding corrosion resistance and biocompatibility [10,50]. Ti-based brackets and tubes reduce bonding failure to enamel, whereas TMA wires are ideal for certain (but not all) orthodontic situations due to the right balance of mechanical properties (e.g., low stiffness and high stringback and formability) and weldability [10,37,42]. The low elastic modulus supports the selection of β-Ti and/or NiTi alloys (Table 2) for orthodontic wires. High manufacturing cost is the most negative drawback [37].

Particular attention should be given to additional Ti-based alloys, such as NiTi and Cu–NiTi alloys, due to its high Ni nominal content. Nitinol®—which stands for "Nickel Titanium Naval Ordnance Laboratory", with near-equiatomic Ni and Ti concentrations—revolutionized orthodontics since its introduction into clinical practice in 1972 [37,51,52]. Due to its distinct mechanical properties, such as shape memory (shape memory alloy—SMA) and superelasticity behavior, this class of metals is now extensively used for the manufacture of orthodontic wires. [42,49,53]. In fact, the initial leveling stage of the orthodontic treatment (Section 1) usually involves NiTi archwires [41,42]. While the shape memory effect allows for the spontaneous recovery of the component form after being subjected to deformation higher than its elastic limit (by heating), the superelasticity tolerates a constant stress as the strain increases. After the initial elastic stress region and the stress/strain release, the NiTi alloy springs back to its original shape. Thus, high elasticity, spring back, and stored energy (Table 2) enable low-force delivery, even when malocclusions involve extreme teeth crowding. To further increase the alloys' strength and reduce energy loss, NiTi alloys have been chemically modified by copper addition (5–6% Cu)—the Cu–NiTi alloys—by acquiring a thermally activated behavior [54–56]. These wires yield lower loads on the teeth and also on deformation percentage; thus, teeth movement proceeds in a more physiological manner, preventing necrosis, hyalinization areas, and the probability of root resorption [57,58]. Other elements, such as Fe and Cr, are also added to Ni–Ti-based SMA alloys to modify their mechanical properties [59]. All Ti-based alloys spontaneously passivate by generating a titanium oxide protective film that provides good oral corrosion resistance.

Other metallic alloys can be found in fixed orthodontic appliances, but with a lower application range. The use of gold (Au)—precious metal-based alloys—for instance, was widespread before 1950 due to its higher corrosion resistance compared with alternative alloys at that time [10,60]. However, high cost and poor mechanical properties (low hardness) undermine its use, even though Au-based and Au-coated aesthetic components are still available today [10,60–64].

2.2. Intraoral Environment

The human body is an extreme environment for any metallic biomaterial [65], and the mouth is its "portal entry" [66]—an "open ecosystem" [67] in which variations in intraoral parameters are frequent and complex, leading to a unique corrosion-promoting medium.

Human saliva—99.5% water, 0.3% proteins, and 0.2% organic compounds—plays multiple important physiological functions, not only in taste, digestion, and speech but also in teeth and tissue lubrication/protection, pH buffering, and microbiological control [68–71]. The main functions of saliva and its constituents are presented in Table 4. This summary intends to reflect the saliva complexity, which is further exacerbated by other factors mentioned during this review.

Table 4. Main functions assigned to saliva and its constituents [69–72].

Function	Description	Agents
Tissue lubrication, repairing, and protection	Seromucous covering of the oral tissues. Barrier against irritants. Lubrication of hard and soft tissues, and prosthesis. Mastication, speech, and deglutition aid due to lubrication. Selective modulation of microbial adhesion to oral tissues. Modulation of dental plaque metabolism. Faster tissue repair.	Mucins and other proteins.
Clearance and pH maintenance	Acids neutralization (e.g., bicarbonate buffer). Alkalinization of dental plaque's pH through urea metabolization by its microbiome. pH modulation to prevent reaching optimal conditions for oral colonization by pathogens.	Bicarbonate, phosphate, urea, amphoteric proteins, and enzymes.
Maintenance of dental integrity	Modulation of pathogens activity to control the progression of caries and enamel damage. Maintenance of the enamel mineralization/demineralization equilibrium. The presence of fluoride in saliva enhances mineralization and forms a fluorapatite-like coating, which is more resistant to caries than the original teeth material.	Calcium, phosphate, fluoride, and several proteins (including statherin, histatins, cystatins, and proline-rich proteins).
Antibacterial activity	Selective action of protein-based immunological and non-immunological agents, allowing the growth of non-cariogenic microorganisms. Among other mechanisms, the non-immunological action involves the adhesion inhibition of colonizers to the oral tissues, namely by aggregation (clumping).	Immunoglobulins, enzymes, and other proteins (including glycoproteins, staherins, agglutinins, histidine-rich proteins, and proline-rich proteins).
Digestion, taste, and smell	Besides lubricating food and tissues, saliva starts the chemical oral digestion, namely by the initial action of the α-amylase (converting complex carbohydrates into simple sugars). The hypotonicity of saliva (low sodium, glycose, bicarbonate, and urea levels) regarding plasma, which enhances the dissolution of the substances. The presence of proteins (such as gustin) is necessary to the growth of gustatory buds.	α-amylase, gustin, lipases and other proteins.

Proteins and glycoproteins from saliva rapidly adhere to teeth enamel and any other surface placed inside the oral cavity to form a thin layer (70–100 nm), making them an important natural lubricant and oral protective film [69,73,74]. The most relevant chemical components of saliva include inorganic ions (e.g., N^+, K^+, Cl^-, F^-, CHO_3^-, PO_4^{3-}, ...), antimicrobial factors, nitrogenous compounds, enzymes, immunoglobulins, albumin and other proteins, and glucose, among others [66,69,71,72,75]. Moreover, the chemical composition, temperature, and pH of human saliva vary between individuals and along the course of the day (circadian rhythms), also depending on the person's lifestyle, diet, and health/disease conditions [72,73,76–80].

Intraoral mean temperature usually ranges around 33–37 °C [76], but abrupt variations up to 65 °C can occur (e.g., drinking a hot coffee after eating an ice cream or drinking a glass of ice water) [77,81]. The pH of non-stimulated saliva—i.e., without consuming food or drinks—usually varies between 6 and 7, but may also oscillate from 5.3 to 7.8 [66,76,82]. A pH value below 5.5 facilitates the development of dental caries [82–84]. An acidic diet can also reduce intraoral pH to 3, for instance, due to acidic soft drinks and fruit juices (pH from 1 to 6) [68,82,85,86]. Another possible contributor to salivary pH fluctuation is regurgitated stomach acid, which has a typical pH value of 1.2 [68]—one of the intraoral problems of bulimic people and oncological patients.

The oral environment is additionally ideal for the inevitable proliferation of microorganisms. So far, over 700 bacterial species have been identified, as well as numerous fungi and viruses [66,67,73,87]. The oral microbiota co-evolved with humans in a mutualistic or even symbiotic manner: While the host provides excellent physiochemical and nutritional conditions, microorganisms (especially bacteria) play important physiological roles, including digestion, oral mucosa cell differentiation, and protection against exogenous pathogens [88,89]. Table 5 provides a summary of the main recognized oral microbiomes.

Table 5. Oral microbiome (reproduced from [73]).

	Oral Bacteria Microbiome
Saliva	*Actinobacteria, Bacteroides, Firmicutes, Fusobacteria, Proteobacteria, Spirochaetes, TM7* (The Human Microbiome Consortium)
Dental plaque	*Firmicutes, Actinobacteria*
Oral mucosa	*Streptococcus salivarius, Rothia mucilaginosa, Eubacterium strain FTB41*
	Oral Bacteria Related to Oral Diseases
Dental caries	*Streptococcus, Veillonella, Actinomyces, Granulicatella, Leptotrichia, Thiomonas, Bifidobacterium, Prevotella, Lactobacillus, Propionibacterium, Pseudoramibacter, Selenomonas*
Periapical infections (periapical periodontitis, root canal infection)	*Proteobacteria, Firmicutes, Bacteroidetes, Fusobacteria, Actinobacteria, Olsenella uli, Prevotella baroniae, Porphyromonas endodontalis, Fusobacterium nucleatum, Tannerella forsythia, Propionibacterium propionicum, Porphyromonas gingivalis, Prevotella intermedia, Prevotella oralis, Parvimonas micra, Porphyromonas endodontalis, Fusobacterium nucleatum, Tannerella forsythia*
Periodontal diseases (gingivitis, periodontitis)	*Actinomycetes, Capnocytophaga, Campylobacter, Eikenella, Fusobacterium, Prevotella, Porphyromonas gingivalis, Treponema denticola, Tannerella forsythia, Bacteroidetes* spp., *Eubacterium saphenum, Porphyromonas endodontalis, Prevotella denticola, Parvimonas micra, Peptostreptococcus* spp., *Filifactor alocis, Desulfobulbus* spp., *Dialister* spp., *Synergistetes*
Halitosis	*Solobacterium moorei, Atopobium parvulum, Eubacterium sulci*

Planktonic (i.e., non-attached, free-floating) bacteria are 1000 times more vulnerable to antimicrobials than when aggregated. Therefore, some species—the primary colonizers—soon physically associate with and then adhere to the glycoprotein-based film over the teeth and biomaterials' surfaces. The mechanism of oral biofilm adhesion is described in Figure 4. Other bacterial species adhere and proliferate along with primary colonizers, forming microcolonies imbedded in an extracellular polysaccharide matrix. At this point, an oral biofilm (the dental plaque) grows: Complex groups of microcolonies positively interact with each other and even form a "primitive circulatory system" [67]. The grown (mature) oral biofilm is therefore advantageous to its inhabitants by providing nutrients to and protecting both aerobic and anaerobic colonizers—even against drugs, antimicrobial factors from saliva, and phagocytic cells [67,89,90].

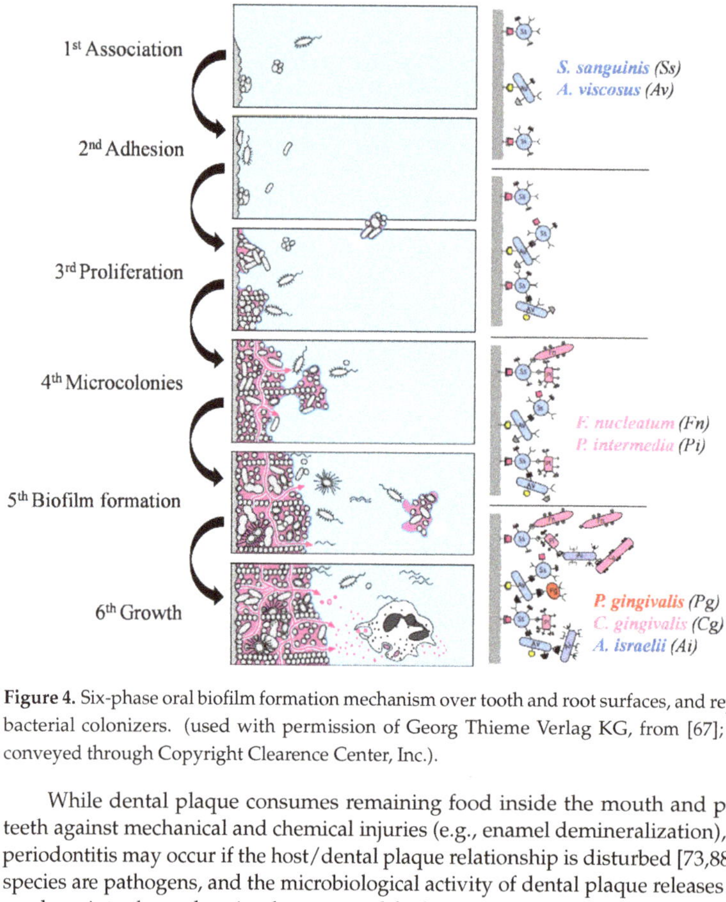

Figure 4. Six-phase oral biofilm formation mechanism over tooth and root surfaces, and representative bacterial colonizers. (used with permission of Georg Thieme Verlag KG, from [67]; permission conveyed through Copyright Clearence Center, Inc.).

While dental plaque consumes remaining food inside the mouth and protects the teeth against mechanical and chemical injuries (e.g., enamel demineralization), caries and periodontitis may occur if the host/dental plaque relationship is disturbed [73,88,89]. Some species are pathogens, and the microbiological activity of dental plaque releases several by-products into the oral cavity that can modify the chemical composition, oxygenation, and oral pH values [73,88,89]. Saliva and self-cleansing by the cheeks and tongue can naturally control biofilm growth to a certain extent. Nevertheless, oral hygiene procedures are crucial for removing dental plaque, including mechanical brushing with fluoride-containing toothpastes and mouth rinsing with fluorinated mouthwashes and elixirs [78,88,89,91,92].

In short, the intraoral environment is a highly dynamic and complex system—an ultimate degradation-promoting scenario for any biometallic material. Corrosion is a necessary (but not sufficient) condition for causing adverse biologic effects during the use of fixed orthodontic appliances.

2.3. Corrosion of Metallic Alloys

Metallic corrosion can be expressed as a "*physicochemical interaction between a metal and the environment that results in changes in the properties of the metal, and which may lead to significant impairment of the function of the metal, the environment, or the technical system, of which these form a part*" [93].

In an aqueous environment, such as the intraoral cavity, corrosion initiates through electrochemical reactions in the metal/solution interface, involving the anodic dissolution of the metal, Me to Me^{2+}_{aq} (oxidation, Equation (1)), and the cathodic reduction of an oxidant from the solution, Ox_{aq} to $Red\left(e^-_{redox}\right)_{aq}$ (reduction, Equation (2)), that is [94]:

$$Me \rightarrow Me^{2+}_{aq} + 2e^-_{metal} \tag{1}$$

$$2Ox_{aq} + 2e^-_{metal} \rightarrow 2Red\left(e^-_{redox}\right)_{aq} \tag{2}$$

The general charge-transfer reaction for a divalent metal can be written according to Equation (3) [94]:

$$Me + 2Ox_{aq} \rightarrow Me^{2+}_{aq} + 2Red\left(e^-_{redox}\right)_{aq} \tag{3}$$

Dissolved oxygen is usually the cathodic reactant—e.g., according to Equation (4)—with the production of hydroxide ions. However, other mechanisms could be more relevant under acidic conditions since protons may accept electrons produced during the anodic reaction. Typical cathodic reactions under low pH include both Equations (5) and (6), producing water or hydrogen gas, respectively [26,42,90].

$$\frac{1}{2}O_2 + H_2O + 2e^- \rightarrow 2OH^- \tag{4}$$

$$O_2 + 4H^+ + 4e^- \rightarrow 2H_2O \tag{5}$$

$$2H^+ + 2e^- \rightarrow H_2 \tag{6}$$

The end result of corrosion is the release of metallic ions into the electrolyte, whose extent depends on the electrolyte's nature, including its chemical composition, dissolved oxygen, and pH [42,95]. Moreover, the type of metal or alloy, its manufacturing process, and surface finishing will also influence corrosion [28,39,95–98].

Most bioalloys—mainly those containing Cr and Ti—rely on the spontaneous formation of a surface protective film. This thin oxide-based layer (some nm thick) may act as a barrier to the movement of ions—a passive film—protecting the metallic substrate against additional electrochemical corrosion [42,93,94]. Passivation is well established, being a spontaneous equilibrium of precipitation and dissolution of ions, with both active and passive films in contact with the electrolyte [99,100].

A simple passivation model [97] is presented in Figure 5. After surface hydroxylation, both metallic ions (from the alloy) and oxide ions (from the aqueous saliva) migrate through the forming film to the electrolyte and to the metal/oxide layer interface, respectively (reactions A and B, Figure 5). The electrons produced both from passivation and corrosion processes will be consumed by oxidants through redox mechanisms (reaction C, Figure 5). Therefore, electronic conduction will be essential to maintain the passivation equilibrium and the oxide film thickness. In other words, the dissolution rate must be lower than the precipitation rate [97,99]. In general, the anodic reaction of oxidation responsible for the passive film formation can be represented by Equation (7):

$$nMe + mH_2O \rightarrow Me_nO_m + 2mH^+ + 2me^- \tag{7}$$

where Me_nO_m is the top metallic oxide [100].

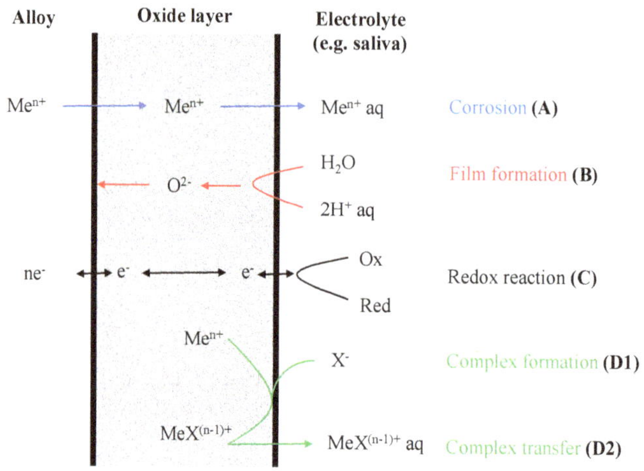

Figure 5. Passivation model of a metallic surface in an electrolyte. Me: metal element; X: complexing anions, e.g., Cl^- and F^-. (used with permission of John Wiley & Sons—Books, from [97]; permission conveyed through Copyright Clearance Center, Inc.).

It is well recognized that some elements—such as manganese (Mn), Mo, Co, and Ni—considering the normal composition of metallic alloys do not participate in passive film formation and are preferentially released into aqueous medium. If those metallic ions bond with biomolecules in vivo, they may present a certain degree of toxicity [99]. Others, however, such as Cr and Ti, are responsible for the growth of oxide films and are most likely to be less toxic [99]. The high corrosion resistance of Fe-based alloys containing Cr (16 < Cr < 25%) is due to the formation of a Cr (III) oxide-hydroxide passive layer [37,101]. The insoluble Cr_2O_3 therefore prevents iron dissolution. Similarly, TiO_2-based passive layers spontaneously form on surface Ti and its alloys, being well recognized for their excellent stability [10,26,102].

Unfortunately, passive layers are fallible [95], that is, top oxide film dissolution and/or disruption may occur, driven by several environmental parameters, including fluctuations in the electrolyte chemical composition. Complexing anions can capture metallic ions from the oxide surface region (reaction D1, in Figure 5), facilitate their migration, dissolve the passive film, and release metallic ions into the saliva (reaction D2, in Figure 5) [97,103]. Aggressive ions such as chlorine (Cl^-), fluoride (F^-) [29,94,100,104], or sulphide (S^{2-}) [105], as well as mechanical damages [26,39,73,106], can disrupt the external passive oxide layer, exposing the active metallic surface for corrosion reactions to proceed. The deleterious effect of the fluoride ions—common in fluorinated prophylactic gels and mouthwash elixirs—on both SS- and Ti-based alloy passivation can be described by the following reactions, Equations (8) and (9), respectively [107].

$$Cr_2O_2 + 12F^- + 8H^+ \rightarrow 12CrHF_6^{2-} + 3H_2O \tag{8}$$

$$TiO_2 + 6F^- + 4H^+ \rightarrow TiF_6^{2-} + 2H_2O \tag{9}$$

Furthermore, both the chemical composition and the presence of defects and impurities [100] influence the thickness, stability, and composition of the external passive films [94]. Those variables, in turn, are determined by the manufacturing process and the corresponding surface finishing—mechanical/physicochemical (polishing, electro-erosion, plasma, acid/alkaline treatments, electrochemical deposition) or biochemical treatments (biological molecules to mimic biointerfaces). In the end, the balance between passive layer damage and its reconstitution will dictate the overall final corrosion resistance behavior of the biometallic alloys [94,100].

2.4. Oral Corrosion Forms

Intraoral degradation of metallic appliances is inevitable. Until today, nine basic corrosion types have been reported: uniform, galvanic, crevice, pitting, intergranular, selective leaching, stress, erosion, and microbiologically induced corrosion (MIC), which will be briefly addressed as follows.

2.4.1. Uniform Attack

Uniform corrosion is recognized as general corrosion. This degradation form seems to occur uniformly over the entire surface due to surface electrochemical reactions, almost at the same rate [81,93]. It is the most common type of oral corrosion, affecting all metallic materials at different rates [95], but sometimes it is difficult to detect—only when a significant amount of material is dissolved. All parts of a metallurgical and compositionally uniform surface should be accessible to the electrolyte (saliva) [29].

2.4.2. Galvanic Corrosion

Galvanic corrosion is observed when two different metallic surfaces—with different corrosion potentials—are joined or sufficiently close in an electrolyte solution by establishing a galvanic coupling. The more electropositive (less noble) metal or alloy becomes the anode and preferentially corrodes [26,30,81,93,95].

In orthodontics, contact between dissimilar metallic surfaces might occur in two situations: By simple contact [29,108] or through bonding processes [10,29,109]. In the first case, bracket/wire interactions are inevitable during orthodontic treatment with fixed appliances. Predictably, in certain combinations, such as in the so-common NiTi wire/SS bracket, galvanic corrosion is susceptible to occurring [29], especially in a fluoride-rich environment [108]. Recent research work [60] reported no evidence of galvanic coupling between SS lingual brackets and SS archwires, but the authors suggested caution when using fluoride-containing products during fixed orthodontic treatment with SS brackets and NiTi archwires.

Different parts of brackets or posted archwires are often made of dissimilar alloys, leading to galvanic corrosion [110]. Furthermore, brazing alloys can be used during the manufacturing of orthodontic components [10,29]. Mechanically active welding joints [95] may be reactive, increasing galvanic corrosion susceptibility accompanied by toxic metallic ion release, particularly for silver (Ag)-, copper (Cu)-, and zinc (Zn)-based welding materials [29,111,112]. A recent in vitro study [113] demonstrated that Ag ion release from Ag-soldered SS bands was an order of magnitude higher than other non-soldered SS orthodontic appliances. The authors assigned this effect to the manufacturing process used (welding).

It is generally accepted that galvanic cells can also occur in different locations of the same metallic surface due to non-uniform surface finish (e.g., roughness and chemical composition) and mechanical properties (e.g., work hardening) or even dissimilar properties of the electrolyte (pH and chemical composition) [95]. In the oral cavity, saliva is the main electrolyte, but extracellular fluids such as blood or gingival fluid are also present. Galvanic currents may take place due to the contact of the metallic surface to different biological fluids [73].

2.4.3. Crevice Corrosion

Crevice corrosion is a localized attack occurring in or near constricted places (crevices) formed by two surfaces, of which at least one is metallic [93]. This leads to a local enhancement of aggressive species and depletion of oxygen, in addition to the consequent acidification of the crevice solution due to the hydrolysis of the dissolving metal ions. Generally, metallic materials that show an affinity to pitting also suffer from crevice corrosion. The main causes include differences in metallic ions, fluoride concentration, or oxygenation between the crevice and its surroundings [28,81], associated with a pH decrease and a chloride ion concentration increase [10,29], which deteriorate the protective passive layer—especially on SS alloys [114].

Elastomeric or metallic ligatures are frequently used to fix orthodontic archwires to brackets, establishing ideal sites for crevice attack on brackets [109] (including on 316L SS alloy): Deep craters, fissures, and pores have been detected after long intraoral exposure, as well as extensive deterioration and perforation of the resin-fixed bracket base [28,30,81]. Daems et al. [106] also noticed this type of corrosion at regions of bracket/archwire contact or with plaque and food remnants. Other factors that reportedly cause crevice corrosion comprise the recycling process of the components [29]—not recommended in several countries—surface defects or irregularities [95,106]—including those caused by handling the components by the orthodontist during treatment (Figure 6)—and the presence of welding areas [50,115].

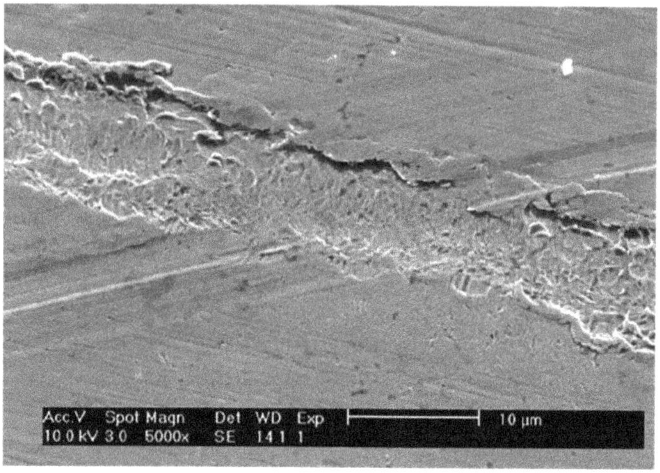

Figure 6. Crevice corrosion detected in a groove on a retrieved archwire (used for at least 6 months) caused by orthodontic pliers (reproduced by permission of Oxford University Press on behalf of the European Orthodontic Society, from [106]).

2.4.4. Pitting Corrosion

Pitting corrosion is a localized type of corrosion that initiates on metallic surfaces when the protective passive film disrupts due to mechanical and/or electrochemical attack [93,100], leading to the formation of pit holes and/or cavities. This attack has been associated with other corrosion types [114,116], such as the one caused by the well-adherent biofilm that forms during orthodontic treatments (Figure 7)—discussed further in Section 2.4.9. Aggressive ions in saliva, such as chlorine and fluoride [29,42,117]—especially under acidic conditions [101,104,118]—and food additives, such as certain spices [119,120], effectively damage surface protective oxides. Manufacturing defects on orthodontic metallic components may also increase pitting corrosion susceptibility for both SS and NiTi alloys [28–30].

The main strategy to improve the pitting resistance of SS alloys is to increase the Cr and Mo nominal content. However, the presence of non-metallic inclusions, such as manganese sulphide (MnS), is of major importance since pits usually initiate at these precipitates [121–123]. Usually, the CoCr-based alloys are resistant to pitting; the dissolution of the protective Cr_2O_3 layer into soluble ions (CrO_4^{2-}) takes place by oxidation at potentials below the oxygen evolution range [124].

2.4.5. Intergranular Corrosion

As the name suggests, intergranular corrosion occurs in microstructural planar defects along grain boundaries or in the immediate near zones, with minimal or no attack on the alloy grain itself [93,114]. The net result is an alloy fracture along these grain boundaries. SS alloys—used for manufacturing orthodontic brackets and archwires—are particularly vulnerable to this corrosion form, leading to surface staining, weakening the mechanical behavior (strength and ductility), or even failure [112,120]. Special attention should be given to heat treatment of steels [37] (or brazing/welding [95,115,125]—termed weld decay). For a prolonged period above the sensitization temperature [28,29], the formation of small precipitate particles of chromium carbide ($Cr_{23}C_6$) occurs [114]. Two major consequences arise: the SS brittleness increases and its corrosion resistance decreases, both due to the Cr-depleted zone adjacent to the grain boundary [37,95].

2.4.6. Selective Leaching

Selective leaching or dealloying is found in solid solution alloys, such as Ni–Cr-based or binary alloys containing calcium (Ca) and zinc (Zn) [26], occurring when one element is preferentially removed during the corrosion process [81,114]. This preferential release of a more reactive element from an alloy, regardless of its chemical composition [93], can occur in vivo [26,28]. Still, the effect of selective leaching seems negligible in dentistry [26,81].

2.4.7. Stress Corrosion

Stress corrosion, sometimes termed stress corrosion cracking (SCC), develops due to the influence of both applied tensile stress and a corrosive environment [28,95]. Some alloys that are virtually inert in a particular corrosive medium can become susceptible to this type of corrosion when under loads. This can seriously compromise the mechanical integrity of the material, and failure may eventually occur under low stress levels (compared with alloys in non-corrosive environments) [81,93,114]. Nitinol archwires bonded to brackets are exposed to compressive and tensile stress and might fracture during orthodontic treatment [29,126].

2.4.8. Erosion Corrosion

Erosion corrosion refers to the deterioration of a metallic material due to mechanical abrasion or wear with the combined action of the chemical attack of the corrodent fluid motion. Three subtypes are well known: Erosion, cavitation, and fretting [114]. In orthodontics, fretting corrosion is the most relevant form [81], due to the slight relative motion (vibration and slip) of two contacting metallic surfaces under load [93,114]. Surfaces of both archwires and bracket slots experience load and may undergo a cold-welding phenomenon. In addition, the required small displacements could disrupt the passive films and, consequently, increase corrosion susceptibility (e.g., by pitting) [28,29,95].

2.4.9. Microbiologically Induced Corrosion

As previously mentioned, oral microorganisms can directly or indirectly degrade metallic materials in vivo, either by metabolizing metal from the surface or by modifying the surrounding electrolyte with their metabolic by-products, respectively [95]. This form of corrosion is known as microbiologically induced corrosion (MIC) [73,114,127]. Zarasvand and Rai [90] extensively studied the MIC mechanisms, while Mystkowska et al. [73] described the intraoral process. Accordingly, oral biofilms create differential concentration cells on the metallic surfaces of three main types: oxygen concentration cells, metal concentration cells, and active–passive cells (Figure 7).

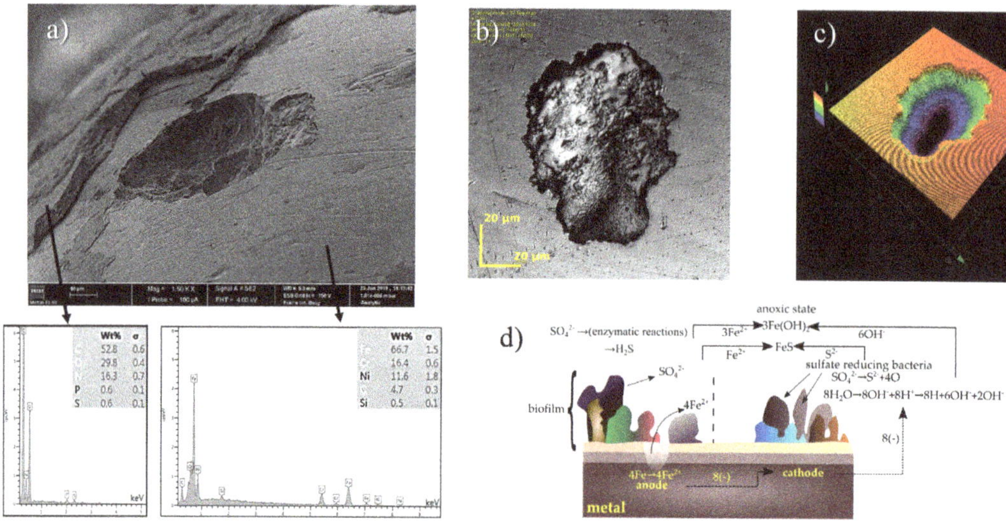

Figure 7. (**a**) SS brackets after two years of intraoral use. Scanning electron microscopy (SEM) and optical micrographs coupled to energy dispersive (EDS) chemical compositions of two distinct zones: SS alloy and pitting morphology underneath a biofilm layer. Adapted from [116]; (**b**,**c**) pitting corrosion induced by sulphate-reducing bacteria: confocal scanning microscope images after a 56-day in vitro test using *D. nigrificans* bacteria on 316LV steel; and (**d**) representation of the proposed corrosion mechanism by [128] (reproduced from [73]).

Oxygenation cells appear due to a non-uniform biofilm layer—in terms of thickness, ratio of aerobic (oxygen-consuming)/anaerobic microorganisms, or due to the presence of layers of corrosion products—that cause differences in oxygenation throughout the surface. Regions with high oxygen concentrations favor cathodic reactions (Equation (2) and Figure 8a), and the metallic surface below becomes the cathode. Conversely, in a poorly oxygenated environment, the anodic reaction is enhanced (Equation (1)), and the surface becomes the anode and corrodes. Differences in metallic ion concentration on different sites (Figure 8b) also occur due to the nature of the extracellular matrix, which has diverse composition and functional groups with different affinities to metallic ions. Under biofilm regions with low affinity to metallic ions, cathodic reactions further progress, whereas anodic dissolution of the metal increases under high-affinity biofilm sites. Finally, if a dense biofilm layer is mechanically or chemically disrupted (Figure 8c), the exposed metallic surface corrodes (the anode), while biofilm-covered regions behave as cathodes [73,90]—active–passive cells [93].

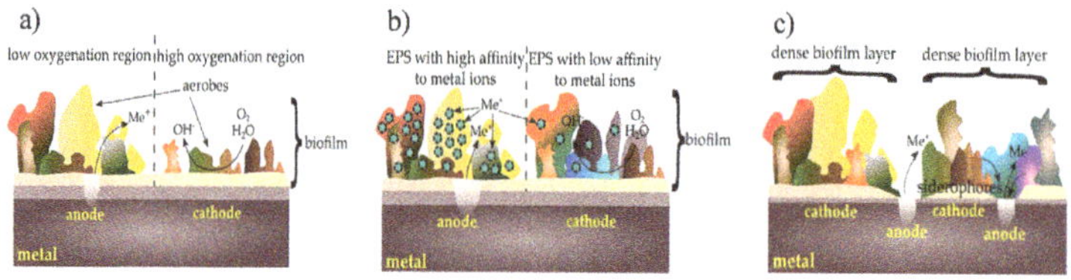

Figure 8. MIC mechanisms on a metallic surface in the oral cavity: (**a**) cells with varying degrees of oxygenation; (**b**) cells with different concentration of metal ions; and (**c**) passive–active cells. EPS: extracellular polymeric substances (reproduced from [73]).

Certain anaerobic microorganisms, such as sulfur-reducing bacteria (SRB), release corrosive metabolic products that degrade metallic alloys. The SRB can produce hydrogen gas (H_2), hydrogen sulfide (H_2S), and sulfur difluoride (F_2S, a strong local cathode), while other Gram-negative bacteria release butyric acid ($C_4H_8O_2$) and carbon dioxide

(CO_2) [73,90,114]. Besides weakening and retarding the passivation mechanism of the metallic surface, H_2S is highly toxic to cells [129] and reacts with metals to form metal sulfides and atomic hydrogen. Metal sulfides may precipitate on the surface, generating new active–passive cells, while released atomic hydrogen can cause SCC [73,90,93].

Recovered SS brackets [110] and archwires [106] unequivocally revealed highly adherent biofilm deposits (Figure 9a), under which pitting or crevice corrosion occurred [106,116], as presented in Figure 7 for brackets after two years of intraoral use. Calcium-precipitating bacteria on SS orthodontic wires cause surface calcium deposits, into which chlorine ions penetrate and induce pitting [130].

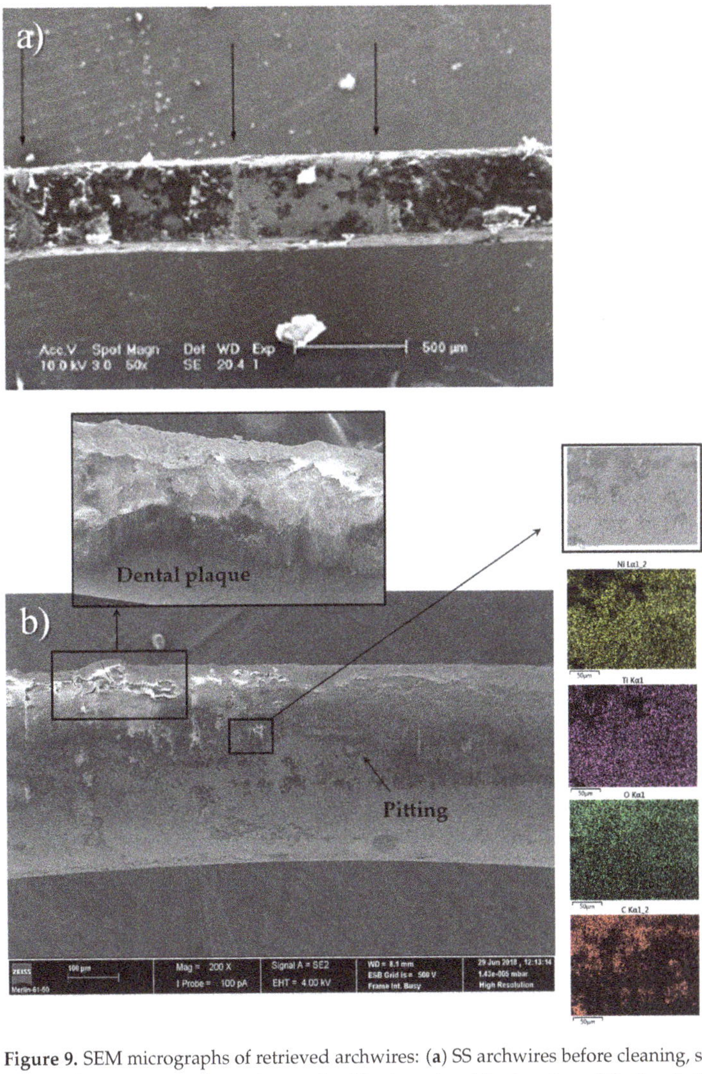

Figure 9. SEM micrographs of retrieved archwires: (**a**) SS archwires before cleaning, showing plaque and food remnants on the surface and evident marks of the location of the bracket borders (black arrows) (reproduced by permission of Oxford University Press on behalf of the European Orthodontic Society, from [106]); (**b**) SEM micrographs and EDS elemental distribution maps of an NiTi archwire after 17 weeks of intraoral use (reproduced from [116]).

Abundant biofilm layers were also observed on NiTi archwires [116] (Figure 9b), in spite of the good resistance against MIC corrosion [131,132] of the Ti-based alloys. Still, dental plaque accumulates and might enhance crevice corrosion in the welding gaps of Ti-based brackets [50,98]. Some research works [105,133] suggest that the corrosion susceptibility of Ti and its alloys decreases in the presence of bacteria, namely by inducing oxygenation cells due to bacterial metabolism.

3. Harmful Effects and Clinical Implications

The main consequences of intraoral aging of orthodontic metallic alloys are briefly presented in this chapter, namely the release of metallic ions into the oral cavity, the friction effect between components, and the consequences of using fluoride-based products during treatment with fixed orthodontic appliances.

Aging of metallic alloys is an important issue in orthodontics since both structural and morphological modifications can occur and thus negatively affect the normal clinical treatment progression. Corrosion of metallic surfaces and the presence of biofilm promote metallic ion release and roughness, which may increase friction between brackets and archwires [134–138] and extend the treatment time. Another pointed aging implication is the friction enhancement between the appliance and the mucosa, which causes oral mucosa lesions (from minor wounds to large ulcers), resulting in patients' pain and discomfort [68,139–141]. Moreover, aging decreases the resistance to fracture of metallic alloys under repeated cyclic loading [30,136]—fatigue—and could lead to premature failure of archwires [95,142] ruining the in vivo function of the biomaterial.

3.1. Release of Metallic Ions

Corrosion processes ultimately cause the release of metallic ions and particles into the oral cavity [28,99,143,144], which may interact with oral tissues and move to the gastrointestinal tract; even so, their impact on health is not yet fully understood [42,73,95,145]. Biocompatibility concerns raised among clinicians and researchers as hazardous species such as nickel, chromium, cobalt, copper, and vanadium (Ni, Cr, Co, Cu, and V, respectively) can be released from metallic appliances [26,29,144,146–148]. Back in 1975, Samitz and Katz [149]—who reviewed data related to Ni released from implanted prostheses—concluded that solubilized metal was found in tissues near implants in laboratory animals.

Multiple researchers have been trying to quantify the release of metallic ions from orthodontic appliances to assess if the concentrations can reach toxic levels for humans, both in vitro and in vivo [144,150–152]. Table 6 compiles an update of the 59 studies found in the literature regarding the in vivo measurements of metallic ions released from orthodontic appliances, sorted out by publication year.

The first study found dates back to 1991, by Gjerdet et al. [153], and measured the Ni and Fe contents in patients' saliva up to 3 months of usage. The authors found an initially higher salivary metal content that decreased over treatment time, but values were small when compared with those from dietary intake. Nonetheless, they were already alerted to the large interindividual variability found, as well as to Ni-sensitive patients [153].

Through time, Ni and Cr concentrations are almost always focused, but most studies concluded that the salivary metallic ion concentration is well below toxicity levels. In fact, dietary studies conducted in different countries obtained a daily intake of nickel between 100 and 300 μg/day from food and drinking water. Consuming Ni-enriched food (e.g., processed food) may increase this value up to 900 μg/day [95,154–160]. Haber et al. [161] estimated a toxicity reference value for Ni-sensitized populations of 4 μg Ni/kg of body weight per day, in addition to Ni in food. Concerning chromium, an average daily intake of 50–280 μg has been proposed [95,162]. However, some authors who analyzed different matrixes (oral mucosa cells, dental plaque, bone, gingiva, hair, and internal organs) found evidence of bioaccumulation that may provoke toxic effects, including DNA damage. In fact, Eliades and Athanasios [28] argued that in vivo studies measuring urinary or serum concentrations of metallic ions in orthodontic patients may give falsely lower Ni levels due to its accumulation in an organ. Further research should therefore persist.

Among the metallic ions released into the oral cavity, Ni raises special health concerns and has been systematically studied [26,154,158,159,161,163–169], including in orthodontics [143,144,168,170–182]. The European Union (EU) currently forbids the use of Ni [183]:

1. "in any post-assemblies which are inserted into pierced ears and other pierced parts of the human body unless the rate of Nickel release from such post-assemblies is less than 0.2 $\mu g/cm^2/week$ (migration limit)";
2. "in articles intended to come into direct and prolonged contact with the skin (...) if the rate of Nickel release from the parts of these articles coming into direct and prolonged contact with the skin is greater than 0.5 $\mu g/cm^2/week$";
3. "in articles referred to in point 2 where these have a non-nickel coating unless such coating is sufficient to ensure that the rate of nickel release from those parts of such articles coming into direct and prolonged contact with the skin will not exceed 0.5 $\mu g/cm^2/week$ for a period of at least two years of normal use of the article" [183].

Unfortunately, biometallic alloys lie outside of this EU regulation regarding this matter. The *American Academy of Pediatrics* also expressed concerns regarding the use of Ni-containing alloys, urging the adoption of regulations similar to the EU nickel directive [168]. Dental biomaterials must still comply with several standards and regulations [184].

This transition metal (Ni) is a well-known allergen [185,186], a strong immunologic sensitizer capable of inducing delayed hypersensitive reactions [169,187], triggering cytotoxic, carcinogenic, and mutagenic effects [144,164,188,189], and affecting several cellular functions by long-term exposure to a small amount [144]. Moreover, emphasis has been given to Ni-induced genetic effects, including DNA damage and the inhibition of enzymes involved in DNA reparation [144,180]. The International Agency for Research on Cancer (IARC) classifies Ni (II) and its compounds as carcinogenic or potentially carcinogenic to humans [165].

Chromium is another well-known toxic element. Between the two most stable oxidation states, Cr(III) and Cr(VI), its hexavalent form is toxic and exhibits mutagenic, cytotoxic, and carcinogenic effects in humans [144]. Reportedly, both oxidation states were found in vitro after the corrosion of SS orthodontic brackets in artificial saliva [190].

Ni carcinogenicity, genotoxicity, and allergy are controversial in orthodontics [95,191–197]. Nonetheless, released Ni from orthodontic components can accumulate in the oral mucosa cells (see studies in Table 6) and decrease cell viability [180], while systemic toxicity should not be ignored [198]. Moreover, Kochanowska et al. [199] showed the in vivo effect of long-term exposure to metal orthodontic appliances on both the metallothionein gene expression and the induction of protein synthesis by using animal models (pigs).

Several subtle to severe intra- and/or extra-oral symptoms of allergic reactions to nickel have been reported due to the use of metallic appliances [31,147,192,200–206], as exemplified in Figure 10. Symptoms include burning sensation, stomatitis, angioedema, severe gingivitis without dental plaque, gingival hyperplasia, generalized urticaria, and widespread eczema [31,147,203,206–209]. Besides discomfort and pain for patients, orthodontists may need to replace high Ni-containing components, interrupt the treatment, and/or refer the patient to an allergologist or other specialist for further examination [201,203–205].

Ni allergy—namely extreme hypersensitive reactions—is (fortunately) rare in orthodontics [32,191,206], but may be ineffectively diagnosed: Subtle signs are easily misinterpreted as mimicking mechanical injuries or microbiologic activity [95,192,202,210]. Schuster and colleagues [192] reported allergy symptoms related to the presence of fixed appliances during treatment without intraoral signs. Corrosion products induce enamel demineralization, metallic ion incorporation, and color change [211], as well as pain and swelling of oral soft tissues, leading to secondary infections [29]. Pazzini et al. [212] concluded that patients treated with Ni-"free" (0.5–4% Ni) appliances had better gingival health and smaller blood changes when compared with those wearing conventional metallic components (13% Ni). Another possible negative effect is the increase in antibiotic resistance of some bacteria exposed to metals and their potential transfer to medically relevant pathogens [95].

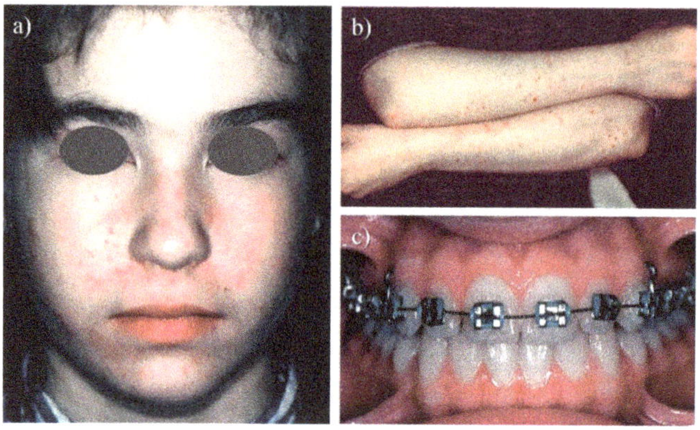

Figure 10. Examples of allergic symptoms related to the use of fixed orthodontic appliances: (**a–c**) extraoral allergic reaction (face and forearms) after placement of fixed appliances, without intraoral symptoms (reproduced from [192] with permission from SNCSC).

For further comprehension of the toxic effects of metallic ions released during a fixed orthodontic treatment, the reading of the outstanding review works conducted by Martín-Caméan and colleagues [144,213], and by Downarowicz and Mikulewicz [214], is recommended.

3.2. Friction in Orthodontics

Resistance to sliding is present when two surfaces come into contact with each other (e.g., bracket/wire and wire/ligature) [12,37], which is clinically relevant in orthodontics since reduced resistance to sliding can decrease treatment time [215]. Kusy and Whitley [216,217] partitioned resistance to sliding into three components: (i) Friction, *"a force that opposes every action that an orthodontist takes to move the teeth"*; (ii) binding, when the angle between the bracket slot and the archwire is high enough to promote contact between the bracket corners and the archwire; and (iii) notching, when a permanent deformation of the wire (or bracket) occurs [215–219]. Figure 11 schematizes these three contributions to resistance to sliding in a bracket/archwire contact.

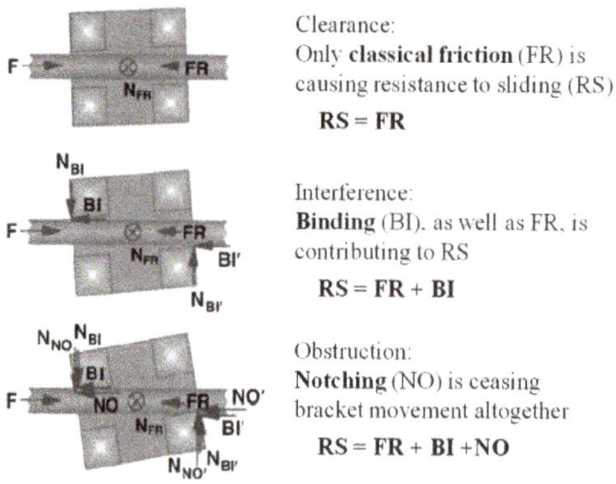

Figure 11. Schematic representations of the resistance to sliding components in a bracket/archwire contact (reproduced by permission of Oxford University Press on behalf of the European Orthodontic Society, from [218]).

Table 6. Overview of in vivo studies reporting the quantification of metallic ions release from orthodontic appliances. DL: detection limit; d.h.m.: dry hair mass; AAS: atomic absorption spectroscopy; ICP-AES/ICP-OES: inductively coupled plasma atomic/optical emission spectroscopy; ICP-MS: inductively coupled plasma mass spectroscopy; XRF: X-ray fluorescence.

Sample Size	Appliances	Matrix and Sampling	Elements and Detection Mode	Mean/Median Concentrations	Main Results	Reference
34 patients	SS brackets and bands; NiTi archwires.	Saliva. Sampling before treatment, right away or more than 3 weeks after application, and 3–5 weeks after removal.	Ni and Fe. AAS.	Without appliances: Ni: 8.2 ppb, Fe: 148 ppb. Immediately after placement: Ni: 67.6 ppb, Fe: 488 ppb. More than 3 weeks after placement: Ni: 7.8 ppb, Fe: 172 ppb.	Significant increase in Ni and Fe concentrations and absolute masses right after placement, but not after 3 months of usage.	[153] Gjerdet et al. (1991)
31 patients	SS brackets and archwires; NiTi archwires.	Blood. Sampling before treatment, and after 3 months (with NiTi archwires) and 4–5 months (with SS archwires).	Ni. AAS.		No significant or consistent increase in Ni blood level during orthodontic treatment.	[220] Bishara et al. (1993)
47 patients	Brackets, bands and archwires.	Saliva. Sampling before and during treatment (1–2 days, 1 week, and 1 month).	Ni and Cr. AAS.	Before treatment: Ni: 68 ppb, Cr: 68 ppb. During treatment: Ni: 55–74 ppb, Cr: 69–90 ppb.	Ni and Cr concentrations were not significantly affected.	[221] Kerosuo et al. (1997)
45 patients + 15 controls	Metallic brackets, tubes, and bands; NiTi archwires.	Saliva. Sampling before and during treatment (1 week, 1 month, and 2 months).	Ni and Cr. AAS.	Before treatment: Controls: Ni: 1.16 μg/mL, Cr: 2.20 μg/mL. Patients: Ni: 0.53–0.54 μg/mL, Cr: 1.35–1.41 μg/mL. During treatment: Controls: Ni: 1.33–1.46 μg/mL, Cr: 2.50–3.43 μg/mL. Patients: Ni: 0.49–0.67 μg/mL, Cr: 0.49–1.98 μg/mL.	Fixed orthodontic appliances do not seem to significantly affect Ni and Cr concentrations in saliva.	[177] Kocadereli et al. (2000)

Table 6. *Cont.*

Sample Size	Appliances	Matrix and Sampling	Elements and Detection Mode	Mean/Median Concentrations	Main Results	Reference
100 patients	SS brackets, bands and archwires; NiTi archwires.	Saliva and Serum. Sampling before and during treatment (1 week, 1 month, 1 year and 2 years).	Ni and Cr. AAS.	Saliva: Before treatment: Ni: 4.45 ppb, Cr: 0.75 ppb. During treatment: Ni: 4.12–11.53 ppb, Cr: 0.53–1.53 ppb. Serum: Before treatment: Ni: 8.36 ppb, Cr: 6.21 ppb. During treatment: Ni: 7.87–10.27 ppb, Cr: 6.16–10.98 ppb.	The maximum levels of Ni and Cr in saliva were recorded 1 month after starting the orthodontic treatment. Below toxicity levels. The maximum levels of Ni and Cr in serum were recorded 2 months after starting the orthodontic treatment. Below toxicity levels.	[172] Ağaoğlu et al. (2001)
17 patients + 7 controls	SS brackets and archwires.	Saliva. Before and after rinsing with distilled water.	Ni, Cr and Fe. ICP-AES.	Controls before: Ni: 18 ppb, Cr: 20 ppb, Fe: 21 ppb. Controls after: Ni: 11 ppb, Cr and Fe: < DL. Patients before: Ni and Cr: < DL, Fe: 14. Patients after: Ni: 10 ppb, Cr: 27 ppb, Fe: 17 ppb.	No significant difference between controls and patients. Below toxic levels.	[222] Eliades et al. (2003)
55 patients + 30 controls	SS brackets, bands and archwires; NiTi and CoCr archwires.	Oral mucosa epithelial cells.	Ni, Co. ICP-MS.	Controls: Ni: 0.725 ppb, Co: 0.202 ppb. Patients: Ni: 2.521 ppb, Co: 0.568 ppb.	Statistically significant differences between controls and patients. DNA damage (Comet assays).	[223] Faccioni et al. (2003)

Table 6. *Cont.*

Sample Size	Appliances	Matrix and Sampling	Elements and Detection Mode	Mean/Median Concentrations	Main Results	Reference
24 patients + 24 controls	SS brackets, bands and archwires; NiTi archwires.	Saliva and dental plaque (on enamel and metallic surfaces). Sampling at approximately 16 months after starting the orthodontic treatment.	Ni. AAS.	Controls: Filtered saliva Ni: 0.004 µg/g, Saliva sediment Ni: 14.85 µg/g. Dental plaque (all tested surfaces) Ni: 0.380–0.875 µg/g. Patients: Filtered saliva Ni: 0.005 µg/g Saliva sediment Ni: 25.25 µg/g Dental plaque (all tested surfaces) Ni: 0.685–2.690 µg/g	Statistically significant Ni increase in both filter-retained fraction (saliva sediments) and dental plaque.	[178] Fors and Persson (2006)
15 patients + controls	Orthodontic appliances (not specified).	Hair. During the orthodontic treatment (not specified).	Ni. AAS.	Controls: Ni: 0.64 µg/g. Patients: Ni: 0.50 µg/g,	No statistically significant different between patients and controls.	[224] Levrini et al. (2006)
10 patients	Brackets, bands, and tubes; NiTi and SS archwires.	Hyperplastic and healthy gingiva. After 2–4 year-treatments with fixed orthodontic appliances.	Ni. AAS.	Ni in healthy gingiva: 1.81 µg/g; Ni in hyperplastic gingiva: 1.32 µg/g.	Non-significant differences between Ni content in healthy and hyperplastic gingiva, but histological differences (toxicity).	[210] Gursoy et al. (2007)
21 patients	SS brackets and bands	Urine. Before and 2 months after placement	Ni. AAS	Before placement: Ni: 17.67 µg/L. After placement: Ni: 19.89 µg/L.	Statistically significant increase in urinary Ni 2 months after placement.	[225] Menezes et al. (2007)
10 patients	SS brackets, bands, and ligatures; NiTi archwires.	Saliva. Sampling before and during treatment (1 and 3 weeks).	Ni and Cr. AAS.	Before treatment: Ni: 31.62 ppb, Cr: 38.82 ppb. After 1 week: Ni: 113.20 ppb, Cr: 83.15 ppb. After 3 weeks: Ni: 65.24 ppb, Cr: 64.21 ppb.	Statistically significant increase in salivary Ni and Cr concentrations during the orthodontic treatment when compared with the basal levels.	[181] Singh et al. (2008)

Table 6. *Cont.*

Sample Size	Appliances	Matrix and Sampling	Elements and Detection Mode	Mean/Median Concentrations	Main Results	Reference
30 patients + 30 controls	SS brackets, bands and archwires; NiTi archwires,	Saliva and mucosa cells. 1 sampling,	Ni, Cr, and Co. AAS.	Controls: Ni: 12.26 ppb, Cr: 3.46 ppb, Co: 0.44 ppb. Patients: Ni: 21.74 ppb, Cr: 4.24 ppb, Co: 0.84 ppb.	Significantly higher Ni concentrations in patients when compared with controls. No differences regarding Cr and Co levels.	[226] Amini et al. (2008)
30 patients	SS brackets	Saliva. Sampling before and after 10 min, 24 h, 7 days, 30 days, and 60 days of usage.	Ni, Cr and Fe. AAS.	Before treatment: Ni: 5.25 µg/L, Cr: 0.64 µg/L, Fe: 94.03 µg/L. During treatment: Ni: 1.69–16.01 µg/L, Cr: 0.52–1.72 µg/L, Fe: 28.31–103.58 µg/L.	Concentration peak for Ni and Cr 10 min after placing the orthodontic appliances, but no significant variations for all metals throughout the study time.	[171] De Souza and Menezes (2008)
18 patients	SS brackets and bands; NiTi archwires.	Saliva. Before and during the orthodontic treatment (immediately after placing the SS components; immediately before and after placing the NiTi archwires; and after 4 and 8 weeks).	Ni. ICP-MS.	Before placing any component: Ni: 34 µg/L. After placing SS components: Ni: 78 µg/L. After placing the archwires: Ni: 56 µg/L. During the remaining study time points: Ni: 28–34 µg/L.	Statistically significant increase in salivary Ni immediately after placing the SS and NiTi components.	[143] Petoumeno et al. (2009) and [131] Petoumeno et al. (2008)
15 patients, divided according to the bracket type	SS, Ti, and Ni-free brackets and tubes.	Oral mucosa cells. Sampling before and 30 days after placing the appliances.	Ni, Cr, Fe, Ti, Co, Mn, and Mo. ICP-MS.	Before treatment: Ni: 3.44 µg/L, Cr: 0.00 µg/L, Fe: 1.95 µg/L, Ti: 0.98 µg/L, Co: 0.00 µg/L, Mn: 0.32 µg/L, Mo: 0.13 µg/L. After 30 days: Ni: 0.00–0.04 µg/L, Cr: 0.00–0.34 µg/L, Fe: 1.24–5.36 µg/L, Ti: 0.82–3.04 µg/L, Co: 0.00 µg/L, Mn: 0.58–1.08 µg/L, Mo: 0.00 µg/L.	Increased Ti and Mn in cells exposed to the SS components. Higher Cr and Fe detected in cells exposed to Ni-free components. Increased Mn in cells exposed to Ti components. Ti components are the most biocompatible.	[227,227] Fernández-Miñano et al. (2011)

Table 6. *Cont.*

Sample Size	Appliances	Matrix and Sampling	Elements and Detection Mode	Mean/Median Concentrations	Main Results	Reference
20 patients + 20 controls	SS brackets and bands; SS and NiTi archwires.	Oral mucosa cells. Sampling after debonding (removal) of the orthodontic appliances, and 30 days later. Minimum treatment time of 18 months.	Ni and Cr. ICP-MS.	Immediately after debonding: Controls Ni: 3.86 ppb, Controls Cr: 2.71 ppb, Patients Ni: 4.09 ppb, Patients Cr: 3.63 ppb. 30 days after debonding: Controls Ni: 3.48 ppb, Controls Cr: 2.26 ppb, Patients Ni: 3.83 ppb, Patients Cr: 2.94 ppb.	No statistically significant differences between the groups regarding Ni and Cr concentrations. Genotoxic damage during orthodontic treatment reverted.	[228] Natarajan et al. (2011)
28 patients divided in 4 groups according to the brackets/archwires type combination+ 18 controls	SS brackets and bands; SS and Ti brackets; SS and NiTi archwires.	Oral mucosa cells. Sampling before and during treatment (3 and 6 months after placement).	Ni and Cr. AAS.	Before treatment: Ni: 0.52 ppb, Cr: 0.31 ppb. After 3 months: Ni: 0.68 ppb, Cr: 0.41 ppb. After 6 months: Ni: 0.78 ppb, Cr: 0.78 ppb.	Statistically significant increase in Cr content fat 3 months, as well as in Ni and Cr content after 6 months. DNA damage. SS brackets/SS archwire combination shows higher biocompatibility, whereas the Ti brackets/NiTi archwires are the less biocompatible.	[180] Hafez et al. (2011)
20 patients divided according to the bracket type	SS conventional or self-ligating brackets and bands; NiTi archwires.	Saliva. Sampling before and during treatment (1, 7 and 30 days after placement).	Ni and Cr. AAS.	Before treatment: Ni: 0.68 µg/L, Cr: 5.19–6.06 µg/L. After 1 day: Conventional—Ni: 1.95 µg/L, Cr: 21.78 µg/L. Self-ligating—Ni: 2.72 µg/L, Cr: 10.65 µg/L. After 7 days: Conventional—Ni: 2.89 µg/L, Cr: 36.69 µg/L. Self-ligating—Ni: 4.95 µg/L; Cr: 14.34 µg/L. After 30 days: Conventional—Ni: 1.18 µg/L; Cr: 8.98 µg/L. Self-ligating—Ni: 1.12 µg/L; Ce: 6.31 µg/L	Non-significant increase in Ni and Cr concentrations during the orthodontic treatment. Below dietary levels.	[170] Sahoo et al. (2011)

Table 6. *Cont.*

Sample Size	Appliances	Matrix and Sampling	Elements and Detection Mode	Mean/Median Concentrations	Main Results	Reference
30 patients + 30 controls	Hyrax appliances (which include 4 orthodontic bands with Ag-based soldering).	Saliva. Sampling before and after placement (10 min, 24 h, 7 days, 30 days, and 60 days).	Cd, Cu, Zn and Ag. AAS.	Controls: Cd: 0.15–0.18 μg/L, Cu: 3.81–6.54 μg/L, Zn: 0.01 μg/L, Ag: < DL. Patients before placement: Cd: 0.14 μg/L, Cu: 16.98 μg/L, Zn: 0.05 μg/L, Ag: 0.80 μg/L. Patients after placement: Cd: 0.18–0.71 μg/L, Cu: 12.63–70.60 μg/L, Zn: 0.05–0.20 μg/L, Ag: 2.01–11.53 μg/L.	All ions showed significant increases 10 min after placing the orthodontic appliances.	[229] Freitas et al. (2011)
28 patients + 18 controls	Orthodontic appliances (non-specified).	Hair. 1.5–2 years orthodontic treatment.	Ni, Cr, Mn, and Fe. ICP-OES.	Controls: Ni: 0.3642 μg/g d.h.m, Cr: 0.1298 μg/g d.h.m, Mn: 0.4850 μg/g d.h.m, Fe: 11.74 μg/g d.h.m. Patients: Ni: 0.5073 μg/g d.h.m, Cr: 0.1331 μg/g d.h.m, Mn: 0.5739 μg/g d.h.m, Fe: 12.22 μg/g d.h.m.	Non-significant differences in the hair metal contents between controls and patients.	[230] Mikulewicz et al. (2011c)
28 patients + 28 controls	SS brackets, and bands archwires.	Saliva. 12–18 months.	Ni and Cr. AAS.	Patients: Ni: 18.5 ng/mL, Cr: 2.6 ng/mL. Control: Ni: 11.9 ng/mL, Cr: 2.2 ng/mL.	Statistically significant difference for Ni between the two studied groups. Below toxic levels.	[231] Amini et al. (2012)
16 patients	SS brackets and tubes; NiTi archwires.	Saliva. Before the orthodontic treatment, immediately after and 8 weeks after placing the NiTi archwires.	Ni. ICP-MS.	Before treatment: Ni: 32 μg/L. Just after placing the NiTi archwire: Ni: 51 μg/L. 8 weeks after placing the NiTi archwire: Ni: 34 μg/L.	Statistically significant Ni increase just after placing the NiTi archwire. Non-significant difference after 8 weeks.	[232] Ousehal and Lazrak (2012)

Table 6. *Cont.*

Sample Size	Appliances	Matrix and Sampling	Elements and Detection Mode	Mean/Median Concentrations	Main Results	Reference
20 patients	SS brackets and bands; SS and NiTi archwires	Saliva. Sampling before and during the treatment (6 and 12 months).	Ni and Cr. AAS.	Before treatment: Ni: 9.75 μg/L, Cr: 3.86 μg/L. After 6 months: Ni: 10.37 μg/L, Cr: 4.60 μg/L. After 12 months: Ni: 8.32 μg/L, Cr: 2.04 μg/L.	Statistically insignificant change in Ni concentrations throughout the treatment time. Small but significant decrease in salivary Cr after 1 year.	[233] Amini et al. (2012)
40 patients + 50 controls	SS brackets and bands; SS and NiTi archwires.	Saliva. Sampling during the orthodontic treatment (ranging from 1 to 32 months after start, 1 sample per patients).	Ni and Cr. ICP-MS and ICP-OES.	Controls: Ni: 2.29 μg/L, Cr: 3.23 μg/L. Patients: Ni: 4.19 μg/L, Cr: 2.83 μg/L.	Statistically significant increase in salivary Ni and decrease in Cr between patients and controls. Non-toxic levels.	[234] Talic et al. (2013)
32 patients	SS brackets and bands; SS and NiTi archwires.	Saliva. Sampling before and during the treatment (20 days, and 3 and 6 months after starting).	Ni and Cr. AAS.	Before treatment: Ni: 5.76 μg/L, Cr: 2.6 μg/L. After 20 days: Ni: 6.54 μg/L, Cr: 3.68 μg/L. After 3 months: Ni: 5.13 μg/L, Cr: 3.41 μg/L. After 6 months: Ni: 5.61 μg/L, Cr: 3.39 μg/L.	Salivary Ni and Cr concentration did not significantly change.	[174] Yassaei et al. (2013)
24 patients + controls	SS brackets and NiTi archwires.	Hair. Sampling 16 weeks after starting the orthodontic treatment.	Ni. AAS.	Before treatment: Controls Ni: 0.245 μg/g, Patients Ni: 0.350 μg/g. After 16 weeks: Controls Ni: 0.382 μg/g, Patients Ni: 0.637 μg/g.	Statistically significant differences between controls and patients after 16 weeks.	[235] Abtahi et al. (2013)
30 patients	SS brackets and bands; NiTi archwires.	Saliva. Before placing the orthodontic treatment, 3 months before introducing stress; and 15 and 30 min after introducing stress.	Ni, Cr. AAS.	Before stress: Ni: 11.9–12.4 μg/L, Cr: 4.1–4.4 μg/L. After stress: Ni: 1.6–14.4 μg/L, Cr: 4.8–5.1 μg/L.	Significant increase in Ni concentrations after stress. No significant alteration in Cr concentrations.	[236] Amini et al. (2013)

Table 6. *Cont.*

Sample Size	Appliances	Matrix and Sampling	Elements and Detection Mode	Mean/Median Concentrations	Main Results	Reference
20 patients + 20 controls	SS brackets, bands, tubes, and ligatures; NiTi and SS archwires	Oral mucosa cells. Sampling between 13–15 months after starting the orthodontic treatment.	Ti, V, and Zr. ICP-MS.	Controls: Ti: 5.14 ng/g, Zr: < DL. Patients: Ti: 5.23 ng/g, Zr: 0.54 ng/g.	Non-significant differences in both groups.	[237] Martín-Cameán et al. (2014)
20 patients + 20 controls	SS brackets, bands, tubes, and ligatures; NiTi and SS archwires,	Oral mucosa cells. Sampling between 13–15 months after starting the orthodontic treatment.	Ni, Cr, Co, and Cu. ICP-MS.	Controls: Ni: 4.3 μg/L, Cr: 2.3 μg/L, Co: 0.6 μg/L, Cu: 4.9 μg/L. Patients: Ni: 24.8 μg/L, Cr: 17.5 μg/L, Co: 11.6 μg/L, Cu: 8.5 μg/L.	Significantly higher values for all metals for patients when compared with controls.	[238] Martín-Cameán et al. (2014)
70 patients + 56 controls	SS brackets, bands, tubes, and ligatures; NiTi and SS archwires.	Hair. Minimum duration of the orthodontic treatment of 24 months.	Ni, Cr, Fe, Mn, Cu. AAS.	Controls: Ni: 0.36 μg/g, Cr: 0.36 μg/g, Fe: 25.3 μg/g, Cu: 33 μg/g, Mn: 0.23 μg/g. Patients: Ni: 0.33 μg/g, Cr: 0.33 μg/g, Fe: 24.86 μg/g, Cu: 24 μg/g, Mn: 0.42 μg/g.	Significant increase for Mn concentrations only.	[239] Martín-Cameán et al. (2014)
24 pigs (12 controls)	SS plates simulating orthodontic appliances.	Hair, kidneys, liver, lungs, aorta, and oral mucosa. Up to 6 months.	Ni, Cr, Fe, Cd, Co, Cu, Mn, Mo, Si and Zn. ICP-OES.	Higher increase in toxic metals registered in the aorta (for Ni), cheek (for Ni) and hair (for Cr).	Products of corrosion passed into selected tissues of pigs. Below toxicity levels.	[240] Mikulewicz et al. (2014)
47 patients	SS brackets and ligatures; NiTi archwires.	Hair. Sampling before and during the orthodontic treatment (4, 8 and 12 months).	Ni, Cr and Fe. ICP-OES.	Before treatment: Ni: 0.275 mg/kg Cr: 0.0201 mg/kg Fe: 13.2 mg/kg After 1 year: Ni: 0.422 mg/kg Cr: 0.158 mg/kg Fe: 14.2 mg/kg	Statistically significant increase in Cr content only. Below toxicity levels.	[241] Mikulewicz et al. (2015)

Table 6. *Cont.*

Sample Size	Appliances	Matrix and Sampling	Elements and Detection Mode	Mean/Median Concentrations	Main Results	Reference
30 patients, divided in two groups according to the type of brackets	Conventional or MIM SS brackets and tubes; NiTi archwires.	Saliva. Immediately before and 60 days after starting the treatment.	Ni and Cr. AAS.	Before the treatment: Conventional—Ni: 7.12 μg/L, Cr: 0.25 μg/L. MIM—Ni: 8.62 μg/L, Cr: 0.42 μg/L. 60 days after treatment: Conventional—Ni: 12.57 μg/L; Cr: 0.35 μg/L. MIM—Ni: 8.86 μg/L, Cr: 0.26 μg/L.	Significant increase in Ni content in each group. Not significant differences between the groups.	[242] Amini et al. (2015)
24 patients	SS brackets and bands; NiTi and SS archwires.	Hair. Immediately before and 6 months after starting the treatment.	Ni and Cr. AAS.	Before the treatment: Ni: 0.1380 μg/g d.h.m, Cr: 0.1455 μg/g d.h.m. After 6 months: Ni: 0.6715 μg/g d.h.m, Cr: 0.1683 μg/g d.h.m.	Ni and Cr content in hair significantly increased (387 and 16%, respectively).	[243] Amini et al. (2015)
13 patients	SS brackets, bands, tubes, lingual sheath, transpalatal arch and archwires; NiTi archwires.	Saliva. Sampling before and during the orthodontic treatment (1 week, and 1 and 3 months after placement).	Ni and Cr. AAS.	Before treatment: Ni: 1.156 μg/L, Cr: 11.570 μg/L. After 1 week: Ni: 6.841 μg/L, Cr: 70.386 μg/L. After 1 month: Ni: 3.403 μg/L, Cr: 21.254 μg/L. After 3 months: Ni: 3.124 μg/L, Cr: 20.002 μg/L.	Significant increase in salivary Ni and Cr after starting the orthodontic treatment. Peak concentrations 1 week after placement.	[176] Dwivedi et al. (2015)
30 patients	SS brackets and bands; NiTi and SS archwires.	Saliva. Sampling before and during treatment (after aligning phase and 10–12 months after placement).	Ni and Cr. ICP-MS.	Before treatment: Ni: 48.78 ppb, Cr: 69.74 ppb. After aligning stage: Ni: 59.19 ppb, Cr: 102.68 ppb. 10–12 months after start: Ni: 46.33 ppb, Cr: 87.07 ppb.	Significant increase in salivary Ni and Cr after the initial aligning phase.	[182] Nayak et al. (2015)

Table 6. *Cont.*

Sample Size	Appliances	Matrix and Sampling	Elements and Detection Mode	Mean/Median Concentrations	Main Results	Reference
50 patients with 4–6 months of fixed orthodontic treatment	Fixed orthodontic appliances (not specified).	Saliva. Sampling after 1 week without using mobile phone (controls); then after 1 week of regular usage (experimental group).	Ni. ICP-MS.	Experimental group: Ni: 16.22 ng/L. Controls: Ni: 12.84 ng/L.	Statistically significant increase in Ni release for the experimental group when compared with the controls.	[244] Saghiri et al. (2015)
1 patient	Fixed appliances (not specified).	Saliva, alveolar bone, and gingiva. Collection during periodontal cosmetic surgery and exostosis removal.	Ni. AAS.	Ni in saliva: 986.4 ppb. Ni in bone: 779.5 ppb. Ni in gingiva: 620.5 ppb.	High Ni accumulation in each sample type.	[245] Arcila et al. (2015)
30 patients	SS bands and closed coil springs; SS self-ligating brackets with NiTi clip; NiTi archwire and open springs.	Saliva. Sampling before and during the orthodontic treatment (immediately after placing brackets and bands and 2 weeks later; immediately after placing the archwires and 4 and 8 weeks later).	Ni. ICP-MS.	Before treatment: Ni: 21.85 μg/L. Immediately after placing brackets and bands: Ni: 85.34 μg/L. Immediately after placing the NiTi archwires: Ni: 57.74 μg/L. For the remaining sampling times: Ni: 13.73–19.83 μg/L.	Significant increase in salivary Ni concentrations after brackets and bands insertion, as well as after placing the archwire. Return to basal levels after 4 weeks. Below dietary intake.	[246] Gölz et al. (2016)
24 patients	SS brackets; NiTi archwires.	Gingival crevicular fluid. Sampling before and during treatment (1 and 6 months).	Ni and Cr. AAS.	Before treatment: Ni: 3.894 μg/g, Cr: 1.978 μg/g. 1 month after starting: Ni: 5.913 μg/g, Cr: 4.135 μg/g. 6 months after starting: Ni: 19.810 μg/g, Cr: 13.760 μg/g.	Significant increase in Ni and Cr (up to 510 and 700%, respectively) during the treatment, as well as gingival inflammation promotion.	[247] Amini et al. (2016)
42 patients, divided according to the bracket type	MIM tubes and SS brackets; NiTi, Cu-NiTi, or epoxy-coated NiTi archwire.	Saliva. Sampling before and 2 months after starting the orthodontic treatment.	Ni. AAS.	Before treatment: Ni: 10.4571 μg/L. After 2 months: Ni: 11.0799 μg/L.	Statistically significant increase in salivary Ni concentration but depends on the archwire type.	[173] Masjedi et al. (2016)

Table 6. *Cont.*

Sample Size	Appliances	Matrix and Sampling	Elements and Detection Mode	Mean/Median Concentrations	Main Results	Reference
10 patients	SS brackets and ligatures; bands with tubes and lingual sheaths (not specified); NiTi archwires.	Saliva. Sampling before and during treatment (10 days and 1 month after placement).	Ni and Cr. ICP-OES.	Before treatment: Ni: 0.0039 mg/L, Cr: 0.0024 mg/L. After 10 days: Ni: 0.0288 mg/L, Cr: 0.0037 mg/L. After 30 days: Ni: 0.0370 mg/L, Cr: 0.0103 mg/L.	Statistically significant increase in salivary Ni after 10 and 30 days when compared with controls, as well as for salivary Cr between the 10th and the 30th day.	[248] Kumar et al. (2016)
47 patients with different dietary habits (coffee, yoghurt, juice and vinegar consumption)	Fixed orthodontic appliances (not specified).	Human hair. Sampling at the beginning of the treatment, and after 4, 8 and 12 months.	Ni and Cr. ICP-OES.	At the beginning: Ni: 0.131–0.331 mg/kg, Cr: 0.00578–0.0338 mg/kg. After 4 months: Ni: 0.222–0.505 mg/kg, Cr: 0.0620–0.446 mg/kg. After 8 months: Ni: 0.252–0.444 mg/kg, Cr: 0.0862–0.292 mg/kg. After 12 months: Ni: 0.207–0.500 mg/kg, Cr: 0.124–0.191 mg/kg.	Consuming foods and drinks with low pH can intensify metal release of Cr and Ni during the orthodontic treatment.	[249] Wołowiec et al. (2017)
30 patients	Fixed orthodontic appliances (not specified).	Saliva, biofilm, and oral mucosa cells. Before and during orthodontic treatment (1 week and 6 months).	Ni. AAS.	Before treatment: Saliva—Ni: 2.213 ppm, Biofilm—Ni: 4.943 ppm, Oral mucosa—Ni: 3.327 ppm. After 1 week: Saliva—Ni: 2.627 ppm, Biofilm—Ni: 5.75 ppm, Oral mucosa—Ni: 3.683 ppm. After 6 months: Saliva—Ni: 3.03 ppm, Biofilm—Ni: 6.917 ppm, Oral mucosa—Ni: 3.143 ppm.	Significant increase in Ni levels, especially in biofilm samples.	[250] Causado-Vitola et al. (2017)
46 patients, divided according to the bracket type	SS MIM or conventional brackets and tubes; NiTi and SS archwires.	Hair. Sampling before and 6 months after starting the orthodontic treatment.	Ni and Cr. AAS.	Before treatment: Ni: 0.1600 µg/g d.h.m, Cr: 0.1657 µg/g d.h.m. After 6 months: Ni: 0.3199 µg/g d.h.m, Cr: 0.3066 µg/g d.h.m.	Statistically significant Ni and Cr content increase, regardless the bracket type.	[251] Masjedi et al. (2017)

Table 6. *Cont.*

Sample Size	Appliances	Matrix and Sampling	Elements and Detection Mode	Mean/Median Concentrations	Main Results	Reference
37 patients	Metallic Fixed appliances (not specified).	Saliva. Sampling immediately before placing the fixed appliances, and after 1 and 24 weeks.	Ni. AAS.		Increased salivary Ni concentration, probably responsible for modifying the oxidative/antioxidative balance of saliva.	[252] Buckzo et al. (2017)
60 patients + 30 controls	Conventional appliances: SS brackets and bands, and NiTi archwires; aesthetic appliances: polycarbonate brackets and tubes and Rh-coated NiTi archwires.	Saliva. Sampling from patients undergoing an orthodontic treatment for 1–6 months. One collection per patient.	Ni, Cr, Fe, and Cu. Total reflection XRF.	Controls: Ni: 4.14 μg/L, Cr: 10.32 μg/L, Fe: 32.04, Cu: 11.40. Patients with conventional appliances: Ni: 22.20 μg/L, Cr: 89.45 μg/L, Fe: 517.77 μg/L, Cu: 15.10 μg/L.	No significant differences regarding Ni and Cr concentrations between conventional or aesthetical and control groups. Ni and Cr influenced by the type of appliances. No differences in Fe and Cu between groups.	[253,253] Lages et al. (2017)
42 patients, divided in two groups (mobile phone users and nun-users)	Fixed orthodontic appliances (not specified).	Saliva. 6–9 months after placement.	Ni. ICP-OES.	Mobile phone users: Ni: 0.012 ppb. Non-users: Ni: 0.0083 ppb.	Mobile phone radiations can influence Ni ion release, but a statistically non-significant difference was obtained.	[254] Nanjannawar et al. (2017)
30 patients	SS brackets and NiTi archwire	Gingival crevicular fluid. Sampling before and during treatment (1 and 6 month)	Ni and Cr. AAS	Before treatment: Ni: 3.2 μg/g Cr: 4.1 μg/g 1 month after starting: Ni: 4.5 μg/g Cr: 4.9 μg/g 6 months after starting: Ni: 14.2 μg/g Cr: 21.4 μg/g	Significant increase in Ni and Cr concentrations during the orthodontic treatment.	[255] Bhasin et al. (2017)
24 patients + 28 controls		Scalp hair. Sampling before and 1 year after placement.	Ni and Cr. AAS.	Before: Controls—Ni: 0.085 μg/g, Cr: 0.299 μg/g. Patients—Ni: 0.061 μg/g, Cr: 0.304 μg/g. 1 year after: Controls—Ni: 0.086 μg/g, Cr: 0.258 μg/g. Patients—Ni: 0.149 μg/g, Cr: 0.339 μg/g.	Statistically significant differences in both Ni and Cr contents in scalp hair between the groups after 1 year of orthodontic treatment.	[256] Jamshidi et al. (2018)

Table 6. *Cont.*

Sample Size	Appliances	Matrix and Sampling	Elements and Detection Mode	Mean/Median Concentrations	Main Results	Reference
42 patients with metallic brackets + 42 with ceramic brackets	SS or ceramic brackets; NiTi archwires.	Saliva. Sampling before and 6 months after starting the orthodontic treatment.	Ni, Cr, Ti, Co, Cu, and Zn. ICP-MS.	Metallic brackets: Before: Ni: 4.24 μg/L, Cr: 1.95 μg/L, Ti: 1.68 μg/L, Co: 0.46 μg/L, Cu: 23.31; μg/L, Zn: 220.67 μg/L. After 6 months: Ni: 5.04 μg/L, Cr: 1.01 μg/L, Ti: 9.29 μg/L, Co: 0.32 μg/L, Zn: 168.45 μg/L.	Statistically significant increase in salivary Ti, and statistically significant decrease in Cr and Zn. Non-significant differences difference between both groups.	[257] Jurela et al. (2018)
20 healthy patients + 20 periodontal patients	Brackets (not specified) and NiTi archwires.	Saliva. Sampling 2 months after starting the orthodontic treatment.	Ni. ICP-MS.	Healthy patients: Ni: 182.8 ng/mL, Cr: 6.35 ng/mL. Periodontal patients: Ni: 338.2 ng/mL, Cr: 7.45 ng/mL.	Statistically significant differences in salivary Ni concentrations between healthy and periodontal patients. Inconclusive for Cr.	[258] Amini et al. (2019)
60 patients divided into 2 groups according to the oral health products used (fluorinated vs. nonfluorinated)	SS brackets and NiTi archwires.	Gingival crevicular fluid. Sampling before and after 1 week, 1 month and 6 months after placement.	Ni, Cr, Ti, and Mn.	Nonfluorinated: Before—Ni: 0.49 μg/L; Ti: 0.49 μg/L. 7 days—Ni: 0.52 μg/L; Ti: 0.51 μg/L. 30 days—Ni: 13.42 μg/L; Ti: 40.09 μg/L; Cr: 0.50 μg/L; Mn: 0.50 μg/L. 6 months—Ni: 0.51 μg/L; Ti: 4.80 μg/L; Cr: 0.49 μg/L; Mn: 0.49 μg/L. Fluorinated: Before—Ni: 0.51 μg/L; Ti: 0.51 μg/L. 7 days—Ni: 0.52 μg/L; Ti: 0.50 μg/L. 30 days: Ni: 101.78 μg/L; Ti: 64.69 μg/L; Cr: 12.00 μg/L. 6 months—Ni: 0.51 μg/L; Ti: 0.51 μg/L; Cr: 0.53 μg/L; Mn: 0.48 μg/L.	Statistically significant increase in Ni, Cr, and Ti concentrations at 30 days only. Higher metal content in patients using fluoride-containing oral hygiene products, with statistically significant difference for Ni when compared with patients using nonfluorinated products.	[259] Chitra et al. (2019)

Table 6. *Cont.*

Sample Size	Appliances	Matrix and Sampling	Elements and Detection Mode	Mean/Median Concentrations	Main Results	Reference
50 patients + 30 controls	SS brackets and bands; SS and NiTi archwires.	Saliva and serum. Sampling before and during the orthodontic treatment (1 week, 3 months, 1 year, and 1.5 years).	Ni, Cr and Zn. AAS.	Saliva—before treatment: Controls—Ni: 4.33 ppb, Cr: 1.13 ppb, Zn: 10.73 ppb; Patients—Ni: 4.24 ppb, Cr: 1.18 ppb, Zn: 11.8 ppb. Saliva—1.5 years after: Controls—Ni: 5.02 ppb, Cr: 1.27 ppb, Zn: 10.24 ppb; Patients—Ni: 67 ppb, Cr: 30.8 ppb, Zn: 164.7 ppb. Serum—before treatment: Controls—Ni: 8.31 ppb, Cr: 6.18 ppb, Zn: 29.1 ppb; Patients—Ni: 8.46 ppb, Cr: 6.46 ppb, Zn: 28.3 ppb; Serum—1.5 years after: Controls—Ni: 8.47 ppb, Cr: 6.02 ppb, Zn: 30.1 ppb; Patients—Ni: 81.65 ppb, Cr: 35.6 ppb, Zn: 597.16 ppb.	Statistically significant increase in salivary and serum concentrations of Ni, Cr, and Zn between controls and patients. Below toxic levels.	[260] Quadras et al. (2019)
100 patients + 40 controls	Fixed orthodontic appliances (not specified).	Serum. Sampling between 3 weeks and over 18 months.	Ni, Cr, Fe, Cu, Mn, and Zn. ICP-MS.	Controls: Ni: 26.95 μg/L, Cr: 44.45 μg/L, Fe: 200.72 μg/L, Cu: 31.43 μg/L, Mn: 13.75 μg/L, Zn: 32.90 μg/L. Patients: Ni: 61.40 μg/L, Cr: 44.28 μg/L, Fe: 454.92 μg/L, Cu: 55.42 μg/L, Mn: 18.85 μg/L, Zn: 143.70 μg/L.	All ions' concentrations increased in the serum, except for Cr. Ni concentration in serum was dependent on treatment time.	[261] Moghadam et al. (2019)
35 patients	Fixed orthodontic appliances (not specified) involving NiTi and SS archwires.	Saliva and urine. Sampling before and during the orthodontic treatment (3 and 6 months).	Ni and Ti. ICP-OES.		Statistically significant differences in the Ni concentrations in saliva between 3 and 6 months, as well as Ti in urine in the same periods.	[198] Velasco-Ibañez et al. (2020)

Table 6. *Cont.*

Sample Size	Appliances	Matrix and Sampling	Elements and Detection Mode	Mean/Median Concentrations	Main Results	Reference
20 patients + 20 controls	SS brackets, bands and archwires.	Saliva. Sampling before and 6–12 months after placement.	Ni. AAS.	Controls: Before: 9.82 ng/mL, After: 10.21 ng/mL. Patients: Before: 9.90 ng/mL, After: 15.83 ng/mL.	Higher but statistically insignificant increase in Ni concentrations in patients when compared with the control group.	[262] Butt et al. (2020)
43 patients divided in groups according to soldering of the lingual arch + 21 controls	Lingual arches composed of SS bands and wires and soldered or welded SS wires.	Saliva. Sampling before and after placement (7, 15 and 30 days).	Ni, Cr, Fe, Cu, Zn, Ag, Cd, and Sn. ICP-MS.	Controls: Ni: 8.0–6.0 μg/L, Cr: 3.5–3.9 μg/L, Fe: 227.9–289.9 μg/L, Cu: 23.1–34.2 μg/L, Zn: 461.0–499.8 μg/L, Ag: 10.1–18.3 μg/L, Cd: 0.6–1.0 μg/L, Sn: 16.6–25.3 μg/L. With lingual arches: Ni: 5.3–34.5 μg/L, Cr: 3.3–4.2 μg/L, Fe: 201.0–314.8 μg/L, Cu: 28.0–40.7 μg/L, Zn: 384.3–963.4 μg/L, Ag: 3.7–20.8 μg/L, Cd: 0.5–1.5 μg/L, Sn: 11.4–27.7 μg/L.	No statistically significant differences for most metallic ions. Below toxic levels.	[263] Schacher et al. (2020)
40 patients divided into 2 groups according to the toothpaste used (nonfluorinated or fluorinated)	Brackets (non-specified) and NiTi archwires.	Gingival crevicular fluid. Before and after placement (7 and 30 days, and 6 months).	Ni and Cr. ICP-MS.	Nonfluorinated toothpaste [ng/mL]: Before: Ni: 0.49, Cr: 0.48; 7 days: Ni: 0.52, Cr: 0.52; 30 days: Ni: 13.4, Cr: 40.6; 60 days: Ni: 0.54, Cr: 4.9; Fluorinated toothpaste [ng/mL]: Before: Ni: 0.52, Cr: 0.52; 7 days: Ni: 0.54, Cr: 0.53; 30 days: Ni: 100.2, Cr: 62.4; 60 days: Ni: 0.52, Cr: 0.52.	Statistically significant increase in Ni and Cr release for patients with prescribed fluorinated toothpastes.	[264] Pritam et al. (2021)
12 patients with fixed appliances + 15 patients with removal appliances	Fixed appliances (NiCr brackets and NiTi archwires); removal aligners (polyurethane).	Saliva. Sampling before and 3 months after starting the treatment.	From Al (Z = 13) to Y (Z = 39). Total reflection XRF.		No significant alterations regarding metals from metal corrosion and inflammatory reactions in patients under dental plaque control.	[265] Zeffa et al. (2021)

Table 6. *Cont.*

Sample Size	Appliances	Matrix and Sampling	Elements and Detection Mode	Mean/Median Concentrations	Main Results	Reference
20 patients	SS brackets (other components were not specified).	Saliva. Before and during treatment (2 weeks, and 1, 4, and 6 months after start).	Ni, Cr, Fe, Ti, and Cu. ICP-OES.	Before treatment: Ni: 3.94 μg/L, Cr:2.37 μg/L, Fe: 45.13 μg/L, Ti: 48.25 μg/L, Cu: 1.53 μg/L. During treatment: Ni: 9.73–34.22 μg/L, Cr: 8.20–17.70 μg/L, Fe: 56.71–99.96 μg/L, Ti: 42.29–62.53 μg/L, Cu: 8.12–25.31 μg/L.	Maximum ion concentrations obtained 1 month after starting the orthodontic treatment. Increase in saliva pH and flow rate. Kinetic model proposed.	[266] Hamadamin (2022)
17 patients	SS brackets, bands, and tubes; NiTi archwires.	Saliva. Sampling before and during treatment (2 days, and 1, 4, and 12 weeks).	Ni, Cr, and Fe. ICP-OES.	Ni: 132–175 μg/L, Cr: 171–192 μg/L, Fe: 826–1023 μg/L.	No statistically significant variations registered throughout the study time.	[116] Fróis et al. (2022)
60 patients	SS brackets; metallic archwires.	Saliva. Sampling 1,3 and 5 years after starting the treatment.	Ni. ICP-MS.	1 week after starting: Ni: 1.25–1.74 ppb; 2 weeks after starting: Ni: 5.76–6.07 ppb; 3 weeks after starting: Ni: 4.32–4.78 ppb.	Significant increase between the 1st and the 2nd sampling time, reportedly linked to the use of hand-held mobile phones.	[267] Rajendran et al. (2023)

Friction always opposes movement between two sliding surfaces under load and may be divided into two modes: Static—when the applied force is still insufficient to induce relative motion—and dynamic—when surfaces are in relative motion (Figure 12). Despite the focus on kinetic friction, the motion of an archwire is hardly ever continuous; therefore, the most likely relevant friction type in orthodontic tooth movement is static friction, opposing any applied force [37,215]. Factors affecting both friction modes include: Geometry and type of archwires, brackets, and ligatures; surface chemical composition and roughness; cleanliness; and lubrication conditions [12,37,68,216,219,268–270]. Moreover, the manipulation of the components when placing and adjusting the orthodontic appliances during the treatment may lead to plastic deformation, wear tracks, and debris. Figure 13 points out these features in a retrieved orthodontic tube after two years of intraoral use.

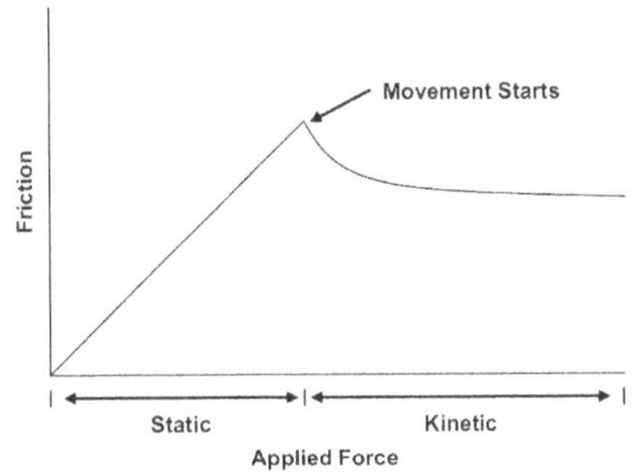

Figure 12. The two types of friction present when a force is applied to an object. Static friction occurs until a certain limit above which the movement starts. Kinetic friction opposes the on-going movement (reprinted from [215], Copyright (2009), with permission from Elsevier).

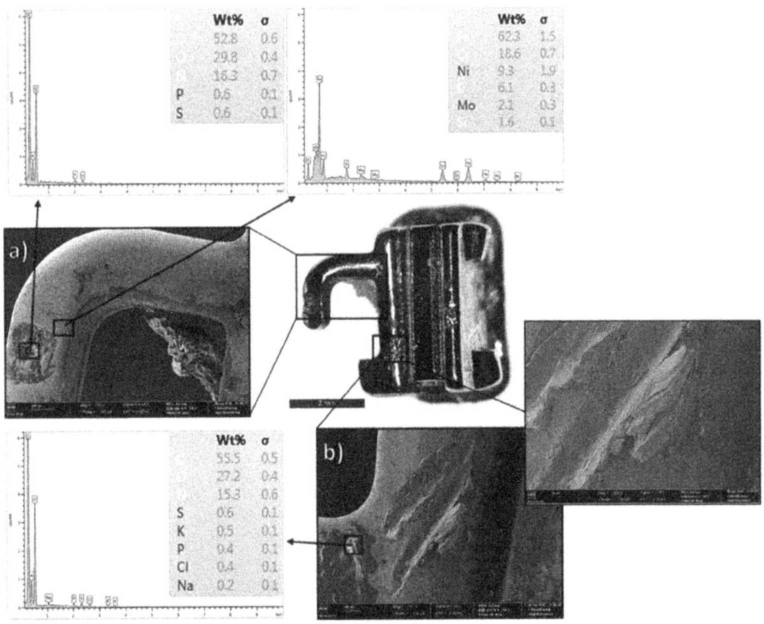

Figure 13. Optical and SEM micrographs of a retrieved SS tube after 2 years of intraoral exposure: (**a**) cavity filled with organic material; (**b**) wear track and debris coupled with the EDS chemical composition (reproduced from [116]).

Saliva is the natural intraoral lubricant by forming a protective pellicle [68–71]—a double layer of proteins [73]—on any material surface and, therefore, reducing the dynamic coefficient of friction [74,271]. While biofilms might have a protective role [66,90,272], microbiological activity most likely contributes to surface degradation by inducing corrosive microcells, rupturing the biofilm, and roughening the appliance surfaces [73]. These effects increase friction, wear, and metallic ions released from the bracket/wire contact pair.

3.3. *Oral Hygiene with Fluoride-Based Products*

Functional and aesthetic success is essential in orthodontics, but patients must comply with proper oral hygiene during treatment to avoid tooth demineralization and white spot lesions [91,108,273]. Fixed appliances make this task difficult, as the number of oral bacteria related to gingivitis increases shortly after their oral placement [273]. In fact, dental plaque accumulates in several regions (see Figures 7, 9 and 13), namely in the gingival areas or behind the archwires (e.g., on the bracket slots) [28,106,116,137,274]. To fight dental plaque, orthodontists prescribe fluoride-containing toothpastes, mouth rinses, gels, and varnishes to further control its accumulation and growth, enhance enamel integrity, and prevent dental and gingival diseases [72,91,275].

The downside of using these fluorides is the increased corrosion susceptibility of metallic alloys [29,42,264,276–282]. Fluoride ions (F^-)—combined with mechanical brushing—easily degrade the protective oxide layers of both SS and Ti-based alloys (see Equations (8) and (9)), increasing localized and general corrosion, promoting metallic ions release [276,283], and negatively impacting their mechanical and surface properties [41,42,277,279,282,284–286]—especially at low pH [104,118] and under the simultaneous presence of chloride ions [287].

Walker et al. [41] reported reduced unloading mechanical properties of SS and β-Ti archwires when exposed to neutral or acidulated prophylactic fluoride gels, which may prolong the orthodontic treatment time. On the other hand, Sufarnap and colleagues [56] reported an increase in both surface roughness and Ni and Cu release from Cu–NiTi archwires in NaF solution in vitro without a significant change in the deflection force.

Corrosion of SS bands and brackets [117,118,288,289] also increases in the presence of fluoride ions. Chantarawaratit and Yanisarapan [289] argued that acidulated phosphate fluoride gel should not be used in patients wearing fixed metal-based orthodontic appliances.

Since these SS components are the support for NiTi or Cu-NiTi archwires, galvanic coupling risk increased during the 1st orthodontic treatment phase (leveling/aligning) with possible mechanical and/or biocompatibility-related adverse consequences [108,112,284]. Figure 14 evidences the risk of galvanic coupling between dissimilar bracket/archwire pairs (concerning their alloys' composition), showing different in vitro metallic ion release profiles, according to the three solutions used. Synthetic F^--free Fusayama–Meyer saliva—widely used in in vitro studies—showed lower aggressiveness than the commercial mouthwashes containing fluoride [108].

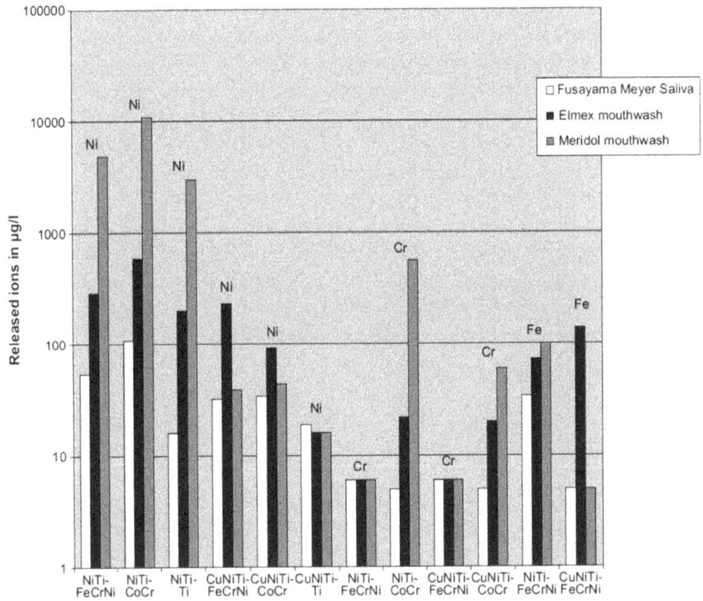

Figure 14. Metal ions (in µg/L) released by wire-bracket combinations in three different solutions (reprinted by permission of Oxford University Press on behalf of the European Orthodontic Society, from [108]; permission conveyed through Copyright Clearance Center, Inc.).

4. Orthodontic Alloys Modification

The main topic of this chapter is to briefly point out the current mitigation strategies to reduce the metal corrosion susceptibility of orthodontic appliances, including existing and emerging approaches for surface and bulk alloy modifications.

4.1. Non-Metallic Components

One way to reduce the release of metallic ions due to intraoral corrosion is to replace metallic-based alloys with non-metallic materials. Several examples of Ni-free substitutes for common orthodontic devices and components are listed in Table 7. However, existing or under development non-metallic alternatives for orthodontic appliances are more focused on satisfying patients' increasing aesthetic demands rather than on clinical concerns. Polymeric brackets—made of polycarbonate (PC), polyurethane (PU), or polyoxymethylene (POM); ceramic brackets—as those made of mono or polycrystalline aluminum oxide (Al_2O_3); non-metallic brackets with and without metallic slots [10,11,253,290–293]; and non-metallic archwires—produced from thermoplastic polymer [294], polyphenylene (PP) [295], poly-ether-ether-ketone (PEEK) resin [296], and fiber-reinforced polymers [34,297,298]—are some examples. A systematic literature review on transparent orthodontic archwires was prepared by Mikulewicz and colleagues [299], while J. Russel [290] reviewed aesthetic orthodontic brackets. Table 8 summarizes some of the current commercially available non-metallic aesthetic brackets and archwires.

Table 7. Ni-containing orthodontic materials and the corresponding Ni-free substitutes (used with permission of The Edward H. Angle Education and Research Foundation, from [28]; permission conveyed through Copyright Clearence Center, Inc.).

Category	Material	Ni-Free Substitute and Modifications
Standard appliances	Brackets.	Ni-free SS, ceramic, plastic, Ti, gold-plated or coated with other precious metals (Pd, Pt) brackets.
Treatment utilities	Bands. SS archwires.	Gold-plated bands. No alternative currently available; development of polymeric wires in progress.
Mechanic helpers	CoCrNi archwires. Sliding yokes, transpalatal. and lingual arches.	No alternative currently available β-Ti (TMA), plastic or inert metal (gold) coatings of wire segments.
Miscellaneous helpers	SS ligatures. Kobayashi hooks. Coil springs.	Teflon-coated ligatures. Teflon-coated Kobayashi hooks; Ni-free brackets with hooks. Elastomeric ligatures.
Fixed expansion appliances	SS appliances (Quad Helix). Rapid palatal expander. SS headgear. NiTi spring screws.	β-Ti (TMA) wires for Quad-Helix. Teflon-coated SS facebow. No alternative currently available.
Removable appliances	SS components of Hawley appliances and variations.	Plastic or elastic retainers; elastic positioners or acrylic splints invisagen™ technique.
Complex therapeutic interventions	Orthognathic surgery lag screws and plates. Distraction osteogenesis apparatus.	Resorbable polylactic-polyglycolic lag screws and plates. No alternative currently available.

Multiple concerns [11,300] still limit clinicians' acceptance and widespread use of alternative aesthetic solutions [290,301]. For instance, polymeric-based brackets may experience hardness reduction [291], undergo discoloration and staining from food dyes and sterilization procedures [302,303], release highly cytotoxic substances (e.g., bisphenol A [BPA]) [304], or have difficulty sliding against metallic archwires [11]. Similarly, ceramic brackets may also stain and discolor [305] during orthodontic treatment, in addition to their brittleness and high susceptibility to fracture [290,306,307], increased friction [308] and more severe archwire notching [218], higher pain during treatment [309], and enamel damage [11]. In a recent in vitro study concerning bracket genotoxicity [194], the authors even argue that ceramic brackets are more genotoxic than conventional SS or CoCr alloy ones. Non-metallic, aesthetical archwires present limitations as well. The foremost drawbacks—crazing formation, plastic deformation, and lower deliverable forces when bended [297,310] in addition to noticeable color change [311]—have limited their clinical widespread use.

Table 8. Non-metallic available aesthetic brackets and archwires [10,11,253,290,293,297–299,310].

	Brackets	Archwires
Polymeric	Polyurethane (PU). Polycarbonate (PC). Polyoxymethylene (POM). Fiberglass-reinforced PC.	Polyethylene terephthalate (PET) polytetrafluoroethylene (PTFE). Polyetheretherketone (PEEK). Fiber-reinforced polymer composites, and 3-layered wire (SiO_2/silicone resin/nylon).
Ceramic	Alumina. Zirconia.	

4.2. Nominal Composition Optimization

Another approach to decreasing intraoral corrosion susceptibility is to adjust the bulk nominal composition of the alloys by adding or subtracting certain elements [95,312]. As already mentioned in Section 2, Ni stabilizes the austenitic phase of SS and increases the alloy's resistance to corrosion and oxidation [10], namely against non-oxidizing acids [127]. However, the abovementioned clinical concerns make these alloys undesirable. Ni-"free" SS alloys (maximum of 2% Ni) were pointed out as viable alternatives to allergic orthodontic patients to this transition element [313]. The main development involves, at least, the presence of another element, such as Mn [10], in the SS nominal composition. Nevertheless, the final microstructural, mechanical, and corrosion resistance properties of the resulting steel can be compromised [36]. Additionally, Ni-"free" brackets/tubes are more cytotoxic than Ti-based ones [312] and may still provoke oral allergic reactions [314], even though conventional SS alloys are less biocompatible [304]. Another alternative approach is appropriately described by Eliades et al. [28] concerning the replacement of the most commonly used bracket SS alloy—316L AISI—by the duplex SS 2205 grade, which contains less Ni nominal content (Table 3).

To mitigate the nickel impact on oral health, the high Ni-containing alloys (TiNi) have been chemically optimized by adding Ag or Cu. The ternary TiNiAg alloys show similar mechanical properties and cytotoxicity effects as the binary TiNi ones and are more effective in reducing bacterial adhesion [315], but up to a certain Ag content limit [316] (reportedly 0.5–1.5 wt.% by Oh et al. [317]). Newer Cu–NiTi archwires [34] provide thermal activation and increase their resistance to permanent deformation. However, such alloys cannot be easily welded, are expensive, have low formability and high friction against SS brackets, and may fracture when bent over sharp edges [10,42,53]. New composite archwires (CAW), produced by NiTi and SS segments laser-welded with Cu, also seem to be promising for orthodontic treatments due to their excellent mechanical properties. Nevertheless, the resulting Cu interlayer could corrode in in vivo conditions, leading to biocompatibility and performance complications [318].

4.3. Manufacturing Processes

Different manufacturing processes lead to different final microstructural and physico-mechanical properties that can influence the overall corrosion resistance of metallic appliances. Brackets, tubes, and bands, for instance, can be made by casting and/or machining (milling) separate metallic pieces (e.g., the base and the wings, as shown in Figure 15). These parts are joined by soldering or welding [319]. Ag-based brazing alloys were popular and are still widely used [320], but these solders may dissolve or induce galvanic coupling inside the oral cavity, releasing cytotoxic ions [111,319,321,322]. Alternatives include the use of Au-brazing alloys, which induce galvanic corrosion of the less-noble SS following the Ag case, or Ni-based brazing alloys, with no available data [98,319] but undesirable due to their toxicity. Laser welding, a more recent and preferable advanced manufacturing technique, allows joining without brazing alloys [95,192,322], with lower corrosion susceptibility [323], even though large gaps in the joints may be a disadvantage for Ti-based brackets [319].

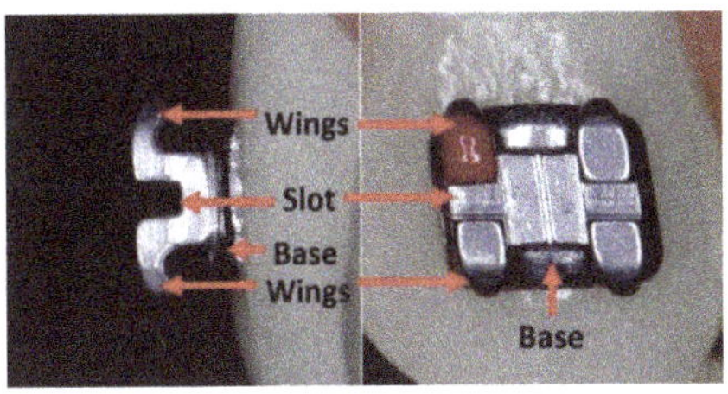

Figure 15. Identification of the bracket parts: wings, slot, and base (reprinted from [319], Copyright (2017), with permission from Elsevier).

Metal injection molding (MIM) is nowadays a popular manufacturing technique [11,319] to provide single-piece metallic orthodontic components with superior surface finishing in a cost-effective manner. Advantages over conventional casting/welding brackets include the absence of galvanic coupling between the base and the wings, the reduction of oral metallic ion release, and lower wear when used with NiTi archwires contacts [98,242,319]. However, internal porosity and possibly different biocompatibility and electrochemical behaviors of well-known alloys are some concerns that motivate further optimization [98,242,319].

Important developments in orthodontic appliance manufacturing include the production of functionally graded archwires (that apply different triggering forces depending on the location, accomplished by heat treatments) [324] and customized pieces by additive manufacturing [325–327], which may lead to less prolonged and more effective orthodontic treatments.

4.4. Surface Modification and Coatings

Another important optimization route for orthodontic alloys is surface modification and/or coatings deposition, preserving their best features, that is, their bulk properties. Corrosion most likely develops on bearing surfaces (Section 2) rather than in the bulk. Hence, different surface finishes can greatly impact the alloys' corrosion performance [95,328]. Mechanical polishing, for example, can reduce surface defects, leaving a smooth topography [329]. According to Hunt et al. [96]—who polished metallic archwires without impacting surface hardness—this process seemed detrimental or indifferent to CoCr, SS, and β-Ti archwires exposed to a corrosive medium (0.9% NaCl solution), but polished NiTi archwires presented higher corrosion resistance than its pristine condition [96]. Electropolishing is a common treatment to enhance both the appearance and corrosion resistance of the components, but galvanic corrosion cells may be produced between polished and unpolished regions [95]. Ion implantation is also a commercially used surface treatment of NiTi archwires [329,330] to reduce friction when compared with uncoated ones. However, Wichelhaus et al. [329] showed that this effect only occurred at the beginning of the orthodontic treatment, alerting them to the risks of recycling these components.

Current progress is also devoted to thin film/coating deposition with new and unique properties—a versatile approach to improving the final surface properties [39]. Within this regard, the flowchart presented in Figure 16 condenses the biomaterial surface properties that can be adjusted by plasma synthesis, as proposed by Chu et al. [331]. Without error, it can be generalized to a variety of methods and technologies available for biomedical applications, including orthodontics.

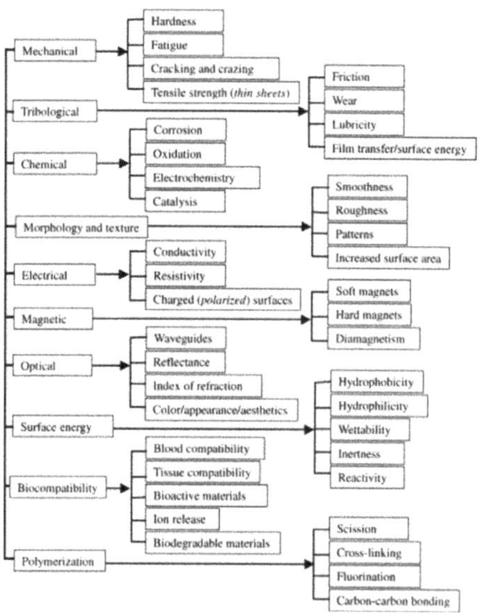

Figure 16. Summary of biomaterials surface properties (reprinted from [331], Copyright (2002), with permission from Elsevier).

Due to the high importance of coatings' deposition in tailoring the surface of orthodontic bioalloys, the authors opted to present and develop this topic separately in the following chapter (Section 5).

5. Protective Coatings in Orthodontics

This chapter aims to present the multitude of coatings available in the literature that have been deposited onto real fixed orthodontic appliances (the widely used archwires, brackets, and bands) to improve features such as corrosion, friction, fretting, biocompatibility, and antibacterial activity. For that, the results of the current review will be presented considering the three main groups: metal-, polymeric- and ceramic-based coatings (Tables 9–11). Several industrial deposition techniques can be used, as reviewed by Arango et al. [39] and presented in Figure 17, some of them which are applied in orthodontics.

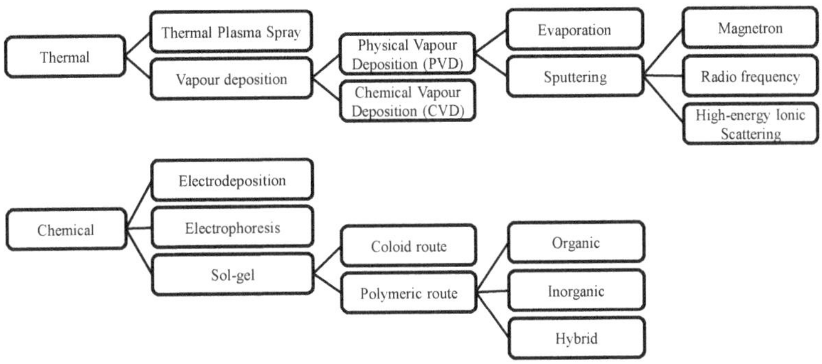

Figure 17. Classification of coating processes used at industrial level (adapted from [39]).

5.1. Metal-Based Coatings

Due to their higher corrosion resistance in addition to their aesthetic and pleasant color, metallic-based coatings have been deposited onto fixed orthodontic appliances, as presented in Table 9. Most pure metal coatings are devoted to rhodium (Rh), gold (Au), and silver (Ag). Titanium (Ti) and zinc (Zn) films were also included in this summary due to their excellent biocompatibility.

5.1.1. Transition Metal-Based Coatings

- Rhodium (Rh):

Rhodium-coated archwires—usually as bimetallic Rh–Au coatings [332–334]—are available on the market due to their pleasant color and superior corrosion resistance, without compromising the mechanical properties of the bulk alloys [253,310,335–338].

Katić and coauthors [332,339–341] compared the corrosion resistance of the uncoated NiTi archwires with those nitrified and coated with Rh. Surprisingly, electrochemical tests showed that corrosion susceptibility increased in artificial saliva, being highest for Rh-coated archwires: The heterogeneous layers induced galvanic coupling between the noble metals (Au and Rh) and the NiTi alloy, causing pitting corrosion and pronounced Ni release, as well as aesthetical appearance degradation [339,340]. Other researchers compared different types of coated and uncoated archwires, finding the worst corrosion resistance for the Rh-coated ones [61,342]. However, when fluoride-based prophylactic agents were weakly added during a 4-week immersion test in artificial saliva, ionic release from Rh-coated archwires decreased with increasing F ionsconcentration [332].

Nsaif et al. [343] found a protective effect of Rh-based coatings against an acidulated phosphate fluoride agent without compromising the overall mechanical properties of the archwires. Mlinaric et al. [279] demonstrated that the alteration of the properties of Rh-coated NiTi archwires depends on the type of oral antiseptics used. In turn, Osmani and colleagues [338] advised that such archwires should be avoided if oral hygiene is compromised due to the higher release of metallic ions.

The use of probiotic supplements—recommended to prevent caries and gingivitis during the fixed orthodontic treatment—also negatively impacts the corrosion resistance of Rh-coated NiTi archwires in artificial saliva: Both general and localized (pitting) corrosion increased, with precipitation of corrosion products on the surfaces [344,345]. However, a simple 1-month immersion test of Rh-coated SS and NiTi archwires in probiotic-containing artificial saliva suggested a decrease in corrosion susceptibility, possibly due to a benign anticorrosion effect of the early-formed biofilms [346]. Since mature biofilms will more likely promote MIC, the in situ higher accumulation of microorganisms for Rh-coated NiTi archwires reported by Lima et al. [347] may be a demerit result. Other researchers disagree [63], reporting that: (i) commercial Rh-coated NiTi archwires can significantly decrease *S. mutans* bacteria adhesion in vitro by altering the apparent surface free energy; and (ii) Au coating on SS archwires did not significantly influence the adhesion of *S. mutans* due to similar apparent surface free energy as uncoated SS [63].

Usui et al. [348] compared the kinetic and static frictional properties of both uncoated and Rh-coated SS archwires, under dry and wet conditions, against a ceramic bracket. In all tests, coated wires generated significantly higher kinetic and static frictional forces than the uncoated ones—more than 46% higher. Still, three-point bending test results indicated equivalent flexural properties. The bending effect can, however, increase the surface roughness of NiTi archwires coated with Rh-based materials [333]. Albawardi and colleagues [349] concluded that Rh-coated β-Ti archwires are not ideal to decrease friction against brackets for angulations up to 10° when compared to uncoated β-Ti or SS wires.

The color stability of commercial Rh-coated NiTi archwires has also been investigated by immersion in a coffee staining solution. While Alsanea and Shehri [335] found just a slight discoloration after 4 weeks, Ramasamy and colleagues [350] reported an extremely marked color change after 3 weeks of wetting.

- Gold (Au) and Platinum (Pt):

Gold and platinum are two noble metals that are widely used in many medical applications due to their bioinertness and notable corrosion resistance [351,352]. Despite their high cost, Au coatings are commercially available in orthodontics [62,64]. Surprisingly, Toy et al. [62] found that Au-plated SS brackets "remarkably decreased cell proliferation of HGFs" (human gingival fibroblasts), possibly related to an increase in in vitro corrosion during the test caused by surface defects of the substrate alloy. Conversely, other tested brackets (uncoated, made of metallic, ceramic, and composite materials) increased cell proliferation [62]. In other research work, Ito et al. [61] evaluated the coating loss of Au-coated SS wires after acid immersion (35% hydrochloric acid, pH = 1.1), wire ending, and mechanical brushing. While the Au coating resisted mechanical brushing and effectively prevented the release of metallic ions during immersion, wire bending showed signs of cracks and delamination of the coated wire [61]. Krishnan et al. [64] studied several coated and uncoated NiTi archwires and found that the breakdown potential of Au-coated ones is only surpassed by PTFE-coated wires. Regarding Pt-based coatings, Khonsari et al. [353] deposited Pt nanoparticles (NPs) on NiTi archwires by an electrochemical method, achieving adherent and homogeneous layers with a suitable tooth-like color.

- Silver (Ag):

Silver is recognized as one of the oldest antibacterial agents known, dating back to ancient times, at least by ~1500 B.C. [354,355]. In orthodontics, Ag-based coatings have been used to take advantage of this feature, that is, to reduce both bacterial adhesion and biofilm growth. Hence, periodontal diseases, tooth decay, and demineralization [356] have been prevented during orthodontic treatments.

Mhaske et al. [357] deposited pure Ag onto two types of archwires (NiTi and SS) by thermal vacuum evaporation. In addition to its antimicrobial effect, the 10 nm thick film effectively reduced the in vitro adhesion of *L. acidophilus* on both wire types, and, therefore, the total weight increase due to biofilm accumulation was lower: 35.5 and 20.5% for uncoated SS and NiTi wires, respectively, and less than 4.5% for Ag-coated wires [357]. Ghasemi et al. [358] reported that 60 and 100 μm thick Ag films deposited by physical vapor deposition (PVD) onto SS brackets were able to reduce the *S. mutans* counts right after 3 h of incubation.

Recently, an in situ study analyzed the initial biofilm formation (up to 48 h) onto an austenitic Ni-free bracket coated with Ag by using 3 distinct production processes: PVD, galvanic, and plasma immersion ion implantation and deposition (PIIID) processes [359]. The results showed statistically significant lower biofilm volume and surface coverage for all thin-film types in comparison to uncoated samples, but not between modified samples. All coatings showed antibacterial properties, with the PIIID-modified bracket displaying the highest increase in the dead/alive cell proportion.

An in vivo study lasting up to 75 days was conducted by Metin-Gürsoy et al. [360] to evaluate the antibacterial properties and the ion release from nano-Ag-coated SS brackets. The PVD-coated brackets placed on the mandibular incisors of Wistar Albino rats effectively inhibited *S. mutans* growth, which reduced caries development on the smooth surfaces. Concerns regarding the biocompatibility of Ag-based coatings for orthodontic applications previously motivated the same authors [356] to perform a 60-day animal histological experiment. Encouraging findings indicated that uncoated and PVD-nano-Ag-coated standard brackets implanted in healthy rats have similar biocompatibility behaviors. Nevertheless, the authors found brown-black silver granules in the adjacent tissues. Further studies are required to assess an eventual negative biological impact.

The research work conducted by Espinosa-Cristóbal et al. [361] described the influence of silver nanoparticles' (Ag-NPs) size on the activity of *S. mutans*. The coatings were chemically deposited on both SS brackets and archwires. The obtained results showed that coatings possessed good antimicrobial and anti-adhesion properties against this bacterial strain, especially for smaller-sized Ag-NPs (diameter of ~8.1 nm). Similarly, other authors [362] reported a significant decrease in the activity of *S. mutans* and *L. acidophilus* after a 24–48 h incubation with Ag-NPs-coated segments of SS orthodontic bands. Ag-NP coatings provide antimicrobial properties to SS brackets [48], bands [363,364] and archwires [365], namely against *S. mutans*, *S. sobrinus*, *C. albicans*, and *L. acidophilus* [48,364,365]. Coated NiTi archwires induced a more than 90% decrease in the adhesion of *S. sanguinis* and *L. salivaris* [366]. Zeidan et al. [367] reported an Ag-NPs coating (deposited on SS brackets by the thermal evaporation method) with an antibacterial effect against *S. mutans* and *L. acidophilus* over a period of 3 months of incubation. According to an interesting work by Anand et al. [368]—who synthesized Ag-NPs from plant extracts—the well-known antibacterial effect of Ag-NP coatings can be further improved by simultaneously adding TiO_2-NPs to functionalize orthodontic archwires. These coatings "*showed an enhanced antibacterial effect against Multi-Drug Resistant (MDR) bacteria*" [368].

Since corrosion and abrasion impact the intraoral performance of coatings, Ryu et al. [369] added Pt (up to 7%) to Ag coatings by PVD. The goal was to increase chemical stability. From the potentiodynamic polarization tests, the resulting AgPt-coated SS substrates showed higher corrosion resistance than the single Ag-coated or uncoated samples. In addition, all coatings diminished *S. mutans* and *A. actinomycetemcomitans* growth by up to 60%. Simultaneously, the cytotoxicity due to silver ion release was statistically insignificant when tested with human gingival fibroblasts. The authors [369] claim that brackets and archwires can be coated with these binary Ag–Pt materials, even though only flat substrates were used in their studies.

Whereas the (micro)biological impact is very important in the oral cavity, the Ag-based coatings must not compromise the surface tribological properties, especially the frictional behavior. Thermal vacuum evaporation-synthesized Ag coatings (~10 nm thick) onto two types of SS archwires did not affect their frictional resistance against SS brackets [370]. Reportedly, friction in the SS bracket/archwire combination decreases when at least the wire is Ag-coated [371]. Nevertheless, thick Ag-electroplated SS brackets (8–10 μm) significantly increased both surface roughness and friction forces against SS and NiTi archwires [372].

Shirakawa, Usui and colleagues [373,374] deposited tooth-colored Ag coatings on SS wires by eletropolishing. After 1 week of in vivo use with metallic brackets, no significant peeling was observed; however, the part of the wire in contact with the bracket slot turned black after 2 days [373].

- Titanium (Ti):

Titanium is scarcely used as a surface material in fixed orthodontics. To the authors' knowledge, only two research works (see Table 9) were found concerning fixed orthodontic appliances [375,376]. Ozeki and colleagues [375,377] coated Nitinol archwires with a 1 μm thick Ti layer by magnetron sputtering. This surface modification effectively reduced the Ni release in physiologically saline immersion tests up to 8 weeks (~5 times lower than uncoated wires). On the other hand, the superelasticity of the NiTi alloys decreased by almost 10.4% in comparison to uncoated wires, and the three-point bending tests (for 2 mm deflection) revealed a load decrease of 39% (from 4.18 to 2.55 N). Despite the decline in mechanical behavior, Ozeki et al. suggested that the performance of the sputter-coated archwires was not clinically compromised. In fact, when used in vivo for 4 weeks, both coated and uncoated archwires showed no inflammation signs, and little coating exfoliation—attributed to the use of pliers—was documented. Yet, the total sample size was very small: Only 5 patients [375]. Anuradha et al. [376] immersed NiTi archwires coated with sputter-deposited Ti (3–5 μm thick) in artificial saliva for 30 days. The results showed a well-adherent and smooth coating that remained mechanically stable over the test duration, without Ni or Ti detection in the artificial saliva solution.

- Zinc (Zn):

Zinc is an essential trace element [378–380], and has been used in medicine and dentistry due to its antibacterial effects, bioactivity/biodegradability, and good mechanical properties [381–384]. Karandish et al. [381] first reported the study of PVD-deposited Zn coatings on SS archwires. According to the obtained results, these coatings can improve both tensile and load-bending strength. In addition, the deposited Zn layers successfully reduced the friction resistance for the bracket/wire angulation of 10° by a factor of ~50%.

5.1.2. Metal-Based Oxides, Nitrides and Carbides Coatings

The addition of interstitial elements—such as oxygen (O), nitrogen (N), carbon (C), or their combinations—to transition metals to form binary, ternary, or quaternary systems is a practical and common method to enhance the hardness performance of protective coatings, mainly those used in mechanical/tribological applications [385]. In orthodontics, Ti-based coatings have a special position. Although the authors did not account for any research work involving the Ti–C system in fixed orthodontic appliances, the same cannot be said concerning those of Ti–O and Ti–N, as can be seen in Table 9. In addition, some research on binary Zr–O, Al–O, Zn–O, W–C/N, and Cr–C systems will also be considered.

- Ti–O System:

The Ti–O system, particularly the titanium oxide phase TiO_2, is nowadays commercially available for coated brackets and archwires [64]. This surface material is well recognized as inert and more stable than the native oxide layer formed onto Ti-based alloys, preventing the degradation of the alloys and the subsequent oral metal release [386]. Electropolishing, followed by sterilization and thermal passivation, not only smooths the surface of bulk metals but also induces the formation of a desirable TiO_2 top layer on the NiTi alloys [386].

Espinar et al. [387] heat-treated NiTi archwires to promote oxygen diffusion, forming a highly-adherent TiO_2 passive film whose thickness was proportional to both treatment temperature and time. The resulting Ni-depleted surface (up to approximately 20 nm depth) exhibited an inferior friction coefficient against 316 SS disks in comparison to untreated archwires: CoF = 0.40 and 1.53, respectively. This oxidation treatment can, therefore, make the bracket/archwire sliding easier and prevent Ni-induced allergy and cytotoxicity since the release of metallic ions decreased.

Horiushi et al. [388] heat treated electrolytically grown TiO_2 on NiTi alloys. They intended to develop a rutile crystalline phase and achieve photolytic activity when exposed to ultraviolet radiation (UV-A light, 315–400 nm), which could provide antibacterial activity. Other researchers explored the photocatalytic effect of TiO_2-based coatings deposited on metallic substrates by sol–gel [389–392] and by magnetron sputtering [393–398] methods. In particular, surface functionalization of SS and NiTi archwires [389,390] and SS brackets [395,397] showed effective decays in bacterial adhesion and survival rates of representative oral microorganisms, namely for *P. gingivalis*, *S. mutans*, *L. acidophilus*, and *Candida albicans*, when exposed to UV-A radiation. Reactive oxygen species (including hydroxyl radicals) decompose surface organic molecules of those microorganisms and damage cell walls, allowing the control of dental plaque growth [395]. On the contrary, Awata et al. [391] found no statistically significant difference in the amount of bacterial adhesion after coating 316L SS disks via sol–gel. Zhang et al. [399] reported that nano-TiO_2 coatings deposited on metallic brackets failed to show antibacterial activity up to 240 min in the dark. However, a nano-Ag–TiO_2 system demonstrated antibacterial activities of ~80–94% after 20 min in the dark, depending on the bacteria strain.

Fatani et al. [398] added silver to sputter-deposited photocatalytic TiO_2 coatings. The results confirmed the beneficial combination effect of this system: When compared with uncoated and Ag or TiO_2-coated metallic SS brackets, Ag–TiO_2 coatings showed lower *S. mutans* and *P. Gingivalis* adhesion and growth and biofilm development without cytotoxicity effects on human gingival fibroblast (HGF) cells. Kielan-Grabowska et al. [400] produced TiO_2 layers containing Ag-NPs via the sol–gel dip-coating method on SS 316L orthodontic wires. The results revealed a significant decrease in the overall corrosion resistance of the coated samples, even though they might possess antimicrobial properties [400,401].

While TiO_2 photocatalytic activity requires UV-A light, magnetron-sputtering-deposited and annealed N-doped Ti oxide films on SS brackets showed similar anti-adherent and antibacterial activity when exposed to visible light [393,394]. A 60-minute irradiation period was enough to prevent in vitro *S. mutans* growth for up to 3 months [393]. Nitrogen addition also seems beneficial to diminish the bracket/archwire friction [402].

Ghasemi et al. [358] deposited 60 and 100 μm thick TiO_2 layers on SS brackets by PVD. Results showed that such oxide coatings effectively improved surface roughness and bacterial growth but failed to reduce friction against archwires. Other authors also reported an inhibition of bacterial adhesion on PVD-deposited TiO_2 layers on SS orthodontic components [403].

Jung and colleagues treated both TMA and NiTi archwires with a plasma electrolytic oxidation (PEO, or micro-arc oxidation—MAO) method [404]. The 20–30 μm thick ceramic coating had a pleasant appearance and a higher biocompatibility and could improve the wear resistance of the NiTi archwire [405]. However, wire mechanical bending tests revealed the formation of defects in the coatings, cracking and chipping, and releasing sharp-edge particles that may harm teeth enamel or soft tissues [404].

Campeol et al. [406] reported that delamination of modified plasma-oxidized NiTi wires only occurs at high processing temperatures (>200 °C). The researchers prevented in vitro Ni release and delamination from substrates with an optimal 75–100 nm thick TiO_2 layer at ~135 °C by using a simplified plasma-assisted method. Moreover, the superelasticity of plasma-treated Nitinol wires increased by 8.6% (at 13% strain) when compared with untreated metallic samples [406].

Supriadi et al. [407] suggested that a previous electropolishing step smooths and cleans the surface of 17–4 PH SS orthodontic brackets and consequently increases the adherence of magnetron-sputtered TiO_2 coatings to the metallic substrates. Sol–gel-synthesized TiO_2 (300–400 nm) directly formed on 316L SS plates by Fu et al. [408] showed long and thin cracks and just slightly increased surface hardness. The solution involved the deposition of a thick TiN interlayer (~8 μm) that prevented the formation of large cracks and significantly increased the hardness (from 280 HV to 810 HV). The main disadvantage reported by the researchers was the decrease in toughness due to surface oxidation, with variable corrosion behavior in a 0.9% saline solution depending on the annealing temperature.

Regarding new composite archwires, Liu et al. [409] used magnetron sputtering to attain TiO_2 and N-doped TiO_2 coatings. They reported an 87% reduction in the antibacterial activity against *S. mutans* of N-doped TiO_2 films (~1.34% of N), in contrast to the undoped films (6% decrease only). The N addition was also beneficial to the in vitro biocompatibility and to the corrosion resistance of bulk CAW, enhancing both performances [409].

For further reading, an interesting meta-analysis was recently published by Solanki et al. [410] concerning in vitro studies using TiO_2 coatings on orthodontic brackets. The notable antibacterial effect with low cytotoxicity on eukaryotic cells was highlighted.

Table 9. Overview of metal-based coatings deposited on different orthodontic appliances and substrates. PVD: physical vapor deposition; PIIID: plasma immersion ion implantation deposition; PEO: plasma electrolytic oxidation; CVD: chemical vapor deposition.

Coatings	Substrate Materials	Deposition Methods	Literature References
Transition Metal-based Coatings			
Rh	NiTi archwires	*Commercial*	[339] Katić et al. (2014)
Rh	NiTi archwires	*Commercial*	[340] Katić et al. (2014)
Rh	NiTi archwires	*Commercial*	[63] Kim et al. (2014)
Rh	NiTi archwires	*Commercial*	[344] Trolic (2017)
Rh	NiTi archwires	*Commercial*	[332] Katić et al. (2017)
Rh	NiTi archwires	*Commercial*	[253] Lages et al. (2017)
Rh	NiTi archwires	*Commercial*	[310] Matias et al. (2018)
Rh	NiTi archwires	*Commercial*	[411] Asiry et al. (2018)
Rh	SS archwires	*Commercial*	[348] Usui et al. (2018)
Rh	NiTi archwires	*Commercial*	[341] Katić et al. (2018)
Rh	NiTi and SS archwires	*Commercial*	[343] Nsaif et al. (2019)
Rh	NiTi archwires	*Commercial*	[335] Alsanea and Shehri (2019)
Rh	NiTi archwires	*Commercial*	[346] Trolic et al. (2019)
Rh	NiTi archwires	*Commercial*	[347] Costa Lima (2019)
Rh	NiTi archwires	*Commercial*	[345] Trolic et al. (2019)
Rh	NiTi archwires	*Commercial*	[279] Mlinaric et al. (2019)
Rh	NiTi archwires	*Commercial*	[337] Batista et al. (2020)
Rh	NiTi archwires	*Commercial*	[336] Pinzan-Vercelino et al. (2020)
Rh	NiTi archwires	*Commercial*	[350] Ramasamy et al. (2020)
Rh	NiTi archwires	*Commercial*	[412] Madasamy et al. (2021)
Rh	NiTi archwires	*Commercial*	[338] Osmani et al. (2022)
Rh	NiTi archwires	*Commercial*	[342] Amorim et al. (2022)
Rh	SS archwires	*Commercial*	[49] Ito et al. (2022)
Rh	β-Ti archwires	*Commercial*	[349] Albawardi et al. (2022)
Rh-Au	NiTi archwires	*Commercial*	[334] Iijima et al. (2012)
Rh-Au	NiTi archwires	*Commercial*	[333] Albuquerque et al. (2017)
Au	SS archwires	*Commercial*	[63] Kim et al. (2014)
Au	NiTi archwires	*Commercial*	[64] Krishnan et al. (2014)
Au	SS brackets	*Commercial*	[62] Toy et al. (2014)
Au	SS archwires	*Commercial*	[49] Ito et al. (2022)
Pt-NPs	NiTi archwires	Electrochemical deposition	[353] Khonsari et al. (2011)
Ag	SS and NiTi archwires	Thermal evaporation	[357] Mhaske et al. (2015)
Ag	SS brackets	Electroplating	[372] Arash et al. (2015)
Ag	SS brackets	Magnetron sputtering	[398] Fatani et al. (2017)
Ag	SS brackets	PVD	[358] Ghasemi et al. (2017)
Ag	SS wires	Electroplating	[374] Usui et al. (2017)
Ag	SS wires	Electroplating	[373] Shirakawa et al. (2017)
Ag	SS archwires	Thermal vacuum evaporation	[370] Shah et al. (2018)
Ag	SS brackets	Galvanic, PVD, and PIIID	[359] Meyer-Kobbe et al. (2019)
Ag	SS brackets and archwires	Thermal vacuum evaporation	[371] Shah et al. (2023)
Nano-Ag	SS brackets	PVD	[356] Metin-Gürsoy et al. (2016)
Nano-Ag	SS brackets	PVD	[360] Metin-Gürsoy et al. (2017)
Nano-Ag	SS brackets and archwires	Chemical deposition	[361] Espinosa-Cristóbal et al. (2018)
Ag-NPs	SS bands	Thermal evaporation	[363] Prabha et al. (2016)
Ag-NPs	SS bands	Thermal evaporation	[362] Bindu et al. (2019)
Ag-NPs	SS archwires	Hydrothermal synthesis	[365] Gonçalves et al. (2020)

Table 9. *Cont.*

Coatings	Substrate Materials	Deposition Methods	Literature References
Ag-NPs	SS and Co–Cr brackets	*Commercial*	[48] Jasso-Ruiz et al. (2020)
Ag-NPs	NiTi archwires	Electrodeposition	[366] Gil et al. (2020)
Ag-NPs	SS brackets	Thermal evaporation	[367] Zeidan et al. (2022)
Ag-NPs	Metallic brackets	Laser ablation	[413] Tawakal et al. (2023)
Ag-NPs	SS bands	Electrostatic spray-assisted vapor deposition	[364] Bahrami et al. (2023)
Ag-NPs	SS archwires	Electrochemical deposition	[368] Anand et al. (2023)
Ag-NPs + TiO_2-NPs	SS archwires	Electrochemical deposition	[368] Anand et al. (2023)
Ti	NiTi archwires	Magnetron sputtering	[375] Ozeki et al. (2003)
Ti	NiTi archwires	Sputtering	[376] Anuradha et al. (2015)
Zn	SS archwires	Thermal evaporation	[381] Karandish et al. (2021)
Metallic Oxides, Nitrides and Carbides Coatings			
TiO_2	SS archwires	Sol–gel dip-coating	[389] Chun et al. (2007)
TiO_2	NiTi archwires	Oxidation treatment	[387] Espinar et al. (2011)
TiO_2	SS brackets	Magnetron sputtering	[395] Shah et al. (2011)
TiO_2	NiTi archwires	Oxidation process	[414] Satiyorini and Pintowantoro (2013)
TiO_2	NiTi archwires	Oxidation process	[415] Pintowantoro and Setiyorini (2013)
TiO_2	SS and NiTi archwires	Sol–gel dip-coating	[390] Chhattani et al. (2014)
TiO_2	NiTi archwires	*Commercial*	[64] Krishnan et al. (2014)
TiO_2	SS archwires	Sol–gel dip-coating	[392] Özyildiz et al. (2014)
TiO_2	metallic brackets	Spin-on deposition	[399] Zhang et al. (2015)
TiO_2	SS brackets	PVD	[358] Ghasemi et al. (2017)
TiO_2	SS brackets	Magnetron sputtering	[397] Baby et al. (2017)
TiO_2	SS brackets	Magnetron sputtering	[398] Fatani et al. (2017)
TiO_2	Composite archwires	Magnetron sputtering	[409] Liu et al. (2017)
TiO_2	SS brackets	Magnetron sputtering	[407] Supriadi et al. (2019)
TiO_2	SS brackets	Magnetron sputtering	[416] Supriadi et al. (2019)
TiO_2	β-Ti and NiTi archwires	PEO	[404] Jung et al. (2019)
TiO_2	SS archwires	PVD	[403] Mollabasci et al. (2020)
TiO_2	NiTi wires	Plasma oxidation	[406] Campeol et al. (2020)
TiO_2	NiTi wires	NH_3 treatments	[417] Kurtoğlu et al. (2020)
TiO_2	SS brackets	Magnetron sputtering	[418] Math et al. (2021)
TiO_2	SS archwires	Sol–gel dip-coating	[400] Kielan-Grabowska et al. (2021)
TiO_2	SS wires	Sol–gel dip-coating	[401] Bącela et al. (2022)
N-doped TiO_2	SS brackets	Magnetron sputtering	[394] Cao et al (2013)
N-doped TiO_2	composite archwires	Magnetron sputtering	[409] Liu et al. (2017)
N-doped TiO_2	SS brackets	Magnetron sputtering	[393] Salehi et al. (2018)
$TiO_{2-x}N_y$	Metallic brackets	Magnetron sputtering	[402] Li et al. (2014)
TiO_xN_y	NiTi wires	NH_3 treatments	[417] Kurtoğlu et al. (2020)
Nano-Ag/TiO_2	metallic brackets	Spin-on deposition	[399] Zhang et al. (2015)
Ag-TiO_2	SS brackets	Magnetron sputtering	[398] Fatani et al. (2017)
TiO_2 + Ag-NPs	SS archwires	Sol–gel dip-coating	[400] Kielan-Grabowska et al. (2021)
TiO_2 + Ag-NPs	SS wires	Sol–gel dip-coating	[401] Bącela et al. (2022)
TiN	NiTi archwires	*Commercial*	[419] Kim and Johnson (1999)
TiN	SS brackets	Ion plating	[420] Kao et al. (2002)
TiN	NiTi and NiTiCu archwires	Nitrogen gas difffusion	[421] Gil et al. (2004)
TiN	NiTi wires	*Commercial*	[422] Iijima et al. (2010)
TiN	SS brackets	Ion plating	[423] Huang et al. (2010)
TiN	SS brackets	Ion plating	[424] Kao et al. (2011)
TiN	SS brackets	*Commercial*	[115] Saporeti et al. (2012)
TiN	NiTi archwires	Chemical deposition	[414] Setiyorini and Pintowantoro (2013)
TiN	NiTi archwires	*Commercial*	[339] Katić et al. (2014)
TiN	NiTi wires	*Commercial*	[332] Katić et al. (2017)
TiN	NiTi archwires	*Commercial*	[340] Katić et al. (2014)
TiN	NiTi archwires	*Commercial*	[64] Krishnan et al. (2014)
TiN	NiTi archwires	*Commercial*	[425] Rongo et al. (2014)
TiN	SS brackets	Ion plating	[426] Zuo et al. (2015)
TiN	NiTi and β-Ti archwires	*Commercial*	[427] Rongo et al. (2016)
TiN	NiTi archwires	*Commercial*	[344] Musa Trolić (2017)
TiN	NiTi archwires	*Commercial*	[341] Katić et al. (2018)
TiN	SS and NiTi archwires	Ion plating	[428] Sugisawa et al. (2018)
TiN	NiTi archwires	*Commercial*	[345] Trolic et al. (2019)
TiN	NiTi archwires	*Commercial*	[279] Mlinaric et al. (2019)
TiN	NiTi archwires	*Commercial*	[346] Musa Trolic et al. (2019)
TiN	NiTi archwires	NH_3 treatments	[417] Kurtoğlu et al. (2020)
TiN	SS brackets and wires; NiTi archwires	Magnetron sputtering	[429] Arici et al. (2021)

Table 9. *Cont.*

Coatings	Substrate Materials	Deposition Methods	Literature References
TiN	SS brackets	Cathodic cage	[430] Teixeira et al. (2021)
TiN	NiTi archwires	*Commercial*	[338] Osmani et al. (2022)
TiN	SS archwires	Ion plating	[49] Ito et al. (2022)
Ti/TiN	NiTi archwires	Magnetron sputtering	[431] Liu et al. (2014)
TiAlN	β-Ti archwires	Cathodic arc PVD	[432] Krishnan et al. (2011)
TiAlN	β-Ti archwires	Cathodic arc PVD	[433] Krishnan et al. (2012)
TiN doped with CaP	SS brackets	Cathodic cage	[430] Teixeira et al. (2021)
ZrO_2	SS, NiTi and β-Ti archwires	Sol–gel	[434] Golshah and Feyli (2022)
Al_2O_3	SS brackets and wires; NiTi archwires	Magnetron sputtering	[429] Arici et al. (2021)
Al-SiO_2	NiTi archwires	Magnetron sputtering	[435] Wu et al. (2022)
Black oxide	NiTi archwires	*Commercial*	[436] Krishnan et al. (2012)
ZnO	SS brackets	Magnetron sputtering	[418] Math et al. (2021)
CrN	SS brackets and wires; NiTi archwires	Magnetron sputtering	[429] Arici et al. (2021)
CrC	SS archwires	Electroplating	[348] Usui et al. (2018)
WC/C	β-Ti archwires	Magnetron sputtering	[432] Krishnan et al. (2011)
WC/C	β-Ti archwires	Magnetron sputtering	[433] Krishnan et al. (2012)
Metal Oxide-Based NPs coatings			
ZnO	SS brackets	Spray pyrolysis	[437] Ramazanzadeh et al. (2015)
ZnO-NPs	SS wires	Chemical solution method	[438] Kachoei et al. (2015)
ZnO	SS archwires	Bath immersion	[439] Behroozian et al. (2016)
ZnO	NiTi archwires	Chemical deposition	[440] Kachoei et al. (2016)
ZnO	NiTi archwires	Electrochemical deposition	[441] Hammad et al. (2020)
ZnO	NiTi archwires	Chemical precipitationCVDSol–gel method	[442] Gholami et al. (2021)
ZnO	SS brackets and archwires	Sol–gel method	[443] Elhelbawy and Ellaithy (2021)
ZnO-NPs	SS bracktes	Thermal evaporation	[367] Zeidan et al. (2022)
ZnO-NPs	NiTi archwires	Hydrothermal method	[444] Palanivel et al. (2022)
ZnO-NPs	SS archwires	Chemical precipitationHydrothermal method	[445] Tanbakuchi et al. (2022)
AlO-NPs	NiTi archwires	Hydrothermal method	[444] Palanivel et al. (2022)
CuO	SS brackets	Spray pyrolysis	[437] Ramazanzadeh et al. (2015)
CuO-NPs	SS brackets	Dip-coating	[446] Ameli et al. (2022)
CuO-NPs	SS brackets	Dip-coating	[447] Ameli et al. (2022)
ZnO-CuO	SS brackets	Spray pyrolysis	[437] Ramazanzadeh et al. (2015)
Ag-NPs+ZnO-NPs	SS brackets	Thermal evaporation	[367] Zeidan et al. (2022)
Ag-HA-NPs	SS brackets	Dip-coating	[446] Ameli et al. (2022)
Ag-HA-NPs	SS brackets	Dip-coating	[447] Ameli et al. (2022)
TiO_2-NPs	NiTi archwires	Magnetron sputtering	[448] Venkatesan et al. (2020)
TiO_2-NPs	SS brackets	Dip-coating	[446] Ameli et al. (2022)
TiO_2-NPs	SS brackets	Dip-coating	[447] Ameli et al. (2022)
TiO_2-NPs	SS archwires	Dip-coating	[449] Silveira et al. (2022)
TiO_2-NPs	SS archwires	Magnetron sputtering	[368] Anand et al. (2023)
TiO_2-NPs	β-Ti and NiTi archwires	Dip-coating	[450] Chaturvedi et al. (2023)
IF-NPs-reinforced Metal-based Coatings			
WS_2-reinforced Ni	SS archwires	Electrochemical co-deposition	[451] Redlich et al. (2008)
WS_2-reinforced Ni	SS archwires	Electrochemical co-deposition	[452] Samorodnitzky-Naveh et al. (2010)
WS_2-reinforced Ni-P	SS archwires	Electroless deposition	[453] Katz et al. (2006)
WS_2-reinforced Ni-P	SS archwires	Electroless deposition	[454] Redlich et al. (2008)
MoS_2-reinforced Ni	SS archwires	Electrochemical co-deposition	[455] Gracco et al. (2019)
WS_2-reinforced Co	NiTi archwires	Co-electrodeposition	[456] Samorodnitzky-Naveh et al. (2009)
WS_2-reinforced Co	NiTi archwires	Co-electrodeposition	[452] Samorodnitzky-Naveh et al. (2010)

- Ti–N System:

The titanium nitride system is commercially accessible as TiN-coated brackets and archwires [64,115,279,332,338,340,341,344–346]. The typically gold-colored TiN coatings may be produced onto orthodontic-compatible metals by several methods, including pulsed high-energy density plasma treatment with Ti electrodes [457], nitrogen gas (N_2) diffusion [421] and ammonia gas (NH_3) treatments [417], plasma immersion ion implantation [422,458], ion beam assisted deposition [459], ion plating [420,423,424,426,428,460], and other PVD methods [396,429,431,461]. As presented below, there is no academic consensus on the corrosion susceptibility of these nitride coatings in orthodontics.

Kao et al. [420] attempted to reduce the corrosion of SS brackets in artificial saliva by TiN ion plating, but no statistically significant difference between coated and uncoated brackets was observed on both metallic ion release and biocompatibility. Later attempts also failed: TiN ion-plated metallic brackets were unable to reduce friction against archwires in several media [423,424]. Opposite results by Zuo et al. [426] showed super low coefficients of friction (CoF < 0.03) under humid wear conditions in artificial saliva, high hardness (14.62 GPa), and improved adhesion of coated SS316L disks and brackets. Such a combination of mechanical and tribological properties is expected to decrease wear and release of toxic elements during orthodontic treatments. In addition, the coated samples presented an appealing golden color.

Jin et al. [460] were also optimistic about their research work: TiN ion-plated NiTi substrates may lead to improved biocompatibility, as fibroblasts' adhesion and proliferation increased, whereas no cracks were reported during the three-point bending test after a 4099 N load. Deposition of the coatings onto SS and NiTi archwires diminished corrosion susceptibility in artificial saliva and friction force against metallic brackets up to a bracket/archwire angulation of 10°, as reported by Saporeti and coauthors [428]. The resulting higher tensile strength and stiffness of the SS wires are particularly useful properties in orthodontics. In turn, Mlinaric et al. [279] argued that, depending on the oral antiseptic used, the mechanical behavior of nitride NiTi (elastic properties and force delivery) can be affected, and Ni release may have been promoted during immersion in both the artificial saliva and the antiseptic.

A single TiN layer and a TiN/TiO_2 multilayer deposited onto AISI 316L substrates by reactive magnetron sputtering effectively reduced the adhesion of calcium-precipitating bacteria, which are responsible for the formation of dental calculus [396]. Wang et al. [462] previously nitrided the surface of commercially pure Ti—a duplex treatment—decreasing ion release by almost 5 times compared to untreated samples when exposed to F-containing artificial saliva. When single Ti was used as a 200 nm thick interlayer for a 1 μm TiN coating on NiTi orthodontic archwires, potentiodynamic tests showed that localized corrosion was prevented in acidulated artificial saliva, both under loading and unloading conditions [431]. Other researchers reported that 1.6 and 3.2 μm thick Ti/TiN multilayer systems (16 bilayers) were more effective in improving the corrosion resistance of 316L SS samples than a simple TiN monolayer. This was demonstrated by using NaCl and Hank's solutions [463]. However, the hardness decreased from ~13 to 10 GPa when using the multilayer system due to the presence of the softer Ti adhesion layers [463].

In opposition to the Ti–O system, no Ag addition to TiN coatings was found in the literature (Table 9). Instead, research work concerning the aluminum (Al) effect on TiN coatings by cathodic arc PVD was carried out [432,433]. The presence of Al (content not specified) decreased through-coating defects and prevented corrosion of the β-Ti archwires but maintained the mechanical properties (namely under unloading process). The results indicated that the TiAlN films effectively protected the wire during potentiodynamic tests in F-containing artificial saliva (no ionic release or coating damage was detected), while offering good biocompatibility without inducing apoptosis of oral epithelial cells [432].

Teixeira et al. [430] attempted to improve the surface properties of SS brackets by depositing single and calcium phosphate (CaPa)-containing TiN coatings via the cathodic cage method. Unfortunately, these coatings were unable to prevent the formation of *S. mutans* biofilms on the surface [430].

Recently, Kurtoğlu et al. [417] presented a simple and low-cost NH_3 surface treatment in He atmosphere at 700 °C and assessed the in vitro corrosion resistance against artificial saliva. Thin TiN and TiO_xN_y films (<1 μm thick) significantly reduced the Ni release from shape-memory NiTi substrates. Nonetheless, the high treatment temperature used may have negatively impacted the mechanical properties of the bulk orthodontic alloys [357, 464].

- Zr–O and Al–O Systems:

Both Zr–O and Zr–Al systems—zirconia (ZrO_2) and alumina (Al_2O_3)—are well recognized as bulk biomaterials due to their attractive combination of high corrosion resistance, low friction, high strength, and high wear resistance [465,466]. As surface material, ZrO_2 coatings (1 μm thick) have been deposited onto AISI 316L substrates by electron-beam PVD [467]. These coatings showed excellent corrosion resistance in NaF-containing Fusayama–Meyer artificial saliva (2% NaF, pH = 3.0), while both *S. mutans* bacterial adhesion and fibroblast cytotoxicity decreased in vitro [467]. More recently, Golshah and Feyli [434] attempted to deposit ZrO_2 on SS, TMA, and NiTi archwires using the sol–gel technique. Unfortunately, the researchers only successfully synthesized the coating on the TMA archwires, without a significant decrease in the static or kinetic friction against SS brackets [434].

Arici et al. [429] studied the binary Al–O system, deposited by magnetron sputtering, on SS brackets and SS and NiTi archwires. For this investigation, the researchers [429] designed a special experimental procedure to simulate the intraoral temperature variation (involving 250 thermal cycles, from −5 to +55 °C) and mechanical toothbrushing with fluoride toothpaste (12 h). The Al_2O_3 coatings presented good thermal, frictional, and brushing behavior. The referred work [429] also intended to study other sputtered coatings, TiN and CrN, which presented worse results in similar test conditions. While the TiN coatings showed small peeling areas, the CrN ones exhibited large detachment areas from the substrate. In fact, the CrN coatings failed the designed thermal, frictional, and brushing tests. The researchers therefore concluded that these Cr-based coatings were unsuitable for decreasing the coefficient of friction (CoF) in orthodontics. The best wear behavior was found for Al_2O_3 (or TiN)-coated archwires against metal brackets (CoF = 0.207 and 0.372, respectively). Wu et al. [435] deposited Al–SiO_2 coatings by magnetron sputtering that smoothed the surface of NiTi and SS archwires and enhanced their overall corrosion resistance with no apparent cytotoxicity.

- Zn–O System:

The Zn–O system, zinc oxide (ZnO), was also deposited as a coating by magnetron sputtering on SS brackets by Math et al. [418]. The coatings underwent thermal oxidation at 500 °C in an open-air furnace for 5 h. The researchers studied both anti-adhesion and antibacterial properties against *S. mutans*. The reported results were disappointing because they showed that ZnO coatings were unable to inhibit adhesion and growth of that bacterial strain.

- W–C, W–N, Cr–C, and Cr–N Systems:

The binary W–C, W–N, Cr–C, and Cr–N systems—included in the well-known protective hard coatings for high-speed cutting operations [385]—have also merited attention in orthodontics. β-Ti archwires were functionalized with sputtered tungsten carbide/carbon (WC/C) coatings by Krishnan et al. [432,433]. The idea was to achieve protection against fluoride-induced corrosion. It was concluded that these coatings were mechanically stable and biocompatible; however, to the identified surface defects could act as initiators of pitting corrosion. In fact, deeper and accentuated cracks were observed in the WC/C coating after fluoride immersion tests [432]. The approach by single W and W–N [468] sputtered coatings was also disappointing on AISI 316L flat samples immersed in artificial saliva immersion. The external passive oxide phase revealed unprotective behavior, accompanied by severe and deep crack formation, as shown in Figure 18.

Usui et al. [348] compared the mechanical, frictional, and aesthetic properties of hard Cr–C-plated SS wires with uncoated, Rh-, and polymeric-coated SS orthodontic archwires. While the three-point bending test revealed similar values of flexural strength and modulus among all wire types, frictional forces significantly decreased against ceramic brackets. Compared with uncoated, Rh-, and polymer-coated samples, both maximum static and kinetic, dry, and wet frictional forces decreased by approximately 15–18, 42–45, and 17–22%, respectively.

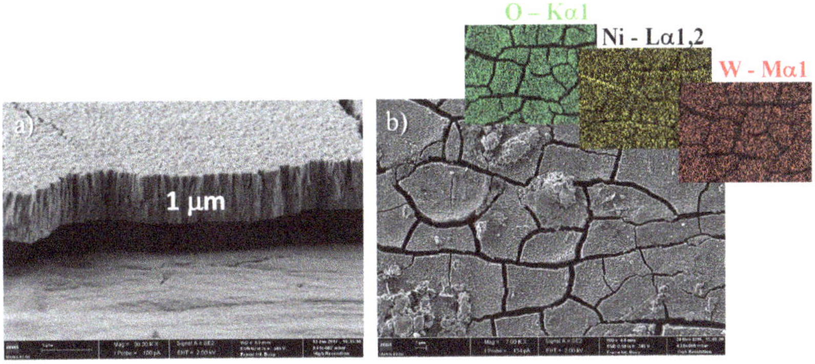

Figure 18. SEM micrographs of as-deposited W–N sputtered coatings (**a**) and its surface morphology after (**b**) Fusayama–Meyer solution aging (37 °C, 30 days, pH 8.3) coupled with EDS elemental maps distribution (adapted from [468], own work).

5.1.3. Metal Oxide-Based Nanoparticles Coatings

Metal oxide-based NPs have been used to improve the performance of biocomponent surfaces, particularly as antimicrobial agents for multiple biomedical applications [469]. According to the literature, the following types of metallic oxide NPs have been considered to cover metallic fixed appliances: ZnO- [367,437–443,445] and/or CuO-NPs [437,446,447], AlO-NPs [444], TiO_2-NPs [446–449], and Ag–HA-NPs [446,447], as presented in Table 9.

ZnO-NPs are novel dental materials with several potential applications in dentistry, including prosthodontic, endodontic, restorative, implantology, periodontal, and orthodontic fields [470]. As protective coatings, the research works concerning ZnO-NPs intend to reduce not only the bacterial adhesion and activity [367,437,440–442], but also the bracket/wire contact wear [439–441,443–445].

Excellent antibacterial results of 93, 96, and 98% against *S. mutans* have been achieved by Gholami et al. [442] with ZnO-NPs deposited by CVD, chemical precipitation and sol–gel methods, respectively, on the surface of NiTi archwires. Hammad et al. [441] also described excellent antibacterial activity for electrochemically deposited ZnO-NPs onto NiTi archwires against different bacterial strains (*S. aureus*, *S. pyogens* and *E. coli*). Other researchers [437] support the antibacterial effect of ZnO-NPs deposited by spray pyrolysis on SS brackets and by a chemical method on NiTi archwires [439]. Nonetheless, a mixture of CuO–ZnO–NPs seems to be more effective than a single ZnO-NPs for SS brackets, by using the spray pyrolysis method. Yet, the color of the surface unsuitably changed to copper (orange-red) [437].

Zeidan et al. [367] compared the antibacterial effect of ZnO- and/or Ag-NPs deposited on SS brackets by thermal evaporation. The researchers found the highest antibacterial effect against *S. mutans* and *L. acidophilus* for the ZnO/Ag-NPs coating. Interestingly, all coatings showed antibacterial effects that persisted for up to 3 months [367].

Concerning the mechanical properties, Elhelbawy and Ellaithy [443] deposited ZnO-NPs coatings via sol–gel on SS brackets and/or archwires, allowing a 64% decrease in the mean friction forces.Other researchers reported more modest reductions for coated NiTi archwires against uncoated SS brackets: ~34% [441,444], ~21% [440] (bracket/wire angulation up to 10°), and ~11% [445]. Palanivel et al. [444] also compared the friction forces of SS brackets against uncoated, ZnO-NPs-, or AlO-NPs-coated NiTi archwires, finding that the latter coating type showed an intermediate performance. ZnO-NPs-coated SS bands showed in vitro biocompatibility with HGF cells [364].

A different approach for reducing bacterial adhesion and subsequent enamel damage consisted of magnetron-sputtered TiO_2 nanoparticles (TiO_2-NPs) [448]. A thin layer (~81 nm) successfully prevented in vivo bacterial adhesion to NiTi archwires for up to 1 month. Coating physical stability was the main downside: Around 60% of the coating showed adhesion failures [448]. Ameli et al. [446,447] showed that TiO_2-NPs and Cu-NPs coatings on SS brackets deposited by a sol–gel method possessed antibacterial properties and could significantly reduce the friction against metallic archwires. Chaturvedi and colleagues [450] dip-coated β-Ti and NiTi archwires with TiO_2-NPs using different concentrations and dipping durations. The authors found an optimum NP concentration range of 1:2–1:6 for 48 h and a decrease in frictional forces for almost all coated wires [450].

5.1.4. Inorganic Fullerene-like Nanoparticles (IF-NPs)-Reinforced Metal-Based Coatings

Metal-based coatings containing inorganic fullerene-like nanoparticles (IF-NPs) are also listed in this review (Table 9). Round-shaped IF-NPS have been embedded into a metallic matrix to achieve composite coatings with self-lubricating properties. The wear mechanism is particularly relevant in bracket/archwires contacts in the salivary environment. Tungsten disulfide (WS_2) was the IF-NPs of choice for nickel (Ni) [451,452,455], nickel–phosphorus (Ni–P) [453,454] and cobalt (Co) [452,456] matrixes, probably due to its excellent dry bulk lubricity.

Samorodnitzky-Naveh et al. [456] reinforced Co coatings with WS_2 IF-NPs by using co-electrodeposition onto NiTi alloys. Successfully, the friction coefficient of the NiTi plates decreased by 66%—from 0.26 to 0.09 (as evaluated by pin-on-disk). Moreover, friction tests of the coated NiTi wires against SS brackets revealed minimum static and kinetic CoF of 0.080 and 0.061, respectively, for a maximum bracket/wire angulation of 5° [456].

Redlich and colleagues [451,453,454] selected Ni and Ni–P matrixes to study the WS_2 IF-NPs effect over SS archwires. Reduced friction forces were claimed, both under dry and wet conditions: 17 and 54% decrease for 0 and 10° bracket/wire angulation, respectively. The 3–5 μm thick Ni-P/[Ni-P + IF-NPs] coatings reduced CoF by half (to ~0.05) after a 50-cycle ball-on-flat test [454].

Recently, Gracco et al. [455] developed and compared molybdenum disulfide (MoS_2)—another solid lubricant—with WS_2 IF-NPs-containing Ni coatings (~20 and 15 μm thick, respectively) on SS orthodontic wires. The coatings were synthesized by electrochemical co-deposition. Overall, both coating types always decreased the in vitro frictional forces against two types of metallic brackets, under dry and wet conditions, for 0 and 5° bracket/wire angulation. The researchers [455] observed no significant damage for minimum bending of the composite Ni/IF-NPs-coated wires, but a complete failure was registered for a maximum bending of 5° angulation.

The use of IF-NPs-containing coatings is, therefore, promising, but further research on their behavior under simulated oral conditions is still required, mainly regarding the effect of sulphur (S) on surfaces' biocompatibility. Based on the reasons mentioned in the previous chapter, the selection of Ni-based matrices is quite undesirable in orthodontics.

Despite the large number of scientific works, further studies on durability, corrosion and wear resistance, and possible cytotoxicity to oral epithelial cells are still required for metals and their oxide, nitride, or carbide phase coatings and thin films.

5.2. Polymer-Based Coatings

Several polymeric coatings are clinically available for fixed orthodontics (Table 10). Polytetrafluoroethylene (PTFE) is, undoubtedly, the most studied thermoplastic polymer for orthodontic applications, followed by epoxy resins. Recently, research on polymeric composite materials has also gained importance, particularly those reinforced by ceramic NPs.

5.2.1. Thermoplastic and Thermoset Coatings

- Polytetrafluoroethylene (PTFE):

Polytetrafluoroethylene, also known as Teflon^{TM}, is widely used in multiple fields, including many medical applications [471,472] for its outstanding biocompatibility and inertness. In orthodontics, PTFE aesthetic coatings for archwires are commercially available due to their suitable white color, similar to the tooth's tone.

A recent study [473] reported that PTFE coatings onto SS, NiTi, and β-Ti wires *"exhibited low coloration, low microbial adhesion, low friction against metallic brackets, and high tolerance of detachment and wear against toothbrushing"*, especially if produced at a low temperature (200 °C) using a two-stage spraying technique. Other researchers agree that PTFE coatings onto metallic archwires are highly stable; significantly improve the corrosion resistance of the substrates in acidulated artificial saliva [330,474,475], in Ringer's solution [64], and in contact with certain food products [476,477]; maintain the surface morphology after cyclic mechanical loading of coated wires [330]; and decrease frictional forces against brackets [478,479]. PTFE coatings can also reduce biofilm adhesion to SS brackets. An in situ study [480] revealed that biofilm accumulation covered more than 22% of the uncoated brackets, in contrast to scarcely 4% of the PTFE-coated brackets. Still, the coated areas under high shear forces (e.g., bracket wings) partially degraded, exposing the metallic substrate.

Other reports, however, concluded that PTFE coatings directly affect load-deflection properties [333,481,482], surface roughness [333,481,483,484] and hardness [335] of NiTi archwires; undergo slight discoloration after 3 to 4 weeks of immersion in a staining solution [335,485]; and suffer from significant to severe coating detachment, both in vitro [486,487] and in vivo [373,488–490] (Figure 19). For instance, Rongo et al. [427] concluded that Teflon™-coated NiTi archwires had similar biocompatibility to uncoated ones. Nevertheless, the mechanical response was different: Friction forces against metallic and aesthetic brackets increased, besides suffering delamination during 1 month of clinical use [425]. Scratch tests performed by Silva et al. [491] with four commercially available archwires coated with a polymer mixture (polytetrafluoroethylene and polyester) revealed high elasticity recoveries of more than 60%, but this mechanical behavior was accompanied by *"delamination, crack propagation, and debris generation along the coatings scratches"*. Lin et al. [492] ascertained that prolonged water immersion (up to 4 weeks) of PTFE-coated archwires had negatively impacted the resistance to sliding against ceramic brackets. Thus, the functionality and durability of such coatings in the oral environment were questioned.

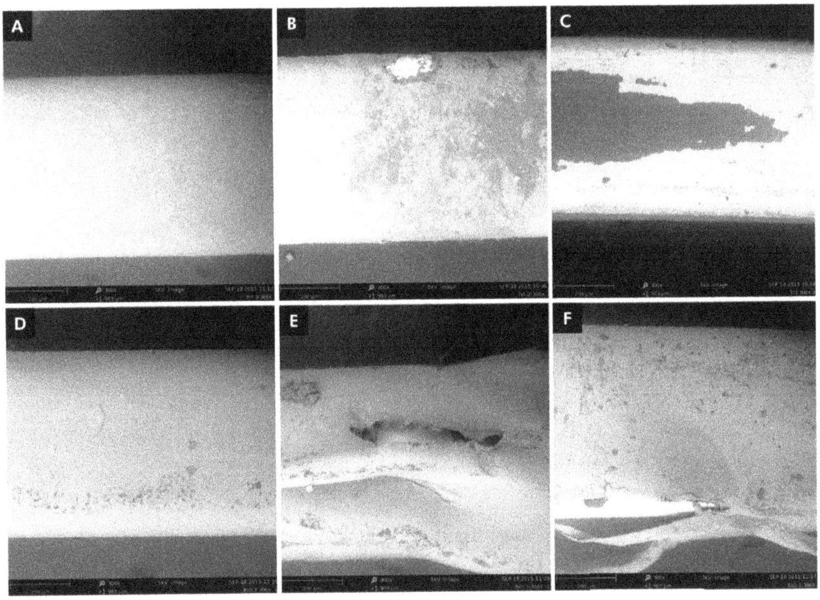

Figure 19. Photomicrographs of coated wires: (**A**) as-received Titanol® Cosmetic, (**B**,**C**) post clinical Titanol® Cosmetic, (**D**) as-received Biocosmetic, (**E**,**F**) post clinical Biocosmetic (reproduced from [488]).

- Epoxy Resins:

Epoxy resins (containing an epoxide group) are another widely used polymer in medicine [493], existing as commercially available aesthetic coatings for archwires with an appealing white color.

Some researchers concluded that epoxy-coated archwires reduced bacterial adhesion [63], prevented dental plaque accumulation [494], and possessed similar mechanical properties (e.g., unloading force) [336] and surface roughness [495] to uncoated wires. Back in 1999, Kim and Johnson [419] recommended the use of epoxy-coated NiTi archwires due to their superior corrosion resistance when compared with uncoated ones. Similarly, Amorim et al. [342] concluded that epoxy resin coatings effectively enhanced the corrosion resistance of NiTi archwires in artificial saliva, for instance, by decreasing Ni release. Nevertheless, several in vitro studies reported major flaws, namely: low coating stability, durability, and color stability [138,348,474,485,486,496,497], lower hardness [335], and loading/unloading forces [497,498], or increased roughness [333,412,481] and friction forces against brackets [499], which may limit the clinical performance.

Abdulkader et al. [474] performed a simple durability test: Coated and uncoated archwires were immersed in artificial saliva at 3.5 and 6.75 pH values for 28 days at 37 °C. After the immersion period, samples were washed with normal saline solution, fixed to brackets with elastomeric ligatures, and subjected to mechanical stress by toothbrushing (210 s). The results revealed *"rupture, roughness, and coatings damage in multiple locations"*. Under acidic conditions, 48% coating loss was reported (Figure 20), while for more neutral pH (6.75), this value decreased to 31%. Alavi and Hosseini [496] immersed coated and uncoated samples in a commercial artificial saliva (at pH 6.7 and 37 °C), refreshed every day. After 3 weeks, they performed 500 thermocycles—10 min at 5 °C, 10 s at room temperature, and 10 min at 55 °C—and then a three-aesthetic bracket bending test. Very recently, Aboalnaga and colleagues [500] tested three commercially available epoxy-coated NiTi archwires and found undesirable surface changes after a one-month immersion test in artificial saliva, including lower surface hardness and higher roughness values.

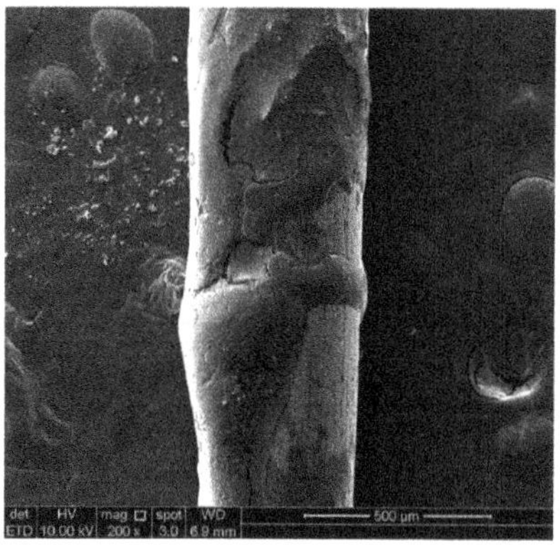

Figure 20. Epoxy coating loss on an archwire when exposed to acidic artificial saliva (reprinted from [474], Copyright (2020), with permission from Elsevier).

Elyayyan et al. [498] obtained lower loading and unloading forces of new coated wires when compared with uncoated wires in a three-point bending test. The same researchers [501] also argued that epoxy-coated NiTi wires *"had low aesthetic value"*, as *"25 % of the coatings was lost within 33 days* in vivo *and surface morphology showed severe deterioration"*. Abdulkhabeer et al. [489] found mean coating losses between 18.3 and 28.6% of epoxy-coated NiTi archwires after 8 weeks of orthodontic treatment. The color stability of these coatings has also been questioned, since an extremely marked change in color was noticed in a 21-day immersion test in coffee staining solutions [350,485].

Shao et al. [502] used a mixture of epoxy resin and PTFE suspension, in different proportions and with white and yellow dies, to dip-coat NiTi archwires. Coatings with the color of human teeth were obtained. The results from the acute toxicity and mucous membrane irritation tests showed no negative signs regarding in vivo biocompatibility [502].

- Polyether-ether-ketone (PEEK):

Polyether-ether-ketone resin is a cheap and non-cytotoxic advanced polymer that meets several chemical, tribological, and mechanical properties for orthodontics. To overcome the low PEEK adhesion to NiTi archwires, Sheiko et al. [503] proposed a new method to increase the adhesion of the polymeric coating. After cleaning the Nitinol archwires, they performed electrochemical polishing to create a thin and smooth TiO_2 interlayer, above which a homogeneous PEEK top layer (~12 μm) grew via dip-coating. The coatings strongly adhered to the metallic substrate without delamination after immersion in Hank's solution for up to 31 days, or 7 million compressing/stretching cycles at 20% strain. Regarding biocompatibility and corrosion properties, the coated archwires showed negligible Ni release during the immersion test and no cytotoxic effect on murine fibroblasts. However, local pressures above 2 GPa may disrupt the PEEK coating and unprotect the NiTi alloy surfaces against corrosion reactions [503].

More recently, Shirakawa et al. [504] followed a different method to cover NiTi archwires with PEEK resin: A simple outer polymeric tube around the arches. Friction tests revealed lower friction forces against brackets when compared to uncoated SS-, CoCr, and NiTi archwires. Since coated wires were able to slide more easily in contact, bracket slots maintained the as-received surface conditions [504].

Table 10. Overview of polymer-based coatings deposited on different orthodontic appliances and substrates. PECVD: plasma-enhanced chemical vapor deposition; n.s.—not specified.

Coatings	Substrates	Deposition Method	Literature References
Thermoplastic and Thermoset Coatings			
PTFE	NiTi and SS archwires	*Commercial*	[330] Neumann et al. (2002)
PTFE	NiTi and SS archwires	*Commercial*	[479] Husman et al. (2002)
PTFE	SS brackets	*n.s.*	[480] Demling et al. (2010)
PTFE	SS and NiTi archwires	*Commercial*	[478] Farronato et al. (2012)
PTFE	NiTi archwire	*Commercial*	[490] Zegan et al. (2012)
PTFE	NiTi archwire	*Commercial*	[64] Krishnan et al. (2014)
PTFE	NiTi archwires	*Commercial*	[425] Rongo et al. (2014)
PTFE	NiTi archwires	*Commercial*	[475] Mareci et al. (2015)
PTFE	NiTi archwires	*Commercial*	[481] Ryu et al. (2015)
PTFE	NiTi archwires	*Commercial*	[484] Choi et al. (2015)
PTFE	NiTi archwires	*Commercial*	[476] Earar et al. (2016)
PTFE	NiTi archwires	*Commercial*	[427] Rongo et al. (2016)
PTFE	NiTi archwires	*Commercial*	[333] Albuquerque et al. (2017)
PTFE	NiTi archwires	*Commercial*	[505] Rego et al. (2017)
PTFE	NiTi archwires	*Commercial*	[477] Matei et al. (2016)
PTFE	NiTi archwires	*Commercial*	[373] Shirakawa et al. (2017)
PTFE	NiTi archwires	*Commercial*	[485] Rego et al. (2017)
PTFE	NiTi archwires	*Commercial*	[310] Matias et al. (2018)
PTFE	NiTi archwires	*Commercial*	[411] Asiry et al. (2018)
PTFE	NiTi archwires	*Commercial*	[483] Dokku et al. (2018)
PTFE	SS archwires	*Commercial*	[487] Shahabi et al. (2018)

Table 10. *Cont.*

Coatings	Substrates	Deposition Method	Literature References
PTFE	NiTi archwires	*Commercial*	[335] Alsanea and Shehri (2019)
PTFE	NiTi archwires	*Commercial*	[347] Costa Lima (2019)
PTFE	SS, NiTi and β-Ti archwires	Thermal spraying	[473] Kameda et al. (2020)
PTFE	NiTi archwires	*Commercial*	[474] Abdulkader et al. (2020)
PTFE	NiTi archwires	*Commercial*	[489] Abdulkhabeer et al. (2020)
PTFE	NiTi archwires	*Commercial*	[486] Jejurikar et al. (2020)
PTFE	NiTi archwires	*Commercial*	[337] Batista et al. (2020)
PTFE	NiTi archwires	*Commercial*	[482] Elsaka et al. (2021)
PTFE	SS archwires	*Commercial*	[492] Lin et al. (2021)
PTFE	β-Ti and SS archwires	Spray treatment	[506] Zhou et al. (2023)
Epoxy	NiTi arcwhires	*Commercial*	[419] Kim and Johnson (1999)
Epoxy	NiTi archwires	*Commercial*	[501] Elayyan et al. (2008)
Epoxy	NiTi archwires	*Commercial*	[498] Elayyan et al. (2010)
Epoxy	NiTi archwires	Electrostatic powder deposition	[138] Bandeira et al. (2011)
Epoxy	NiTi archwires	*Commercial*	[496] Alavi et al. (2012)
Epoxy	NiTi archwires	*Commercial*	[494] Raji et al. (2014)
Epoxy	NiTi archwires	*Commercial*	[64] Krishnan et al. (2014)
Epoxy	NiTi archwires	*Commercial*	[63] Kim et al. (2014)
Epoxy	NiTi archwires	*Commercial*	[497] Pop et al. (2015)
Epoxy	NiTi archwires	*Commercial*	[484] Choi et al. (2015)
Epoxy	NiTi archwires	*Commercial*	[333] Albuquerque et al. (2017)
Epoxy	NiTi archwires	*Commercial*	[505] Rego et al. (2017)
Epoxy	SS archwires	*Commercial*	[374] Usui et al. (2017)
Epoxy	NiTi archwires	*Commercial*	[485] Rego et al. (2017)
Epoxy	SS archwires	*Commercial*	[348] Usui et al. (2018)
Epoxy	NiTi archwires	*Commercial*	[310] Matias et al. (2018)
Epoxy	NiTi archwires	*Commercial*	[411] Asiry et al. (2018)
Epoxy	NiTi archwires	*Commercial*	[483] Dokku et al. (2018)
Epoxy	NiTi archwires	*Commercial*	[335] Alsanea and Shehri (2019)
Epoxy	NiTi archwires	*Commercial*	[499] Dragomirescu et al. (2019)
Epoxy	NiTi archwires	*Commercial*	[495] Shamohammadi et al. (2019)
Epoxy	NiTi archwires	*Commercial*	[474] Abdulkader et al. (2020)
Epoxy	NiTi archwires	*Commercial*	[336] Pinzan-Vercelino et al. (2020)
Epoxy	NiTi archwires	*Commercial*	[489] Abdulkhabeer et al. (2020)
Epoxy	NiTi archwires	*Commercial*	[486] Jejurikar et al. (2020)
Epoxy	NiTi archwires	*Commercial*	[350] Ramasamy et al. (2020)
Epoxy	NiTi archwires	*Commercial*	[412] Madasamy et al. (2021)
Epoxy	SS archwires	*Commercial*	[492] Lin et al. (2021)
Epoxy	NiTi archwires	*Commercial*	[342] Amorim et al. (2022)
Epoxy	β-Ti and SS archwires	Spray treatment	[506] Zhou et al. (2023)
Epoxy	NiTi archwires	*Commercial*	[500] Aboalnaga et al. (2023)
PEEK	Nitinol wires	Dip-coating deposition	[503] Sheiko et al. (2016)
PEEK tubes	SS, NiTi, and CoCr archwires	Tube coverage	[504] Shirakawa et al. (2018)
PE	NiTi archwires	*Commercial*	[330] Neumann et al. (2002)

Table 10. *Cont.*

Coatings	Substrates	Deposition Method	Literature References
PE	NiTi archwires	*Commercial*	[479] Husman et al. (2002)
PEN (PE naphthalate)	SS archwires	*Commercial*	[61] Ito et al. (2022)
Polyamide	NiTi archwires	Dipping treatment	[507] Bravo et al. (2014)
Epoxy + PTFE	NiTi archwires	Dip-coating deposition	[502] Shao et al. (2009)
PTFE + polyester	SS and NiTi archwires	*Commercial*	[491] da Silva et al. (2015)
Other Polymer Coatings			
Lysozyme	Composite archwires	Coating protein deposition	[318] He et al. (2020)
Hexamethyldisiloxane (HMDSO)	SS brackets	PECVD	[508] Tupinambá et al. (2017)
Organosilane	SS brackets	Sol–gel method	[509] Oliveria et al. (2015)
2-methacryloyloxyethyl phosphorylcholine	SS archwires	Chemical deposition	[510] Kunimatsu et al. (2022)
Chitosan	Brackets *(n.s.)*	Freeze-drying	[511] Want et al. (2023)
Chitosan nanoparticles	SS brackets and archwires	Sol–gel method	[443] Elhelbawy and Ellaithy (2021)
Ag–chitosan nanoparticles	Metallic brackets	Laser ablation	[413] Tawakal et al. (2023)
Polyoxazoline + tryptophan	SS brackets	Plasma polymerization/Immobilization	[512] Kumarasinghe et al. (2021)
1H,1H,2H,2H-perfluorodecyltrimethoxysilane(FAS) + bovine serum albumin (BSA)	SS brackets and archwires	Chemical deposition	[513] Liu et al. (2018)
PVA hydrogel	SS archwires	Chemical deposition	[514] MingWen et al. (2023)
Polydopamine + honokiol C-dots	SS brackets	Hydrothermal method/chemical deposition	[515] Wang et al. (2023)
Butyl-3-methylimidazolium chloride	NiTi archwires	Ionic liquid coating	[516] Ahmed et al. (2021)
Parylene	SS archwires	*Commercial*	[492] Lin et al. (2021)
Parylene with Ag–Pt layer	NiTi archwires	*Commercial*	[334] Iijima et al. (2012)
Parylene with Ag–Pt layer	NiTi archwires	*Commercial*	[484] Choi et al. (2015)
Ag/biopolymer bilayer	NiTi archwires	*Commercial*	[63] Kim et al. (2014)
Ag/polymer bilayer	NiTi archwires	*Commercial*	[482] Elsaka et al. (2021)
Silicone	β-Ti and SS archwires	Dip coating	[506] Zhou et al. (2023)
Ceramic-reinforced epoxy composite	β-Ti and SS archwires	Spray deposition	[506] Zhou et al. (2023)
Polymer (?)	NiTi archwires	*Commercial*	[496] Alavi et al. (2012)
Polymer (?)	NiTi archwires	*Commercial*	[425] Rongo et al. (2014)
Polymer (?)	NiTi archwires	*Commercial*	[427] Rongo et al. (2016)
Polymer (?)	NiTi archwires	*Commercial*	[488] Argalji et al. (2017)
Polymer (?)	NiTi archwires	*Commercial*	[499] Dragomirescu et al. (2019)
Polymer (?)	NiTi archwires	*Commercial*	[495] Shamohammadi et al. (2019)
Polymer (?)	NiTi archwires	*Commercial*	[482] Elsaka et al. (2021)
NPs reinforced Polymer Coatings			
TiO_2 reinforced Epoxy	NiTi archwires	Electrophoretic deposition	[517] Xu et al. (2019)
ZnO reinforced PVP	NiTi archwires	Electrospinning	[442] Gholami et al. (2021)
ZnO reinforced PVA	NiTi archwires	Polymer composite coating	[442] Gholami et al. (2021)

- Polyethylene (PE) and Polyethylene Naphthalate (PEN):

Polyethylene is a commercially available coating for NiTi archwires [330,479]. Husman et al. [479] indicated that the plasma-based synthesis process starts by removing micro-contaminants on the wire surface through high-energy oxygen radicals, followed by the formation of titanium oxide, and finally the deposition of polyethylene. The surface is smooth due to the filling of its depressions [330,479]. Commercial PE-coated archwires showed an apparently superior in vitro corrosion resistance than the uncoated ones, even though the rupture potential has been lowered. This may indicate different corrosion processes. Moreover, PE-coated NiTi archwires showed no changes after cyclic mechanical loading [330] and reduced the friction forces against SS brackets [479]. Ito et al. [61] recommended the use of polyethylene naphthalate—also a commercial coating—for patients with metal allergies and when wire bending is required due to the in vitro mechanical stability of coated SS archwires. Even so, only toothbrushes with soft filaments should be used for oral hygiene procedures to prevent damaging the coating and, consequently, leading to corrosion of the metallic substrate [61].

- Polyamides:

Polyamides are additional potential polymeric coatings for NiTi superelastic archwires. According to Bravo et al. [507], the dip-coated wires showed 85% less Ni release than uncoated samples after the 30-day electrochemical corrosion tests. During self-designed friction tests against SS- and Ti6Al4V-brackets, wear rates and both static and dynamic CoF were always inferior in comparison to other alloy archwires types (NiTi, TiMo, commercially pure Ti, Cu–NiTi or SS 304 AISI). In addition, both coated and uncoated wires had very similar superelastic behavior [507].

- Other Polymer Coatings:

Lysozyme coatings were used by He et al. [318] to avoid Cu corrosion from laser-welded NiTi + SS CAW. These coatings, synthesized via liquid-phase deposition, presented antimicrobial properties. *S. aureus* activity decreased from 97 down to 59%, with increased electrochemical corrosion resistance, while cytocompatibility with fibroblast cells was higher for coated samples (79 vs. 83–90%).

Hexamethyldisilane (HMDSO) films were also deposited on conventional and self-ligating SS brackets by PECVD [508]. The plasma-polymerized resulting coatings were only effective in reducing surface roughness and bacterial adhesion in conventional brackets. Their uncoated, smoother surface and more favorable external geometry led to a better film deposition than in self-ligating brackets.

Organosilane-based coatings with super-hydrophobic properties were deposited via sol–gel on SS and ceramic brackets by Oliveira and colleagues [509]. A water contact angle of ~123° was reported, in addition to an exponential decrease in in vitro biofilm accumulation with increasing water contact angle of the surface up to 24 h of incubation (Figure 21).

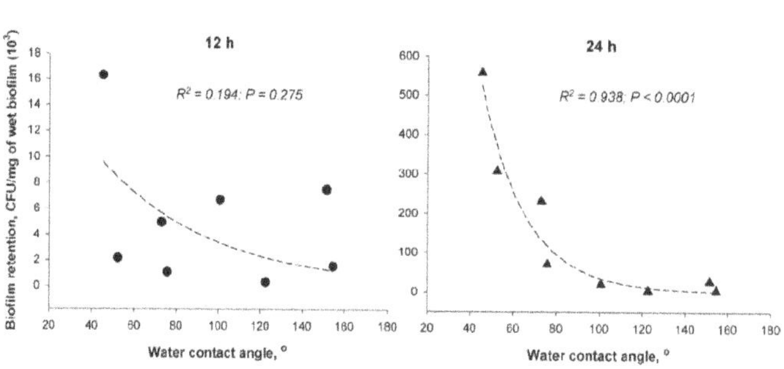

Figure 21. Biofilm retention after 12 and 24 h of incubation as a function of the water contact angle on organosilane coated brackets (used with permission of IOP Publishing, Ltd., from [509]; permission conveyed through Copyright Clearence Center, Inc.).

Chitosan—an antibacterial carbohydrate polymer—was studied as a coating for orthodontic brackets. Non-crosslinked chitosan was deposited by freeze-drying from previously dissolved chitosan in 2% (w/v) acetic acid at different concentrations (1.0, 2.5, 5.0, and

10 mg/mL). Promising results were obtained, as *S. mutans* bacterial adhesion decreased by up to 99.6% for coated brackets [511].

Chitosan-NP materials were also studied as antibacterial surfaces for orthodontics. The deposition process was performed by sol–gel thin film dip coating. A decrease in friction of SS brackets and archwires against coated and/or uncoated counterparts [443] was accessed by the researchers. A significant reduction in friction forces and *S. aureus* activity was reported as well. Ag–chitosan NPs deposited on metallic brackets showed outstanding antibacterial activity against that bacterial strain [413].

MingWen et al. [514] quite recently deposited a toughly-adherent polyvinyl alcohol (PVA) hydrogel onto SS archwires. The authors reported a remarkable decrease in friction in water (CoF as low as 0.005) and a significant decrease in *E. coli* and *S. aureus* adhesion on coated samples.

Further attempts regarding polymer-based coatings deposited onto fixed orthodontic metallic components include: **(i) 2-methacryloyloxyethyl phosphorylcholine**: an hydrophilic and antibacterial phospholipid polymer with low friction property, which reduced the CoF of coated SS wires against brackets, and inhibited the in vitro adhesion of *S. mutans* and *P. aeruginosa* [510]; **(ii) butyl-3-methylimidazolium chloride**, which reduced the in vitro corrosion of NiTi archwires in fluoride- and/or bovine albumin-containing artificial saliva; **(iii) 1H,1H,2H,2H-perfluorodecyltrimethoxysilane** with bovine serum albumin, showing a non-bactericidal but highly antibacterial effect by preventing adhesion through a protein-mediated mechanism [513]; **(iv) polydopamine and carbon dots:** a fluorescent and antibacterial coating, with in vitro cytocompatibility, deposited on SS brackets [515]; **(v) polyoxazoline with tryptophan** coating, which decreased metal release and showed good in vitro biocompatibility to primary human dermal fibroblasts [512]; and **(vi) silicone** coatings on β-Ti and SS archwires [506].

5.2.2. Polymeric-Based Composite Coatings

Parylene polymer matrix reinforced with Ag and Pt is a commercially available coating on SS archwires [334,484,492]. The surface of the coated wires is typically rougher and softer, even though the bulk maintained its mechanical properties when compared with the uncoated ones [334].

Epoxy matrix reinforced with nanoscale TiO_2 particles (~200 nm) was studied by Xu et al. [517] on NiTi sheets and round wires. The researchers concluded that the tooth-like-colored coatings were thinner (~22 μm) and presented better surface quality (smooth, defect-free, and uniform) than commercially available aesthetic epoxy coatings (~50 μm). Stress–strain curves of both coated and uncoated archwires, from loading-unloading tensile tests at 8% pre-strain, almost overlapped, indicating that superelasticity and shape-memory properties remained unchanged. After a 30-day immersion test in Fusayama-Meyer artificial saliva, both coating microhardness and adhesion slightly decreased (yet statistically non-significant), while Ni release was effectively blocked without corrosion signs. Moreover, no cytotoxicity in MG-63 cells was detected. Results are, therefore, optimistic, even though further variables must be added to the experiment to better simulate the oral environment; for instance, a more corrosive medium, bracket/wire contact, and microbiological attack [517]. Zhou et al. spray-coated β-Ti and SS archwires with commercial ceramic-reinforced epoxy coatings [488].

Recently, Gholami et al. [442] successfully reinforced polymeric coatings with ZnO-NPs: (i) ZnO-NPs-containing polyvinyl pyrrolidone (PVP) by electrospinning; and (ii) ZnO-NPs-containing PVA by a polymer composite coating method. However, such coatings displayed less antibacterial efficacy than simple ZnO-NPs coatings (non-composites) even though all surfaces were antibacterial: the reduction percentages regarding *S. mutans* viability with non-composite ZnO-NPs coatings ranged between ~93 and 99%, whereas ZnO-NPs-containing PVP and PVA composite films showed values of ~72 and 90%, respectively [442].

A more complex polymeric-based composite coating system, which consisted of a 3-layered structure covered with Ag nanoparticles onto a flat SS substrate, was developed by Lee et al. [518] and is worth mentioning. The studied architecture can be seen in Figure 22. While the two poly(3,4-ethylenedioxythiophene)-based inner layers enhanced adhesion to the substrates, the outmost dopamine layer decreased bacterial adhesion for both *S. mutans* and *E. coli*. Adsorbed silver (Ag^+ and Ag-NPs) conferred antifouling and antibacterial properties without releasing toxic amounts of Ag species or compromising the in vitro cell viability of human gingival fibroblasts [518].

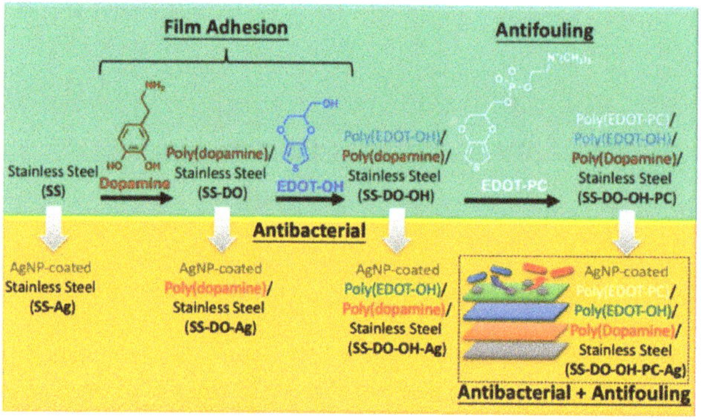

Figure 22. A complex polymeric-based composite deposited onto an SS flat substrate to confer the surface with antibacterial and antifouling properties (reprinted with permission from [518]. Copyright (2020) American Chemical Society).

5.3. Ceramic-Based Coatings

The authors chose to include within this section the studies covering calcium (Ca) and phosphorus (P)-based coatings, comprehending hydroxyapatite (HA) materials, and bioactive glass (Table 11). An overview of covalent C-based materials is also presented, including hydrogen-free amorphous carbon nitride (a-CN_x) and diamond-like carbon (DLC) coatings, as well as silicon-based coatings (SiN, SiC, and SiO_2-NPs).

5.3.1. Hydroxyapatite (HA)

Hydroxyapatite (HA), that is, pentacalcium hydroxyl apatite [$Ca_{10}(PO_4)_6(OH)_2$], is a mineral frequently used in many biomedical applications, namely bone-related conditions [519], due to its excellent biocompatibility and osteoconductive properties. Moreover, HA is now commercially available in oral care products [520], and some attempts were made to use it as a surface material for orthodontic components.

Jiang et al. [521] coated porous SMA–Nitinol alloy with HA by the SHS method, doing a 5-day immersion in Hank's solution after NH_3 and NaOH aqueous solution treatments. The resulting bone-like HA layer reduced, by one order of magnitude, the Ni release after a 50-day body fluid immersion test. Kocijan et al. [522] confirmed the barrier property of the HA coatings: The electrodeposited layer effectively enhanced corrosion resistance in Hank's solution of NiTi samples. Even a thin amorphous calcium phosphate layer (~600 nm thick) deposited by r.f. magnetron sputtering reduced nickel release from NiTi plates [523].

Unfortunately, only two studies were found using real orthodontic components. Setiyorini and Pintowantoro [414] synthesized biomimetic HA on NiTi wires by electrodeposition, revealing the highest cell viability percentage compared to other studied coatings. Dimasruhin et al. [524] deposited a SiO_2-containing HA coating via electrodeposition. The researchers successfully coated NiTi archwires but only performed a morphological characterization.

The evident lack of studies dealing with metallic orthodontic components (brackets, archwires) is probably related to the ceramic-brittle nature of HA.

5.3.2. Bioactive Glasses

Bioactive glasses have also been studied by Kawaguchi et al. by deposition on both SS disks [525] and archwires [526] through an electrophoretic process under 10 or 15 V (direct or alternating current: d.c. or a.c., respectively). Coatings (1–4 μm thick) onto SS disks showed promising aesthetics, non-cytotoxicity, and remineralization ability up to 3 months of acid etching [525]. Archwires coated at 10 V demonstrated significantly higher friction against SS brackets up to 10° of bracket/archwire angulation [526]. This can be due to the lower coating hardness when compared with the metallic surface (<0.5 vs. 6.11 GPa, respectively). On the other hand, coatings deposited at 15 V, with a hardness value of ~2 GPa, provided similar friction as unmodified wires against SS brackets. The main downside reported was the disruption/detachment behavior in the bracket/archwire interface during the friction tests [526].

5.3.3. Silicon-Based Coatings

SiN and SiC-based coatings were deposited on SS brackets and archwires by Rapiejko et al. [527] in an attempt to reduce both friction and wear during the sliding of those orthodontic components. Those authors designed an ex-situ fretting test to allow micro-sliding (simulating the occlusion movements) to superimpose on macro-sliding (which in turn simulates the global displacement of the teeth), both in dry and wet (artificial saliva) environments. The lowest friction and the highest wear were found for the uncoated SS bracket and archwire contact, whereas the opposite was true for the coated pairs. The researchers concluded that due to such a higher CoF (0.50 vs. 0.20 for coated and uncoated pairs, respectively, in artificial saliva), and despite the lower wear, the coatings are unsuitable for medical applications [527]. Silveira et al. [449] also reported a decrease in surface roughness and friction forces in both dry and artificial saliva environments when SS orthodontic wires were coated with an SiO_2-NPs-based film, when compared with uncoated ones, against brackets.

5.3.4. Carbon-Based Coatings

- C and C–H Systems:

Single C and binary C–H systems are very broad groups of materials, encompassing the non-hydrogenated and hydrogenated amorphous carbon—a-C and a-C:H, respectively. Usually, these materials are ambiguously recognized as diamond-like carbon (DLC) in the literature [528,529], despite their variable C=C sp^2 (similar to graphite) and C-C sp^3 (similar to diamond) hybridization bondsand H content. DLC coatings are widely used in the medical field [530–533]. The major advantages of DLC coatings are their outstanding overall biocompatibility, corrosion resistance, achievable high hardness, and low CoF. For further knowledge on the DLC materials, it is recommended to read the well-known review by J. Robertson [534]. To the authors' best knowledge, there is yet no commercial application in fixed orthodontics, and the first attempts to use DLC coatings on real orthodontic components date back to 2005 (Table 11). Studies focused on coating brackets and archwires with DLC are hereafter described.

Kobayashi et al. [535,536] were able to reduce the in vitro corrosion of NiTi archwires. The researchers successfully deposited a DLC coating (presumably a-C:H) with an SiC interlayer by ion beam plating with benzene as the reactive gas. The coating was stable for a 24 h continuous mechanical brushing and significantly decreased the Ni release in physiological saline solution up to 14 days (short-term, 80 °C) and 6 months (long-term, 37 °C) when compared with uncoated NiTi archwires. In all, the DLC coating showed excellent adhesion and mechanical properties to mechanical brushing and prevented corrosion [535]. In two quite similar reports [537,538], the ~1 μm thick a-C:H coatings—deposited by an arc discharge ion plating method with benzene—reduced Ni release from NiTi archwires by 80% in a 5-day static immersion test (physiological saline solution, 85 °C).

Simulating basic intraoral conditions is essential, but the corrosion susceptibility of orthodontic alloys increases in fluoride-acidic environments [29,42,95,104,276,277]—and daily-used mouthwashes and toothpastes include fluoride prophylactic agents in their compositions. Accordingly, the corrosion protection ability of a 100 nm thick DLC (likely a-C) film deposited by mirror-confinement-type electron cyclotron resonance (MCECR) plasma sputtering on NiTi archwires was studied against a NaF-based anticavity dental rinse (pH 4) [539]. All samples were statically immersed in an artificial saliva for 12 weeks at 37 °C, and additionally dipped for 5 min in a mouthwash (0.044% sodium fluoride, in acidulated phosphate solution) three times a day. While the surface roughness of uncoated samples increased by ~50% due to F-induced dimples and cracks, the coatings protected the substrate against F-induced corrosion, with only a slight increase in surface roughness (~8%) and no significant morphological alterations [539].

To reduce resistance to sliding, some researchers deposited a-C:H coatings onto orthodontic brackets [423,539–542] and archwires [535,537–539,542–547], or even both components [548]. Deposition techniques included Plasma-based ion implantation/deposition (PBIID) [540,542,544,547], MCECR plasma sputtering [539,548], arch-discharge ion plating [535,537,538,545], and PECVD [541,543,546]. To promote and evaluate bracket/wire sliding properties, such as frictional forces and friction coefficient, custom-made devices are usually selected.

Tantiwinyupong et al. [547] found significantly different static frictional forces between conventional and DLC-coated NiTi archwires against brackets (dry, room temperature 25 °C), representing a decrease of ~33%. The PBIID-deposited a-C:H coating also improved surface hardness from 1.06 to 11.44 GPa. Similar results were reported by Muguruma et al. [542] for NiTi and SS wires coated by PBIID against uncoated conventional or self-ligating brackets, up to a 10° angulation at room temperature and in a dry environment. The 500 nm thick a-C:H layer increased surface hardness from 11.6 to 17.6, and from 4.7 to 9.1 in the case of SS and NiTi, respectively. Thicker a-C:H coatings (~5–7 μm) produced by the same technique on SS brackets can also reduce both friction forces against SS brackets under dry and wet conditions [540].

Zhang et al. [546] produced a-C:H coatings (~1 μm) on SS wires by PECVD, obtaining a DLC surface smoother and 1.46 times harder than the uncoated wire. However, no statistically significant differences between the static friction coefficient or maximum static friction force were observed under dry conditions—only the kinetic friction coefficient decreased by ~40%. Kang and colleagues [548] also reported a decrease in the kinetic friction between SS wires and brackets. They tested MCECR plasma sputtering-produced, 200 nm thick a-C:H coatings, obtaining a reduction in the kinetic friction coefficient by ~80 and 70% under ambient air and artificial saliva, respectively, when at least the wire was coated. The lowest value of CoF = 0.11 was reported for both components under ambient air. Although kinetic friction seems irrelevant in orthodontics [215], the authors argued that their test involved oscillations with small displacements (±150 μm, 0.5 Hz), which recreated the discontinuous clinical motion (5 mm of distance, 10 mm/min) [548], in opposition to others [423,540,542].

Danisman et al. [549] deposited an 83 nm thick DLC top-coating above a Ti/TiN coating (36/687 nm) film on SS brackets by closed-field magnetron sputtering and measured the static friction force against SS archwires in air (5 mm of distance, 10 mm/min). The researchers reported a 28–39% decrease in the friction forces when coatings were used in comparison with uncoated brackets [549].

S. Huang et al. [539] coated NiTi archwires with a thin 100 nm a-C:H layer by PECVD. Surface roughness did not significantly change, but bracket/wire friction coefficients diminished up to 79.7% and 70.0% in ambient air and artificial saliva, respectively, when compared with uncoated tribo-couples. In a static immersion test simulating fluoride mouth rinse, surface roughness variations—caused by fluoride-induced corrosion—decreased by 91.3%, demonstrating the anti-corrosion behavior of amorphous carbon films [539].

The general tribological improvement by using a-C:H coatings may be related to the expected increase in surface hardness, which leads to a lower wear rate and, therefore, lower frictional forces [542]. Higher surface hardness is advantageous for orthodontics, as it may minimize binding or notching effects [542,547]. Muguruma et al. [544] pointed out the role of H on the a-C:H surface, suggesting that DLC coatings with high H content (>30 at.%) are potential candidates for this application.

Surface wettability may also play another important role in reducing static friction [541,543]. To further explore this idea, Akaike et al. [541] deposited unmodified and F- and Si-modified a-C:H coatings on SS bracket slots by PECVD. Under wet (PBS-sprayed) conditions, all coatings decreased static friction when compared with uncoated bracket slots, up to a bracket/wire angulation of 10°. However, a-C:H:F coatings showed the lowest static friction. This may be due to the higher hydrophobicity of such coating, reflected by the highest PBS and distilled water contact angles (~80 and 90°, respectively) when compared to the SS surface (~60°) and even to the a-C:H film (~70 and 75°). The a-C:H:Si films were more hydrophilic than the other two DLC-based coatings (contact angles of ~35°), and therefore the decrease in static friction was lower [541].

Mechanical brushing with fluorine-containing toothpaste is an indispensable daily hygiene procedure, especially for patients undergoing fixed orthodontic treatment [29,91,108]. Therefore, Oghoe, Kobayashi et al. studied the stability of 1 μm thick a-C:H coatings deposited on orthodontic NiTi archwires by arch-discharge ion plating with [535] and without [537,545] a SiC interlayer. The researchers simulated a 6-month daily procedure by mechanically brushing the coated wires with equipment carrying a toothbrush for 250 min, 350 min, and 24 h at a constant load of 35 g. The coatings protected the substrate that corroded when uncoated; no peeling, cracks, or even microstructural changes were detected, indicating good coatings' adherence and stability against brushing hygiene procedures [535,537,545].

Designing coatings with antibacterial properties—for instance, by reducing bacterial adhesion—can minimize the formation of biofilm layers [550]. Currently, the eventual antibacterial properties of DLC coatings are non-consensual, even though the introduction of antibacterial agents is possible [551]. Furthermore, and to the authors' best knowledge, little investigation has been conducted using amorphous carbon coatings specifically for orthodontics [550,552], and no study was found assessing the antibacterial properties of DLC-coated fixed orthodontic components. Further research is required to assess and enhance the antibacterial properties for orthodontic applications.

- C–N System:

Carbon nitride (CN_x) coatings are well-recognized biocompatible and chemically stable surface materials with attractive mechanical and tribological properties for biomedical applications. Following previous research [553], Wei et al. [554] coated SS orthodontic archwires with CN_x by ion beam assisted deposition (IBAD). Published results show lower and more stable frictional forces and CoF under both ambient air and artificial saliva, especially at higher bracket/wire angulations (up to 15°). Compared with bare or TiN-coated SS304L disks, those IBAD-produced CN_x thin films (<500 nm) presented the best surface properties, that is, a lower and more stable coefficient of friction (CoF < 0.2), biocompatibility with human fibroblasts and effective antibacterial properties against *S. mutans* [459].

Table 11. Overview of ceramic-based coatings deposited on different orthodontic appliances and substrates. CVD: chemical vapor deposition; PECVD: plasma-enhanced CVD; PBIID: plasma-based ion implantation and deposition; IBAD: ion beam-assisted deposition; MCECR: mirror-confinement-type electron cyclotron resonance.

Coatings	Substrate Materials	Deposition Methods	Literature References
Hydroxyapatite and Bioactive Glass Coatings			
Bioactive glass	SS archwires	Electrophoretic deposition	[526] Kawaguchi et al. (2020)
SiO_2-reinforced HA	NiTi archwires	Electrodeposition	[524] Dimasruhin et al. (2014)
HA	NiTi archwires	Electrodeposition	[414] Satiyorini and Pintowantoro (2013)

Table 11. *Cont.*

Coatings	Substrate Materials	Deposition Methods	Literature References
Silicon-based Coatings			
SiN	SS brackets and archwires	PECVD	[527] Rapiejko et al. (2009)
SiC	SS brackets	PECVD	[527] Rapiejko et al. (2009)
SiO_2-NPs	SS archwires	Commercial ceramic paint	[449] Silveira et al. (2022)
Carbon-based Coatings			
DLC (a-C:H)	NiTi archwires	Ion beam plating	[535] Kobayashi et al. (2005)
DLC (a-C:H)	NiTi archwires	CVD	[536] Kobayashi et al. (2005)
DLC (a-C:H)	NiTi archwires	Arch discharge ion plating	[538] Ohgoe et al. (2006)
DLC (a-C:H)	NiTi archwires	Arch discharge ion plating	[537] Ohgoe et al. (2007)
DLC (a-C:H)	NiTi archwires	Arch discharge ion plating	[545] Kobayashi et al. (2007)
DLC (a-C:H)	SS brackets	PECVD	[423] Huang et a. (2010)
DLC (a-C:H)	NiTi and SS archwires	PBIID	[542] Muguruma et al. (2011)
DLC (a-C:H)	SS brackets	PBIID	[540] Muguruma et al. (2013)
DLC (a-C:H)	NiTi archwires	PECVD	[539] Huag et al. (2013)
DLC (a-C)	SS archwires	MCECR plasma sputtering	[548] Kang et al. (2015)
DLC (a-C:H)	SS backet slots	PECVD	[543] Akaike et al. (2015)
DLC (a-C:H)	SS bracket slots	PECVD	[541] Akaike et al. (2016)
DLC-F (a-C:H:F)	SS bracket slots	PECVD	[541] Akaike et al. (2016)
DLC-Si (a-C:H:Si)	SS bracket slots	PECVD	[541] Akaike et al. (2016)
DLC (a-C:H)	SS archwires	PECVD	[546] Zhang et al. (2016)
DLC (a-C:H)	SS archwires	PBIID	[544] Muguruma et al. (2018)
DLC (a-C:H)	NiTi archwires	PBIID	[547] Tantiwinyupong et al. (2019)
DLC (a-C)	SS brackets	Magnetron sputtering	[549] Danisman et al. (2021)
CNx	SS archwires	IBAD	[554] Wei et al. (2011)
Carbon-based Nanocomposite Coatings			
Graphene sheets embedded in carbon	SS archwires	MCECR plasma sputtering	[555] Pan et al. (2022)
Graphene sheets embedded in carbon	SS archwires	MCECR plasma sputtering	[556] Wang et al. (2022)

- Carbon-based Nanocomposite Coatings:

Graphene sheets embedded in carbon (GSEC) matrixes are a recently deposited material on SS archwires by Pan and coauthors [555,556] via MCECR plasma sputtering. The aim was to decrease the CoF against SS brackets in wet conditions (Fusayama–Meyer artificial saliva). The researchers reported the lowest CoF and wear rate of 0.05 and 0.11×10^{-6} mm^3/Nm when sliding against three-row micro-groove-textured SS brackets (10,000 times, 150 μm displacement stroke, 0.5 Hz, and 0.5–2.0 N). They proposed that the excellent friction and wear performances are due to the formation of a tribofilm on the contact interface formed by an adsorbed saliva layer and graphene sheets, coupled with the removal of debris with artificial saliva through the micro-grove (Figure 23) [555,556].

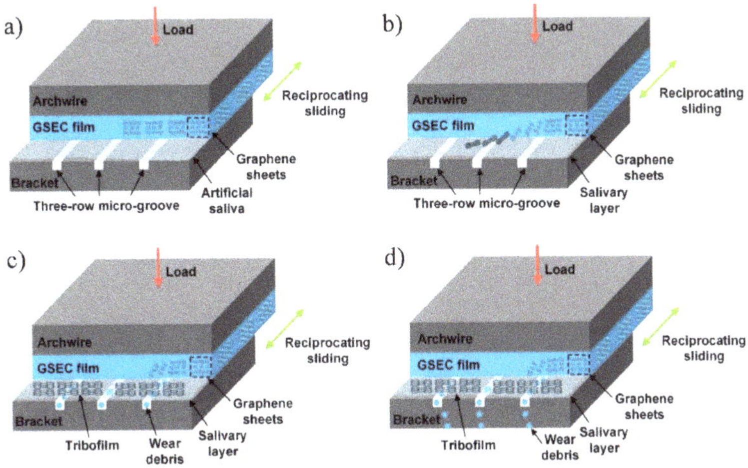

Figure 23. Low friction mechanism of graphene sheets embedded carbon (GSEC) coating on an SS archwire against an SS bracket: (**a**) initial state of the contact combination; (**b**) stable low friction with the formation of graphete-rich tribofilm and salivary adsorbed layer; (**c**) accumulation of wear debris detached from the GSEC film with micro-groove; and (**d**) flow out of wear debris with artificial saliva from the micro-groove (reproduced from [556]).

6. Conclusions

Contemporary orthodontics heavily relies on using fixed orthodontic appliances to treat dental malocclusions. Those devices are commonly manufactured using metallic bioalloys due to their suitable mechanical properties that induce the necessary tooth movement through the treatment time—usually around 2 years. Stainless steels and nickel–titanium stand out due to their widespread use, followed by CoCr- and other Ti-based alloys.

The main downside of using metallic biomaterials in vivo is the inevitable corrosion process in biological environments. In fact, the human mouth is an extreme corrosion-promoting scenario, encompassing frequent and complex variations in chemical composition, temperature, and pH, promoted by many factors such as diet, oral microorganisms, and the use of F^--containing hygiene products. Consequently, several types of corrosion could occur, including pitting, crevice, fretting, and so-called microbiologically induced corrosion.

The main consequences of intraoral corrosion are presented in this overview. The release of metallics is the most frequently studied feature; 67 studies measured the metal content in various matrices. While concentrations of released metal ions are below toxic levels, reported allergic reactions to metals such as Ni during orthodontic treatments with fixed appliances raised flags. This topic is not new, but further clarification is still required. Moreover, corrosion can increase the bracket/archwire pair friction, which may make it difficult to determine the correct treatment progression. The use of fluoride-containing products—unquestionably important to maintain oral health—was also focused, as they can potentially aggravate metallic corrosion susceptibility.

The literature shows several paths that researchers have been following to improve the overall corrosion resistance and, in consequence, the biocompatibility of orthodontic alloys. Those include the replacement of metallic components with non-metallic or composite substitutes; the modification of the alloys' chemical composition; the development of different manufacturing processes; and the application of surface modification techniques by depositing protective coatings. The versatile use of thin films and coatings stands out in this biomedical field and is particularly focused on this review, as it can maintain the crucial properties of the alloys' bulk. Many different deposition technologies and materials have been used in multiple in vivo and in vitro efforts to protect against oral aging, from monolithic to composite architectures and micro- to nano-scale materials, to meet the best and safest oral practice demands. Unfortunately, the challenging oral environment still undermines the development of a fully effective coating, whereas the high variability of research methodologies challenges the direct comparison between the studies.

Data suggest that even existing commercially available alternatives have important drawbacks and are fallible. Further multidisciplinary research should, therefore, persist, namely by applying new coating materials from the surface engineering field to fixed orthodontics and conducting broader studies to mimic the intraoral cavity complexity.

Author Contributions: A.F. performed the literature review and drafted the work; C.S.L. designed the work, participated in drafting, and substantially revised it; A.C.S. substantially revised the work. All authors have read and agreed to the published version of the manuscript.

Funding: This work was supported by national funds through FCT-Fundação para a Ciência e a Tecnologia, under the projects UIDB/00285/2020 and LA/P/0112/2020, and through the PhD Grant SFRH/BD/143905/2019 attributed to A. Fróis; and under the projects COMPETE/FEDER/POCI programs CENTRO-01-0247-FEDER-039880 and PTDC/EMD-EMD/21402020.

Data Availability Statement: Not applicable.

Conflicts of Interest: The authors declare no conflict of interest.

Abbreviations

AAS	Atomic absorption spectrometry
a-CNx	Amorphous carbon nitride
AISI	American Iron and Steel Institute
CAW	Composite archwires
CVD	Chemical vapor deposition
CoF	Coefficient of friction
d.h.m.	Dry hair mass
DL	Detection limit
DLC	Diamond-like carbon
EDS	Energy-dispersive spectroscopy
FCC	Face centered cubic
GSEC	Graphene sheets embedded carbon
HA	Hydroxyapatite
HGF	Human gingival fibroblasts
IARC	International Agency for Research on Cancer
IBAD	Ion beam-assisted deposition
ICP-AES/ICP-OES	Inductively coupled plasma-atomic/optical emission spectrometry
ICP-MS	Inductively coupled plasma mass spectroscopy
IF-NPs	Inorganic fullerene-like nanoparticles
ISO	International organization for standardization
MAO	Micro-arc oxidation
MCECR	Mirror-confinement-type electron cyclotron resonance
MIC	Microbiologically induced corrosion
NPs	Nanoparticles
PC	Polycarbonate
PECVD	Plasma-enhanced chemical vapor deposition
PVD	Physical vapor deposition
PE	Polyethylene
PEEK	Poly-ether-ether-ketone
PEN	Polyethylene naphthalate
PET	Polyethylene terephthalate
PEO	Plasma electrolytic oxidation

PH	Precipitation hardening
PIIID	Plasma immersion ion implantation and deposition
POM	Polyoxymethylene
PP	Polyphenylene
PTFE	Polytetrafluoroethylene
PU	Polyurethane
PVA	Polyvinyl alcohol
PVP	Polyvinylpyrrolidone
SCC	Stress corrosion cracking
SEM	Scanning electron microscopy
SMA	Shape memory alloy
SRB	Sulphur reducing bacteria
SS	Stainless steel
XRF	X-ray fluorescence

References

1. Dental Board of Australia. Dental List of Recognised Specialties, Related Specialist Titles and Definitions. Available online: https://www.dentalboard.gov.au/Registration-Standards.aspx (accessed on 10 December 2022).
2. Guo, L.; Feng, Y.; Guo, H.-G.; Liu, B.-W.; Zhang, Y. Consequences of Orthodontic Treatment in Malocclusion Patients: Clinical and Microbial Effects in Adults and Children. *BMC Oral Health* **2016**, *16*, 112. [CrossRef] [PubMed]
3. Zou, J.; Meng, M.; Law, C.S.; Rao, Y.; Zhou, X. Common Dental Diseases in Children and Malocclusion. *Int. J. Oral Sci.* **2018**, *10*, 7. [CrossRef] [PubMed]
4. Lombardo, G.; Vena, F.; Negri, P.; Pagano, S.; Barilotti, C.; Paglia, L.; Colombo, S.; Orso, M.; Cianetti, S. Worldwide Prevalence of Malocclusion in the Different Stages of Dentition: A Systematic Review and Meta-Analysis. *Eur. J. Paediatr. Dent.* **2020**, *21*, 115–122. [CrossRef]
5. Cenzato, N.; Nobili, A.; Maspero, C. Prevalence of Dental Malocclusions in Different Geographical Areas: Scoping Review. *Dent. J.* **2021**, *9*, 117. [CrossRef] [PubMed]
6. Redzepagic Vrazalica, L.; Ilic, Z.; Laganin, S.; Dzemidzic, V.; Tiro, A. An Epidemiological Study of Malocclusion and Occlusal Traits Related to Different Stages of Dental Development. *S. Eur. J. Orthod. Dentofac. Res.* **2017**, *4*, 9–13. [CrossRef]
7. Mtaya, M.; Brudvik, P.; Astrom, A.N. Prevalence of Malocclusion and Its Relationship with Socio-Demographic Factors, Dental Caries, and Oral Hygiene in 12- to 14-Year-Old Tanzanian Schoolchildren. *Eur. J. Orthod.* **2009**, *31*, 467–476. [CrossRef]
8. Jamilian, A.; Kiaee, B.; Sanayei, S.; Khosravi, S.; Perillo, L. Orthodontic Treatment of Malocclusion and Its Impact on Oral Health-Related Quality of Life. *Open Dent. J.* **2016**, *10*, 236–241. [CrossRef]
9. Proffit, W.R.; Fields, H.W.; Sarver, D.M.; Ackerman, J.L. Malocclusion and Dentofacial Deformity in Contemporary Society. In *Contemporary Orthodontics*; Mosby: St. Louis, MO, USA; Elsevier: Amsterdam, The Netherlands, 2012; pp. 2–18, ISBN 978032308317.
10. Abdallah, M.-N.; Lou, T.; Retrouvey, J.-M.; Suri, S. Biomaterials Used in Orthodontics: Brackets, Archwires, and Clear Aligners. In *Advanced Dental Biomaterials*; Khurshid, Z., Najeeb, S., Zafar, M.S., Sefat, F., Eds.; Elsevier: Amsterdam, The Netherlands, 2019; pp. 541–579, ISBN 978-0-08-102476-8.
11. Proffit, W.R.; Fields, H.W.; Sarver, D.M.; Ackerman, J.L. Contemporary Orthodontic Appliances. In *Contemporary Orthodontics*; Mosby: St. Louis, MO, USA; Elsevier: Amsterdam, The Netherlands, 2012; pp. 347–389, ISBN 978032308317.
12. Proffit, W.R.; Fields, H.W.; Sarver, D.M.; Ackerman, J.L. Mechanical Principles in Orthodontic Force Control. In *Contemporary Orthodontics*; Mosby: St. Louis, MO, USA; Elsevier: Amsterdam, The Netherlands, 2012; pp. 312–346, ISBN 978032308317.
13. Wen, Y.; Niu, Q.; Wang, A.; Yang, H.; Wang, H.; Jin, Z. Clear Aligner Treatment for an Adult with Severe Anterior Open Bite Malocclusion. *AJO-DO Clin. Companion* **2022**, *2*, 409–417. [CrossRef]
14. Puppin Filho, A. Má Oclusão de Classe I Com Biprotrusão e Ausência Dos Primeiros Molares Inferiores. *Dent. Press J. Orthod.* **2011**, *16*, 119–129. [CrossRef]
15. Aljhani, A.S.; Zawawi, K.H. Nonextraction Treatment of Severe Crowding with the Aid of Corticotomy-Assisted Orthodontics. *Case Rep. Dent.* **2012**, *2012*, 694527. [CrossRef]
16. Kaya, D.; Taner, T.U. Management of an Adult with Spaced Dentition, Class III Malocclusion and Open-Bite Tendency. *Eur. J. Dent.* **2011**, *5*, 121–129. [CrossRef]
17. Schroeder, M.A.; Schroeder, D.K.; Júnior, J.C.; da Silva Santos, D.J. Orthodontic Traction of Impacted Maxillary Canines Using Segmented Arch Mechanics. *Dent. Press J. Orthod.* **2019**, *24*, 79–89. [CrossRef] [PubMed]
18. Cobourne, M.T.; DiBiase, A.T.; Seehra, J.; Papageorgiou, S.N. Should We Recommend Early Overjet Reduction to Prevent Dental Trauma? *Br. Dent. J.* **2022**, *233*, 387–390. [CrossRef] [PubMed]
19. Fooladi, B.; MacCarthy, T.; Maloney, T.; Suri, L. Category 4: Class II Division 2 Malocclusion with Deep Overbite. *Am. J. Orthod. Dentofac. Orthop.* **2007**, *132*, 252–259. [CrossRef] [PubMed]
20. Carriére, L. Nonsurgical Correction of Severe Skeletal Class III Malocclusion. *J. Clin. Orthod.* **2016**, *50*, 216–230. [PubMed]

21. de Oliveira, S.R. Class III Malocclusion with Unilateral Posterior Crossbite and Facial Asymmetry. *Dent. Press J. Orthod.* **2010**, *15*, 182–191. [CrossRef]
22. Moro, A.; Borges, S.W.; Spada, P.P.; Morais, N.D.; Correr, G.M.; Chaves, C.M.; Cevidanes, L.H.S. Twenty-Year Clinical Experience with Fixed Functional Appliances. *Dent. Press J. Orthod.* **2018**, *23*, 87–109. [CrossRef]
23. Tsichlaki, A.; Chin, S.Y.; Pandis, N.; Fleming, P.S. How Long Does Treatment with Fixed Orthodontic Appliances Last? A Systematic Review. *Am. J. Orthod. Dentofac. Orthop.* **2016**, *149*, 308–318. [CrossRef]
24. Proffit, W.R.; Fields, H.W.; Sarver, D.M.; Ackerman, J.L. The First Stage of Comprehensive Treatment: Alignment and Leveling. In *Contemporary Orthodontics*; Mosby: St. Louis, MO, USA; Elsevier: Amsterdam, The Netherlands, 2012; pp. 530–555, ISBN 978032308317.
25. Business Wire Inc. Business Wire. Available online: https://www.businesswire.com/news/home/20200728005527/en/Global-Orthodontic-Supplies-Market-Worth-4.6-Billion (accessed on 10 September 2020).
26. Eliaz, N. Corrosion of Metallic Biomaterials: A Review. *Materials* **2019**, *12*, 407. [CrossRef]
27. Proffit, W.R.; Fields, H.W.; Sarver, D.M.; Ackerman, J.L. The Biology Basis of Orthodontic Therapy. In *Contemporary Orthodontics*; Mosby: St. Louis, MO, USA; Elsevier: Amsterdam, The Netherlands, 2012; pp. 278–311, ISBN 978032308317.
28. Eliades, T.; Athanasiou, A.E. In Vivo Aging of Orthodontic Alloys: Implications for Corrosion Potential, Nickel Release, and Biocompatibility. *Angle Orthod.* **2002**, *72*, 222–237.
29. Chaturvedi, T.P.; Upadhayay, S.N. An Overview of Orthodontic Material Degradation in Oral Cavity. *Indian J. Dent. Res.* **2010**, *21*, 275–284. [CrossRef] [PubMed]
30. Sifakakis, I.; Eliades, T. Adverse Reactions to Orthodontic Materials. *Aust. Dent. J.* **2017**, *62*, 20–28. [CrossRef] [PubMed]
31. Shukoor, K.M.; Shaj, F.; Shabeer, N.N.; Jayarajan, J. Nickel Allergies in Orthodontic Treatment. *Int. J. Prev. Clin. Dent. Res.* **2016**, *3*, 143–146. [CrossRef]
32. Agarwal, P.; Upadhyay, U.; Tandon, R.; Kumar, S. Nickel Allergy and Orthodontics. *Asian J. Oral Health Allied Sci.* **2011**, *1*, 61–63.
33. *EN 10088-1*; Stainless Steels—Part 1: List of Stainless Steels. European Committee for Standardization (CEN): Brussels, Belgium, 2014.
34. Malik, N.; Dubey, R.; Kallury, A.; Chauksye, A.; Shrivastav, T.; Kapse, B.R. A Review of Orthodontic Archwires. *J. Orofac. Res.* **2015**, *5*, 6–11. [CrossRef]
35. Olszewska, A.; Hanć, A.; Barałkiewicz, D.; Rzymski, P. Metals and Metalloids Release from Orthodontic Elastomeric and Stainless Steel Ligatures: In Vitro Risk Assessment of Human Exposure. *Biol. Trace Elem. Res.* **2020**, *196*, 646–653. [CrossRef]
36. Brüngger, D.; Koutsoukis, T.; Al Jabbari, Y.S.; Hersberger-Zurfluh, M.; Zinelis, S.; Eliades, T. A Comparison of the Compositional, Microstructural, and Mechanical Characteristics of Ni-Free and Conventional Stainless Steel Orthodontic Wires. *Materials* **2019**, *12*, 3424. [CrossRef]
37. Brantley, W.; Berzins, D.; Iijima, M.; Tufekçi, E.; Cai, Z. Structure/Property Relationships in Orthodontic Alloys. In *Orthodontic Applications of Biomaterials*; Eliades, T., Brantley, W.A., Eds.; Elsevier: Amsterdam, The Netherlands, 2017; pp. 3–38, ISBN 9780081003831.
38. Wendl, B.; Wiltsche, H.; Lankmayr, E.; Winsauer, H.; Walter, A.; Muchitsch, A.; Jakse, N.; Wendl, M.; Wendl, T. Metal Release Profiles of Orthodontic Bands, Brackets, and Wires: An in Vitro Study. *J. Orofac. Orthop. Fortschritte Kieferorthopädie* **2017**, *78*, 494–503. [CrossRef]
39. Arango, S.; Peláez-Vargas, A.; García, C. Coating and Surface Treatments on Orthodontic Metallic Materials. *Coatings* **2013**, *3*, 1–15. [CrossRef]
40. Arango Santander, S.; Luna Ossa, C.M. Stainless Steel: Material Facts for the Orthodontic Practitioner. *Rev. Nac. Odontol.* **2015**, *11*. [CrossRef]
41. Walker, M.P.; Ries, D.; Kula, K.; Ellis, M.; Fricke, B. Mechanical Properties and Surface Characterization of Beta Titanium and Stainless Steel Orthodontic Wire Following Topical Fluoride Treatment. *Angle Orthod.* **2007**, *77*, 342–348. [CrossRef]
42. Castro, S.M.; Ponces, M.J.; Lopes, J.D.; Vasconcelos, M.; Pollmann, M.C.F. Orthodontic Wires and Its Corrosion—The Specific Case of Stainless Steel and Beta-Titanium. *J. Dent. Sci.* **2015**, *10*, 1–7. [CrossRef]
43. Cuy, J.L.; Mann, A.B.; Livi, K.J.; Teaford, M.F.; Weihs, T.P. Nanoindentation Mapping of the Mechanical Properties of Human Molar Tooth Enamel. *Arch. Oral Biol.* **2002**, *47*, 281–291. [CrossRef] [PubMed]
44. Niinomi, M. Mechanical Properties of Biomedical Titanium Alloys. *Mater. Sci. Eng. A* **1998**, *243*, 231–236. [CrossRef]
45. Bauer, S.; Schmuki, P.; Von Der Mark, K.; Park, J. Progress in Materials Science Engineering Biocompatible Implant Surfaces Part I: Materials and Surfaces. *Prog. Mater. Sci.* **2013**, *58*, 261–326. [CrossRef]
46. Tian, K.; Darvell, B.W. Determination of the Flexural Modulus of Elasticity of Orthodontic Archwires. *Dent. Mater.* **2010**, *26*, 821–829. [CrossRef] [PubMed]
47. Anusavice, K.J.; Phillips, R.W. *Phillips' Science of Dental Materials*, 10th ed.; W.B. Saunders: Philadelphia, PA, USA, 1996; ISBN 9780721657417.
48. Jasso-Ruiz, I.; Velazquez-Enriquez, U.; Scougall-Vilchis, R.J.; Morales-Luckie, R.A.; Sawada, T.; Yamaguchi, R. Silver Nanoparticles in Orthodontics, a New Alternative in Bacterial Inhibition: In Vitro Study. *Prog. Orthod.* **2020**, *21*, 24. [CrossRef]
49. Arango-Santander, S.; Ramírez-Vega, C. Titanio: Aspectos Del Material Para Uso En Ortodoncia. *Rev. Nac. Odontol.* **2016**, *12*, 63–71. [CrossRef]

50. Gioka, C.; Bourauel, C.; Zinelis, S.; Eliades, T.; Silikas, N.; Eliades, G. Titanium Orthodontic Brackets: Structure, Composition, Hardness and Ionic Release. *Dent. Mater.* **2004**, *20*, 693–700. [CrossRef]
51. Uysal, I.; Yilmaz, B.; Atilla, A.O.; Evis, Z. Nickel Titanium Alloys as Orthodontic Archwires: A Narrative Review. *Eng. Sci. Technol. Int. J.* **2022**, *36*, 101277. [CrossRef]
52. Wadood, A. Brief Overview on Nitinol as Biomaterial. *Adv. Mater. Sci. Eng.* **2016**, *2016*, 4173138. [CrossRef]
53. Sifakakis, I.; Bourauel, C. Nickel–Titanium Products in Daily Orthodontic Practice. In *Orthodontic Applications of Biomaterials*; Elsevier: Amsterdam, The Netherlands, 2017; pp. 107–127, ISBN 9780081003831.
54. Gravina, M.A.; Canavarro, C.; Elias, C.N.; Chaves, M.D.G.A.M.; Brunharo, I.H.V.P.; Quintão, C.C.A. Mechanical Properties of NiTi and CuNiTi Wires Used in Orthodontic Treatment. Part 2: Microscopic Surface Appraisal and Metallurgical Characteristics. *Dent. Press J. Orthod.* **2014**, *19*, 69–76. [CrossRef] [PubMed]
55. Parvizi, F. The Load/Deflection Characteristics of Thermally Activated Orthodontic Archwires. *Eur. J. Orthod.* **2003**, *25*, 417–421. [CrossRef] [PubMed]
56. Sufarnap, E.; Harahap, K.; Cynthiana, S.; Reza, M. Nickel and Copper Ion Release, Deflection and the Surface Roughness of Copper-Nickel-Titanium Orthodontic Archwire in Sodium Fluoride Solution. *J. Orthod. Sci.* **2023**, *12*, 44. [CrossRef] [PubMed]
57. Seyyed Aghamiri, S.M.; Ahmadabadi, M.N.; Raygan, S. Combined Effects of Different Heat Treatments and Cu Element on Transformation Behavior of NiTi Orthodontic Wires. *J. Mech. Behav. Biomed. Mater.* **2011**, *4*, 298–302. [CrossRef] [PubMed]
58. Sarul, M.; Kawala, B.; Kawala, M.; Antoszewska-Smith, J. Do the NiTi Low and Constant Force Levels Remain Stable in Vivo? *Eur. J. Orthod.* **2015**, *37*, 656–664. [CrossRef] [PubMed]
59. Farzin-Nia, F.; Yoneyama, T. Orthodontic Devices Using Ti-Ni Shape Memory Alloys. In *Shape Memory Alloys for Biomedical Applications*; Elsevier: Amsterdam, The Netherlands, 2008; pp. 257–296, ISBN 9781845693442.
60. Polychronis, G.; Al Jabbari, Y.S.; Eliades, T.; Zinelis, S. Galvanic Coupling of Steel and Gold Alloy Lingual Brackets with Orthodontic Wires: Is Corrosion a Concern? *Angle Orthod.* **2018**, *88*, 450–457. [CrossRef]
61. Ito, A.; Kitaura, H.; Noguchi, T.; Ohori, F.; Mizoguchi, I. Analysis of Coating Loss from Coated Stainless Steel Orthodontic Wire. *Appl. Sci.* **2022**, *12*, 9497. [CrossRef]
62. Toy, E.; Malkoc, S.; Corekci, B.; Bozkurt, B.S.; Hakki, S.S. Real-Time Cell Analysis of the Cytotoxicity of Orthodontic Brackets on Gingival Fibroblasts. *J. Appl. Biomater. Funct. Mater.* **2014**, *12*, 248–255. [CrossRef]
63. Kim, I.-H.; Park, H.-S.; Kim, Y.K.; Kim, K.-H.; Kwon, T.-Y. Comparative Short-Term in Vitro Analysis of Mutans Streptococci Adhesion on Esthetic, Nickel-Titanium, and Stainless-Steel Arch Wires. *Angle Orthod.* **2014**, *84*, 680–686. [CrossRef]
64. Krishnan, M.; Seema, S.; Kumar, A.V.; Varthini, N.P.; Sukumaran, K.; Pawar, V.R.; Arora, V. Corrosion Resistance of Surface Modified Nickel Titanium Archwires. *Angle Orthod.* **2014**, *84*, 358–367. [CrossRef]
65. Hansen, D.C. Metal Corrosion in the Human Body: The Ultimate Bio-Corrosion Scenario. *Electrochem. Soc. Interface* **2008**, *17*, 31–34. [CrossRef]
66. Maruthamuthu, S.; Rajasekar, A.; Sathiyanarayanan, S.; Muthukumar, N.; Palaniswamy, N. Electrochemical Behaviour of Microbes on Orthodontic Wires. *Curr. Sci.* **2005**, *89*, 988–996.
67. Wolf, H.F.; Hassell, T.M. Biofilm—Plaque Formation on Tooth and Root Surfaces. In *Color Atlas of Dental Hygene-Periodontology*; Thieme: Stuttgart, Germany; New York, NY, USA, 2006; p. 24, ISBN 9783131417619.
68. Zhou, Z.R.; Zheng, J. Tribology of Dental Materials: A Review. *J. Phys. D Appl. Phys.* **2008**, *41*, 113001. [CrossRef]
69. Dawes, C.; Pedersen, A.M.L.; Villa, A.; Ekström, J.; Proctor, G.B.; Vissink, A.; Aframian, D.; McGowan, R.; Aliko, A.; Narayana, N.; et al. The Functions of Human Saliva: A Review Sponsored by the World Workshop on Oral Medicine VI. *Arch. Oral Biol.* **2015**, *60*, 863–874. [CrossRef] [PubMed]
70. Mosca, A.C.; Chen, J. Food-Saliva Interactions: Mechanisms and Implications. *Trends Food Sci. Technol.* **2017**, *66*, 125–134. [CrossRef]
71. de Almeida, P.D.V.; Grégio, A.M.T.; Machado, M.A.N.; de Lima, A.A.S.; Azevedo, L.R. Saliva Composition and Functions: A Comprehensive Review. *J. Contemp. Dent. Pract.* **2008**, *9*, 72–80.
72. Humphrey, S.P.; Williamson, R.T. A Review of Saliva: Normal Composition, Flow, and Function. *J. Prosthet. Dent.* **2001**, *85*, 162–169. [CrossRef]
73. Mystkowska, J.; Niemirowicz-Laskowska, K.; Łysik, D.; Tokajuk, G.; Dąbrowski, J.R.; Bucki, R. The Role of Oral Cavity Biofilm on Metallic Biomaterial Surface Destruction–Corrosion and Friction Aspects. *Int. J. Mol. Sci.* **2018**, *19*, 743. [CrossRef]
74. Yakubov, G.E.; Macakova, L.; Wilson, S.; Windust, J.H.C.; Stokes, J.R. Aqueous Lubrication by Fractionated Salivary Proteins: Synergistic Interaction of Mucin Polymer Brush with Low Molecular Weight Macromolecules. *Tribol. Int.* **2015**, *89*, 34–45. [CrossRef]
75. Castagnola, M.; Cabras, T.; Vitali, A.; Sanna, M.T.; Messana, I. Biotechnological Implications of the Salivary Proteome. *Trends Biotechnol.* **2011**, *29*, 409–418. [CrossRef] [PubMed]
76. Choi, J.E.; Lyons, K.M.; Kieser, J.A.; Waddell, N.J. Diurnal Variation of Intraoral PH and Temperature. *BDJ Open* **2017**, *3*, 17015. [CrossRef] [PubMed]
77. Moore, R.J.; Watts, J.T.F.; Hood, J.A.A.; Burritt, D.J. Intra-Oral Temperature Variation over 24 Hours. *Eur. J. Orthod.* **1999**, *21*, 249–261. [CrossRef] [PubMed]
78. Kwak, D.Y.; Kim, N.Y.; Kim, H.J.; Yang, S.Y.; Yoon, J.E.; Hyun, I.A.; Nam, S.H. Changes in the Oral Environment after Tooth Brushing and Oral Gargling. *Biomed. Res.* **2017**, *28*, 7093–7097.

79. Neyraud, E.; Palicki, O.; Schwartz, C.; Nicklaus, S.; Feron, G. Variability of Human Saliva Composition: Possible Relationships with Fat Perception and Liking. *Arch. Oral Biol.* **2012**, *57*, 556–566. [CrossRef] [PubMed]
80. Poles, A.A.; Balcão, V.M.; Chaud, M.V.; Vila, M.M.D.C.; Aranha, N.; Yoshida, V.M.H.; Oliveira, J.M. Study of the Elemental Composition of Saliva of Smokers and Nonsmokers by X-Ray Fluorescence. *Appl. Radiat. Isot.* **2016**, *118*, 221–227. [CrossRef]
81. Upadhyay, D.; Panchal, M.A.; Dubey, R.S.; Srivastava, V.K. Corrosion of Alloys Used in Dentistry: A Review. *Mater. Sci. Eng. A* **2006**, *432*, 1–11. [CrossRef]
82. Hans, R.; Thomas, S.; Garla, B.; Dagli, R.J.; Hans, M.K. Effect of Various Sugary Beverages on Salivary PH, Flow Rate, and Oral Clearance Rate amongst Adults. *Scientifica* **2016**, *2016*, 5027283. [CrossRef]
83. Dental Caries. *Essentials of Oral Pathology and Oral Medicine*; Cawson, R., Odell, E., Eds.; Churchill Livingstone: London, UK, 2008; pp. 40–59, ISBN 978-0443-10125-0.
84. Walsh, L.J. Dental Plaque Fermentation and Its Role in Caries Risk Assessment. *Int. Dent. S. Afr.* **2006**, *8*, 34–40.
85. Goel, I.; Navit, S.; Mayall, S.S.; Rallan, M.; Navit, P.; Chandra, S. Effects of Carbonated Drink & Fruit Juice on Salivary PH of Children: An in Vivo Study. *Int. J. Sci. Study* **2013**, *1*, 60.
86. Lubis, H.F.; Simamora, G.H. Release of Nickel Ions and Changes in Surface Microstructure of Stainless Steel Archwire after Immersion in Tomato and Orange Juice. *IOP Conf. Ser. Earth Environ. Sci.* **2021**, *912*, 012018. [CrossRef]
87. He, J.; Li, Y.; Cao, Y.; Xue, J.; Zhou, X. The Oral Microbiome Diversity and Its Relation to Human Diseases. *Folia Microbiol.* **2015**, *60*, 69–80. [CrossRef]
88. Kilian, M.; Chapple, I.L.C.; Hannig, M.; Marsh, P.D.; Meuric, V.; Pedersen, A.M.L.; Tonetti, M.S.; Wade, W.G.; Zaura, E. The Oral Microbiome—An Update for Oral Healthcare Professionals. *Br. Dent. J.* **2016**, *221*, 657–666. [CrossRef]
89. Marsh, P.D.; Head, D.A.; Devine, D.A. Dental Plaque as a Biofilm and a Microbial Community—Implications for Treatment. *J. Oral Biosci.* **2015**, *57*, 185–191. [CrossRef]
90. Alasvand Zarasvand, K.; Rai, V.R. Microorganisms: Induction and Inhibition of Corrosion in Metals. *Int. Biodeterior. Biodegrad.* **2014**, *87*, 66–74. [CrossRef]
91. Øgaard, B. White Spot Lesions During Orthodontic Treatment: Mechanisms and Fluoride Preventive Aspects. *Semin. Orthod.* **2008**, *14*, 183–193. [CrossRef]
92. Weyant, R.J.; Tracy, S.L.; Anselmo, T.T.; Beltrán-Aguilar, E.D.; Donly, K.J.; Frese, W.A.; Hujoel, P.P.; Iafolla, T.; Kohn, W.; Kumar, J.; et al. Topical Fluoride for Caries Prevention. *J. Am. Dent. Assoc.* **2013**, *144*, 1279–1291. [CrossRef]
93. *ISO 8044:2015*; Corrosion of Metals and Alloys—Basic Terms and Definitions. ISO (International Organization for Standardization): Geneva, Switzerland, 2015. Available online: https://www.iso.org/obp/ui/#iso:std:iso:8044:ed-4:v1:en (accessed on 18 May 2018).
94. Sato, N. Basics of Corrosion Chemistry. In *Green Corrosion Chemistry and Engineering: Opportunities and Challenges*; Wiley: Hoboken, NJ, USA, 2011; pp. 1–32, ISBN 9783527329304.
95. House, K.; Sernetz, F.; Dymock, D.; Sandy, J.R.; Ireland, A.J. Corrosion of Orthodontic Appliances-Should We Care? *Am. J. Orthod. Dentofac. Orthop.* **2008**, *133*, 584–592. [CrossRef]
96. Hunt, N.; Cunningham, S.; Golden, C.; Sheriff, M. An Investigation into the Effects of Polishing on Surface Hardness and Corrosion of Orthodontic Archwires. *Angle Orthod.* **1999**, *69*, 433–440. [CrossRef]
97. Strehblow, H.-H. Phenomenological and Electrochemical Fundamentals of Corrosion. In *Materials Science and Technology, Corrosion and Environmental Degradation, Vol. I.*; Schütze, M., Cahn, R., Haasen, P., Kramer, E., Eds.; Wiley-VCH: Weinheim, Germany, 2000; pp. 1–66, ISBN 3-527-29505-4.
98. Eliades, T.; Zinelis, S.; Bourauel, C.; Eliades, G. Manufacturing of Orthodontic Brackets: A Review of Metallurgical Perspectives and Applications. *Recent Pat. Mater. Sci.* **2008**, *1*, 135–139. [CrossRef]
99. Hanawa, T. Metal Ion Release from Metal Implants. *Mater. Sci. Eng. C* **2004**, *24*, 745–752. [CrossRef]
100. Marcus, P.; Maurice, V. Passivity of Metals and Alloys. In *Materials Science and Technology, Corrosion and Environmental Degradation, Vol. I.*; Schütze, M., Cahn, R., Haasen, P., Kramer, E., Eds.; Wiley-VCH: Weinheim, Germany, 2000; pp. 131–169, ISBN 3-527-29505-4.
101. Hedberg, Y.S.; Odnevall Wallinder, I. Metal Release from Stainless Steel in Biological Environments: A Review. *Biointerphases* **2016**, *11*, 018901. [CrossRef]
102. Prasad, S.; Ehrensberger, M.; Gibson, M.P.; Kim, H.; Monaco, E.A. Biomaterial Properties of Titanium in Dentistry. *J. Oral Biosci.* **2015**, *57*, 192–199. [CrossRef]
103. Mazinanian, N.; Hedberg, Y.S. Metal Release Mechanisms for Passive Stainless Steel in Citric Acid at Weakly Acidic PH. *J. Electrochem. Soc.* **2016**, *163*, C686–C693. [CrossRef]
104. Walker, M.P.; White, R.J.; Kula, K.S. Effect of Fluoride Prophylactic Agents on the Mechanical Properties of Nickel-Titanium-Based Orthodontic Wires. *Am. J. Orthod. Dentofac. Orthop.* **2005**, *127*, 662–669. [CrossRef]
105. Mystkowska, J. Biocorrosion of Dental Alloys Due to Desulfotomaculum Nigrificans Bacteria. *Acta Bioeng. Biomech.* **2016**, *18*, 87–96. [CrossRef]
106. Daems, J.; Celis, J.-P.; Willems, G. Morphological Characterization of As-Received and in Vivo Orthodontic Stainless Steel Archwires. *Eur. J. Orthod.* **2009**, *31*, 260–265. [CrossRef]
107. Porcayo-Calderon, J.; Casales-Diaz, M.; Salinas-Bravo, V.M.; Martinez-Gomez, L. Corrosion Performance of Fe-Cr-Ni Alloys in Artificial Saliva and Mouthwash Solution. *Bioinorg. Chem. Appl.* **2015**, *2015*, 930802. [CrossRef]

108. Schiff, N. Galvanic Corrosion between Orthodontic Wires and Brackets in Fluoride Mouthwashes. *Eur. J. Orthod.* **2006**, *28*, 298–304. [CrossRef]
109. Oh, K.T.; Choo, S.U.; Kim, K.M.; Kim, K.N. A Stainless Steel Bracket for Orthodontic Application. *Eur. J. Orthod.* **2005**, *27*, 237–244. [CrossRef]
110. Mendes, B.D.A.B.; Ferreira, R.A.N.; Pithon, M.M.; Horta, M.C.R.; Oliveira, D.D. Physical and Chemical Properties of Orthodontic Brackets after 12 and 24 Months: In Situ Study. *J. Appl. Oral Sci.* **2014**, *22*, 194–203. [CrossRef]
111. Jacoby, L.S.; Junior, V.D.S.R.; Campos, M.M.; de Menezes, L.M. Cytotoxic Outcomes of Orthodontic Bands with and without Silver Solder in Different Cell Lineages. *Am. J. Orthod. Dentofac. Orthop.* **2017**, *151*, 957–963. [CrossRef]
112. Tahmasbi, S.; Ghorbani, M.; Masudrad, M. Galvanic Corrosion of and Ion Release from Various Orthodontic Brackets and Wires in a Fluoride-Containing Mouthwash. *J. Dent. Res. Dent. Clin. Dent. Prospects* **2015**, *9*, 159–165. [CrossRef]
113. Petković Didović, M.; Jelovica Badovinac, I.; Fiket, Ž.; Žigon, J.; Rinčić Mlinarić, M.; Čanadi Jurešić, G. Cytotoxicity of Metal Ions Released from NiTi and Stainless Steel Orthodontic Appliances, Part 1: Surface Morphology and Ion Release Variations. *Materials* **2023**, *16*, 4156. [CrossRef]
114. Schweitzer, P.A. *Fundamentals of Metallic Corrosion*, 2nd ed.; CRC Press: Boca Raton, FL, USA, 2006; ISBN 9780429127137.
115. Saporeti, M.P.; Mazzieiro, E.T.; Sales, W.F. In Vitro Corrosion of Metallic Orthodontic Brackets: Influence of Artificial Saliva with and without Fluorides. *Dent. Press J. Orthod.* **2012**, *17*, 24e1–24e7. [CrossRef]
116. Fróis, A.; Mendes, A.R.; Pereira, S.A.; Louro, C.S. Metal Release and Surface Degradation of Fixed Orthodontic Appliances during the Dental Levelling and Aligning Phase: A 12-Week Study. *Coatings* **2022**, *12*, 554. [CrossRef]
117. Zhang, Y. Corrosion Resistance of Passive Films on Orthodontic Bands in Fluoride-Containing Artificial Saliva. *Int. J. Electrochem. Sci.* **2017**, *12*, 292–304. [CrossRef]
118. Kao, C.-T.; Huang, T.-H. Variations in Surface Characteristics and Corrosion Behaviour of Metal Brackets and Wires in Different Electrolyte Solutions. *Eur. J. Orthod.* **2010**, *32*, 555–560. [CrossRef]
119. Mahato, N.; Sharma, M.R.; Chaturvedi, T.P.; Singh, M.M. Effect of Dietary Spices on the Pitting Behavior of Stainless Steel Orthodontic Bands. *Mater. Lett.* **2011**, *65*, 2241–2244. [CrossRef]
120. Chaturvedi, T. Corrosion of Orthodontic Brackets in Different Spices:In Vitro Study. *Indian J. Dent. Res.* **2014**, *25*, 630. [CrossRef]
121. Chiba, A.; Muto, I.; Sugawara, Y.; Hara, N. Pit Initiation Mechanism at MnS Inclusions in Stainless Steel: Synergistic Effect of Elemental Sulfur and Chloride Ions. *J. Electrochem. Soc.* **2013**, *160*, C511–C520. [CrossRef]
122. Yang, S.; Zhao, M.; Feng, J.; Li, J.; Liu, C. Induced-Pitting Behaviors of MnS Inclusions in Steel. *High Temp. Mater. Process.* **2018**, *37*, 1007–1016. [CrossRef]
123. Alnajjar, M.; Christien, F.; Barnier, V.; Bosch, C.; Wolski, K.; Fortes, A.D.; Telling, M. Influence of Microstructure and Manganese Sulfides on Corrosion Resistance of Selective Laser Melted 17-4 PH Stainless Steel in Acidic Chloride Medium. *Corros. Sci.* **2020**, *168*, 108585. [CrossRef]
124. Hodgson, A.W.E.; Kurz, S.; Virtanen, S.; Fervel, V.; Olsson, C.-O.A.; Mischler, S. Passive and Transpassive Behaviour of CoCrMo in Simulated Biological Solutions. *Electrochim. Acta* **2004**, *49*, 2167–2178. [CrossRef]
125. Bagatin, C.R.; Ito, I.Y.; Andrucioli, M.C.D.; Nelson-Filho, P.; Ferreira, J.T.L. Corrosion in Haas Expanders with and without Use of an Antimicrobial Agent: An in Situ Study. *J. Appl. Oral Sci.* **2011**, *19*, 662–667. [CrossRef]
126. Wang, J.; Li, N.; Rao, G.; Han, E.; Ke, W. Stress Corrosion Cracking of NiTi in Artificial Saliva. *Dent. Mater.* **2007**, *23*, 133–137. [CrossRef]
127. Kameda, T.; Oda, H.; Ohkuma, K.; Sano, N.; Batbayar, N.; Terashima, Y.; Sato, S.; TeradaA, K. Microbiologically Influenced Corrosion of Orthodontic Metallic Appliances. *Dent. Mater. J.* **2014**, *33*, 187–195. [CrossRef]
128. Muyzer, G.; Stams, A.J.M. The Ecology and Biotechnology of Sulphate-Reducing Bacteria. *Nat. Rev. Microbiol.* **2008**, *6*, 441–454. [CrossRef]
129. Jiang, J.; Chan, A.; Ali, S.; Saha, A.; Haushalter, K.J.; Lam, W.-L.M.; Glasheen, M.; Parker, J.; Brenner, M.; Mahon, S.B.; et al. Hydrogen Sulfide—Mechanisms of Toxicity and Development of an Antidote. *Sci. Rep.* **2016**, *6*, 20831. [CrossRef]
130. Dhandapani, P.; MuraliKannan, M.; Anandkumar, B.; Maruthamuthu, S.; Manoharan, S.P. Electrochemistry of Calcium Precipitating Bacteria in Orthodontic Wire. *Oral Sci. Int.* **2014**, *11*, 22–29. [CrossRef]
131. Petoumeno, E.; Kislyuk, M.; Hoederath, H.; Keilig, L.; Bourauel, C.; Jäger, A. Corrosion Susceptibility and Nickel Release of Nickel Titanium Wires during Clinical Application. *J. Orofac. Orthop.* **2008**, *69*, 411–423. [CrossRef]
132. Harada, R.; Kokubu, E.; Kinoshita, H.; Yoshinari, M.; Ishihara, K.; Kawada, E.; Takemoto, S. Corrosion Behavior of Titanium in Response to Sulfides Produced by Porphyromonas Gingivalis. *Dent. Mater.* **2018**, *34*, 183–191. [CrossRef] [PubMed]
133. Fukushima, A.; Mayanagi, G.; Nakajo, K.; Sasaki, K.; Takahashi, N. Microbiologically Induced Corrosive Properties of the Titanium Surface. *J. Dent. Res.* **2014**, *93*, 525–529. [CrossRef] [PubMed]
134. Kumar, A.; Khanam, A.; Ghafoor, H. Effects of Intraoral Aging of Arch-Wires on Frictional Forces: An Ex Vivo Study. *J. Orthod. Sci.* **2016**, *5*, 109. [CrossRef] [PubMed]
135. Cury, S.; Aliaga-Del Castillo, A.; Pinzan, A.; Sakoda, K.; Bellini-Pereira, S.; Janson, G. Orthodontic Brackets Friction Changes after Clinical Use: A Systematic Review. *J. Clin. Exp. Dent.* **2019**, *11*, e482–e490. [CrossRef] [PubMed]
136. Eliades, T.; Bourauel, C. Intraoral Aging of Orthodontic Materials: The Picture We Miss and Its Clinical Relevance. *Am. J. Orthod. Dentofac. Orthop.* **2005**, *127*, 403–412. [CrossRef]

137. Regis, S.; Soares, P.; Camargo, E.S.; Guariza Filho, O.; Tanaka, O.; Maruo, H. Biodegradation of Orthodontic Metallic Brackets and Associated Implications for Friction. *Am. J. Orthod. Dentofac. Orthop.* **2011**, *140*, 501–509. [CrossRef]
138. Bandeira, A.M.B.; dos Santos, M.P.A.; Pulitini, G.; Elias, C.N.; da Costa, M.F. Influence of Thermal or Chemical Degradation on the Frictional Force of an Experimental Coated NiTi Wire. *Angle Orthod.* **2011**, *81*, 484–489. [CrossRef]
139. Lima, A.A.S.; de Grégio, A.M.T.; Tanaka, O.; Machado, M.Â.N.; França, B.H.S. Tratamento Das Ulcerações Traumáticas Bucais Causadas Por Aparelhos Ortodônticos. *Rev. Dent. Press Ortod. Ortop. Facial* **2005**, *10*, 30–36. [CrossRef]
140. Pires, L.P.B.; de Oliveira, A.H.A.; da Silva, H.F.; de Oliveira, P.T.; dos Santos, P.B.D.; Pinheiro, F.H. de S.L. Can Shielded Brackets Reduce Mucosa Alteration and Increase Comfort Perception in Orthodontic Patients in the First 3 Days of Treatment? A Single-Blind Randomized Controlled Trial. *Am. J. Orthod. Dentofac. Orthop.* **2015**, *148*, 956–966. [CrossRef]
141. Kluemper, G.T.; Hiser, D.G.; Rayens, M.K.; Jay, M.J. Efficacy of a Wax Containing Benzocaine in the Relief of Oral Mucosal Pain Caused by Orthodontic Appliances. *Am. J. Orthod. Dentofac. Orthop.* **2002**, *122*, 359–365. [CrossRef] [PubMed]
142. Bourauel, C.; Scharold, W.; Jäger, A.; Eliades, T. Fatigue Failure of As-Received and Retrieved NiTi Orthodontic Archwires. *Dent. Mater.* **2008**, *24*, 1095–1101. [CrossRef]
143. Petoumeno, E.; Arndt, M.; Keilig, L.; Reimann, S.; Hoederath, H.; Eliades, T.; Jäger, A.; Bourauel, C. Nickel Concentration in the Saliva of Patients with Nickel-Titanium Orthodontic Appliances. *Am. J. Orthod. Dentofac. Orthop.* **2009**, *135*, 59–65. [CrossRef] [PubMed]
144. Martín-Cameán, A.; Jos, Á.; Mellado-García, P.; Iglesias-Linares, A.; Solano, E.; Cameán, A.M. In Vitro and in Vivo Evidence of the Cytotoxic and Genotoxic Effects of Metal Ions Released by Orthodontic Appliances: A Review. *Environ. Toxicol. Pharmacol.* **2015**, *40*, 86–113. [CrossRef] [PubMed]
145. Wang, J.J.; Sanderson, B.J.S.; Wang, H. Cyto- and Genotoxicity of Ultrafine TiO_2 Particles in Cultured Human Lymphoblastoid Cells. *Mutat. Res. Toxicol. Environ. Mutagen.* **2007**, *628*, 99–106. [CrossRef] [PubMed]
146. Messer, R.L.W.; Bishop, S.; Lucas, L.C. Effects of Metallic Ion Toxicity on Human Gingival Fibroblasts Morphology. *Biomaterials* **1999**, *20*, 1647–1657. [CrossRef] [PubMed]
147. Chakravarthi, S.; Chitharanjan, A.; Padmanabhan, S. Allergy and Orthodontics. *J. Orthod. Sci.* **2012**, *1*, 83. [CrossRef] [PubMed]
148. Primožič, J.; Poljšak, B.; Jamnik, P.; Kovač, V.; Čanadi Jurešić, G.; Spalj, S. Risk Assessment of Oxidative Stress Induced by Metal Ions Released from Fixed Orthodontic Appliances during Treatment and Indications for Supportive Antioxidant Therapy: A Narrative Review. *Antioxidants* **2021**, *10*, 1359. [CrossRef]
149. Samitz, M.H.; Katz, S.A. Nickel Dermatitis Hazards from Prostheses: In Vivo and in Vitro Stabilization Studies. *Br. J. Dermatol.* **1975**, *92*, 287–290. [CrossRef]
150. Mikulewicz, M.; Chojnacka, K. Release of Metal Ions from Orthodontic Appliances by In Vitro Studies: A Systematic Literature Review. *Biol. Trace Elem. Res.* **2011**, *139*, 241–256. [CrossRef]
151. Macedo de Menezes, L.; Cardoso Abdo Quintão, C. The Release of Ions from Metallic Orthodontic Appliances. *Semin. Orthod.* **2010**, *16*, 282–292. [CrossRef]
152. Urbutytė, K.; Barčiūtė, A.; Lopatienė, K. The Changes in Nickel and Chromium Ion Levels in Saliva with Fixed Orthodontic Appliances: A Systematic Review. *Appl. Sci.* **2023**, *13*, 4739. [CrossRef]
153. Gjerdet, N.R.; Erichsen, E.S.; Remlo, H.E.; Evjen, G. Nickel and Iron in Saliva of Patients with Fixed Orthodontic Appliances. *Acta Odontol. Scand.* **1991**, *49*, 73–78. [CrossRef] [PubMed]
154. Cempel, G.N.M. Nickel: A Review of Its Sources and Environmental Toxicolog. *Pol. J. Environ. Stud.* **2006**, *15*, 372–382.
155. Smart, G.A.; Sherlock, J.C. Nickel in Foods and the Diet. *Food Addit. Contam.* **1987**, *4*, 61–71. [CrossRef] [PubMed]
156. WHO (World Health Organization). Nickel in Drinking Water. In *Background Document for Development of WHO Guidelines for Drinking-Water Quality*; (WHO/SDE/WSH/04.08/55); World Health Organization: Geneva, Switzerland, 2005.
157. Becker, W.; Kumpulainen, J. Contents of Essential and Toxic Mineral Elements in Swedish Market-Basket Diets in 1987. *Br. J. Nutr.* **1991**, *66*, 151–160. [CrossRef] [PubMed]
158. CCME (Canadian Council of Ministers of the Environment). *Scientific Criteria Document for Canadian Soil Quality Guidelines for the Protection of Environmental and Human Health: Nickel*; Canadian Council of Ministers of the Environment: Winnipeg, MB, Canada, 2015.
159. PHE (Public Health England). Nickel. In *Toxicological Overview*; Version 1; Toxicology Department, CRCE, Public Health England: London, UK, 2009.
160. WHO (World Health Organization). Nickel. In *Air Quality Guidelines for Europe*; Regional Office for Europe, Ed.; World Health Organization: Copenhagen, Denmark, 2000; ISBN 9789289013581.
161. Haber, L.T.; Bates, H.K.; Allen, B.C.; Vincent, M.J.; Oller, A.R. Derivation of an Oral Toxicity Reference Value for Nickel. *Regul. Toxicol. Pharmacol.* **2017**, *87*, S1–S18. [CrossRef]
162. WHO (World Health Organization). Chromium. In *Air Quality Guidelines for Europe*; Regional Office for Europe, Ed.; World Health Organization: Copenhagen, Denmark, 2000; ISBN 9789289013581.
163. Duda-Chodak, A.; Blaszczyk, U. The Impact of Nickel on Human Health. *J. Elem.* **2008**, *13*, 685–696.
164. Forgacs, Z.; Massányi, P.; Lukac, N.; Somosy, Z. Reproductive Toxicology of Nickel—Review. *J. Environ. Sci. Health Part A* **2012**, *47*, 1249–1260. [CrossRef]
165. IARC (International Agency for Research on Cancer). *Nickel and Nickel Compounds*; Academic Press: New York, NY, USA, 2011; Volume 100C, pp. 169–218. [CrossRef]

166. World Health Organization. *Guidelines for Drinking-Water Quality*, 4th ed.; World Health Organization: Valletta, Malta, 2011.
167. Genchi, G.; Carocci, A.; Lauria, G.; Sinicropi, M.S.; Catalano, A. Nickel: Human Health and Environmental Toxicology. *Int. J. Environ. Res. Public Health* **2020**, *17*, 679. [CrossRef]
168. Silverberg, N.B.; Pelletier, J.L.; Jacob, S.E.; Schneider, L.C.; Cohen, B.; Horii, K.A.; Kristal, C.L.; Maguiness, S.M.; Tollefson, M.M.; Weinstein, M.G.; et al. Nickel Allergic Contact Dermatitis: Identification, Treatment, and Prevention. *Pediatrics* **2020**, *145*, e20200628. [CrossRef] [PubMed]
169. Das, K.K.; Das, S.N.; Dhundasi, S.A. Nickel, Its Adverse Health Effects & Oxidative Stress. *Indian J. Med. Res.* **2008**, *128*, 412–425. [PubMed]
170. Sahoo, N.; Kailasam, V.; Padmanabhan, S.; Chitharanjan, A.B. In-Vivo Evaluation of Salivary Nickel and Chromium Levels in Conventional and Self-Ligating Brackets. *Am. J. Orthod. Dentofac. Orthop.* **2011**, *140*, 340–345. [CrossRef] [PubMed]
171. De Souza, R.M.; De Menezes, L.M. Nickel, Chromium and Iron Levels in the Saliva of Patients with Simulated Fixed Orthodontic Appliances. *Angle Orthod.* **2008**, *78*, 345–350. [CrossRef] [PubMed]
172. Ağaoğlu, G.; Arun, T.; Izgü, B.; Yarat, A. Nickel and Chromium Levels in the Saliva and Serum of Patients with Fixed Orthodontic Appliances. *Angle Orthod.* **2001**, *71*, 375–379. [CrossRef] [PubMed]
173. Masjedi, M.; Niknam, O.; Haghighat Jahromi, N.; Javidi, P.; Rakhshan, V. Effects of Fixed Orthodontic Treatment Using Conventional, Copper-Included, and Epoxy-Coated Nickel-Titanium Archwires on Salivary Nickel Levels: A Double-Blind Randomized Clinical Trial. *Biol. Trace Elem. Res.* **2016**, *174*, 27–31. [CrossRef]
174. Yassaei, S.; Dadfarnia, S.; Ahadian, H.; Moradi, F. Nickel and Chromium Levels in the Saliva of Patients with Fixed Orthodontic Appliances. *Orthodontics* **2013**, *14*, e76–e81. [CrossRef]
175. Haleem, R.; Ahmad Shafiai, N.; Mohd Noor, S. Perspective on Metal Leachables from Orthodontic Appliances: A Scoping Review. *J. Int. Oral Health* **2021**, *13*, 539–548.
176. Dwivedi, A.; Tikku, T.; Khanna, R.; Maurya, R.P.; Verma, G.; Murthy, R.C. Release of Nickel and Chromium Ions in the Saliva of Patients with Fixed Orthodontic Appliance: An in-Vivo Study. *Natl. J. Maxillofac. Surg.* **2015**, *6*, 62–66. [CrossRef]
177. Kocadereli, L.; Ataç, A.; Kale, S.; Özer, D. Salivary Nickel and Chromium in Patients with Fixed Orthodontic Appliances. *Angle Orthod.* **2000**, *70*, 431–434. [CrossRef]
178. Fors, R.; Persson, M. Nickel in Dental Plaque and Saliva in Patients with and without Orthodontic Appliances. *Eur. J. Orthod.* **2006**, *28*, 292–297. [CrossRef] [PubMed]
179. Olms, C.; Yahiaoui-Doktor, M.; Remmerbach, T.W. Contact Allergies to Dental Materials. *Swiss Dent. J.* **2019**, *129*, 571–579. [PubMed]
180. Hafez, H.S.; Selim, E.M.N.; Kamel Eid, F.H.; Tawfik, W.A.; Al-Ashkar, E.A.; Mostafa, Y.A. Cytotoxicity, Genotoxicity, and Metal Release in Patients with Fixed Orthodontic Appliances: A Longitudinal in-Vivo Study. *Am. J. Orthod. Dentofac. Orthop.* **2011**, *140*, 298–308. [CrossRef]
181. Singh, D.P.; Sehgal, V.; Pradhan, K.L.; Chandna, A.; Gupta, R. Estimation of Nickel and Chromium in Saliva of Patients with Fixed Orthodontic Appliances. *World J. Orthod.* **2008**, *9*, 196–202.
182. Nayak, R.S.; Khanna, B.; Pasha, A.; Vinay, K.; Narayan, A.; Chaitra, K. Evaluation of Nickel and Chromium Ion Release During Fixed Orthodontic Treatment Using Inductively Coupled Plasma-Mass Spectrometer: An In Vivo Study. *J. Int. Oral Health* **2015**, *7*, 14–20.
183. European Comission Regulation (EC). 1907/2006 of the European Parliament and of the Council of 18 December 2006—REACH. Available online: http://eur-lex.europa.eu/legal-content/EN/TXT/PDF/?uri=CELEX:32006R1907&from=en (accessed on 5 June 2023).
184. Mikulewicz, M.; Chojnacka, K. Human Exposure to Trace Elements from Dental Biomaterials. In *Recent Advances in Trace Elements*; Wiley: Hoboken, NJ, USA, 2018; pp. 469–479, ISBN 9781119133780.
185. Büdinger, L.; Hertl, M. Immunologic Mechanisms in Hypersensitivity Reactions to Metal Ions: An Overview. *Allergy Eur. J. Allergy Clin. Immunol.* **2000**, *55*, 108–115. [CrossRef]
186. Saito, M.; Arakaki, R.; Yamada, A.; Tsunematsu, T.; Kudo, Y.; Ishimaru, N. Molecular Mechanisms of Nickel Allergy. *Int. J. Mol. Sci.* **2016**, *17*, 202. [CrossRef]
187. Peltonen, L. Nickel Sensitivity. *Int. J. Dermatol.* **2008**, *20*, 352–353. [CrossRef]
188. Zambelli, B.; Uversky, V.N.; Ciurli, S. Nickel Impact on Human Health: An Intrinsic Disorder Perspective. *Biochim. Biophys. Acta Proteins Proteom.* **2016**, *1864*, 1714–1731. [CrossRef]
189. Buczko, P.; Szarmach, I.; Grycz, M.; Kasacka, I. Caspase-3 as an Important Factor in the Early Cytotoxic Effect of Nickel on Oral Mucosa Cells in Patients Treated Orthodontically. *Folia Histochem. Cytobiol.* **2017**, *55*, 37–42. [CrossRef] [PubMed]
190. Luz, M.; Souza, A.; Haddad, A.; Tartomano, A.; Oliveira, P. In Vitro Cr(VI) Speciation in Synthetic Saliva after Releasing from Orthodontic Brackets Using Silica-Aptes Separation and GF AAS Determination. *Quim. Nova* **2016**, *39*, 951–955. [CrossRef]
191. Setcos, J.C.; Babaei-Mahani, A.; Di Silvio, L.; Mjör, I.A.; Wilson, N.H.F. The Safety of Nickel Containing Dental Alloys. *Dent. Mater.* **2006**, *22*, 1163–1168. [CrossRef]
192. Schuster, G.; Reichle, R.; Bauer, R.R.; Schopf, P.M. Allergies Induced by Orthodontic Alloys: Incidence and Impact on Treatment. *J. Orofac. Orthop.* **2004**, *65*, 48–59. [CrossRef]

193. Flores-Bracho, M.G.; Takahashi, C.S.; Castillo, W.O.; Saraiva, M.C.P.; Küchler, E.C.; Matsumoto, M.A.N.; Ferreira, J.T.L.; Nelson-Filho, P.; Romano, F.L. Genotoxic Effects in Oral Mucosal Cells Caused by the Use of Orthodontic Fixed Appliances in Patients after Short and Long Periods of Treatment. *Clin. Oral Investig.* **2019**, *23*, 2913–2919. [CrossRef]
194. Loyola-Rodríguez, J.P.; Lastra-Corso, I.; García-Cortés, J.O.; Loyola-Leyva, A.; Domínguez-Pérez, R.A.; Avila-Arizmendi, D.; Contreras-Palma, G.; González-Calixto, C. In Vitro Determination of Genotoxicity Induced by Brackets Alloys in Cultures of Human Gingival Fibroblasts. *J. Toxicol.* **2020**, *2020*, 1467456. [CrossRef] [PubMed]
195. Bass, J.K.; Fine, H.; Cisneros, G.J. Nickel Hypersensitivity in the Orthodontic Patient. *Am. J. Orthod. Dentofac. Orthop.* **1993**, *103*, 280–285. [CrossRef] [PubMed]
196. Kerosuo, H.; Kullaa, A.; Kerosuo, E.; Kanerva, L.; Hensten-Pettersen, A. Nickel Allergy in Adolescents in Relation to Orthodontic Treatment and Piercing of Ears. *Am. J. Orthod. Dentofac. Orthop.* **1996**, *109*, 148–154. [CrossRef] [PubMed]
197. Staerkjaer, L.; Menne, T. Nickel Allergy and Orthodontic Treatment. *Eur. J. Orthod.* **1990**, *12*, 284–289. [CrossRef] [PubMed]
198. Velasco-Ibáñez, R.; Lara-Carrillo, E.; Morales-Luckie, R.A.; Romero-Guzmán, E.T.; Toral-Rizo, V.H.; Ramírez-Cardona, M.; García-Hernández, V.; Medina-Solís, C.E. Evaluation of the Release of Nickel and Titanium under Orthodontic Treatment. *Sci. Rep.* **2020**, *10*, 22280. [CrossRef]
199. Kochanowska, I.E.; Chojnacka, K.; Pawlak-Adamska, E.; Mikulewicz, M. Metallic Orthodontic Materials Induce Gene Expression and Protein Synthesis of Metallothioneins. *Materials* **2021**, *14*, 1922. [CrossRef] [PubMed]
200. Muris, J.; Feilzer, A.J. Micro Analysis of Metals in Dental Restorations as Part of a Diagnostic Approach in Metal Allergies. *Neuro Endocrinol. Lett.* **2006**, *27* (Suppl. 1), 49–52.
201. Dunlap, C.L.; Vincent, S.K.; Barker, B.F. Allergic Reaction to Orthodontic Wire: Report of Case. *J. Am. Dent. Assoc.* **1989**, *118*, 449–450. [CrossRef] [PubMed]
202. Ellis, P.E.; Benson, P.E. Potential Hazards of Orthodontic Treatment—What Your Patient Should Know. *Dent. Update* **2002**, *29*, 492–496. [CrossRef] [PubMed]
203. Kolokitha, O.E.; Chatzistavrou, E. A Severe Reaction to Ni-Containing Orthodontic Appliances. *Angle Orthod.* **2009**, *79*, 186–192. [CrossRef] [PubMed]
204. Noble, J.; Ahing, S.I.; Karaiskos, N.E.; Wiltshire, W.A. Nickel Allergy and Orthodontics, a Review and Report of Two Cases. *Br. Dent. J.* **2008**, *204*, 297–300. [CrossRef] [PubMed]
205. Ehrnrooth, M.; Kerosuo, H. Face and Neck Dermatitis from a Stainless Steel Orthodontic Appliance. *Angle Orthod.* **2009**, *79*, 1194–1196. [CrossRef]
206. Navarro-Triviño, F.J.; Ruiz-Villaverde, R. Contact Urticaria/Angioedema Caused by Nickel from Metal Dental Braces. *Contact Dermat.* **2020**, *83*, 425–427. [CrossRef] [PubMed]
207. Maheshwari, S.; Verma, S.; Dhiman, S. Metal Hypersensitivity in Orthodontic Patients. *J. Dent. Mater. Tech.* **2015**, *4*, 111–114.
208. Rahilly, G.; Price, N. Nickel Allergy and Orthodontics. *J. Orthod.* **2003**, *30*, 171–174. [CrossRef]
209. Kolokitha, O.E.G.; Chatzistavrou, E. Allergic Reactions to Nickel-Containing Orthodontic Appliances: Clinical Signs and Treatment Alternatives. *World J. Orthod.* **2008**, *9*, 399–406.
210. Gursoy, U.K.; Sokucu, O.; Uitto, V.J.; Aydin, A.; Demirer, S.; Toker, H.; Erdem, O.; Sayal, A. The Role of Nickel Accumulation and Epithelial Cell Proliferation in Orthodontic Treatment-Induced Gingival Overgrowth. *Eur. J. Orthod.* **2007**, *29*, 555–558. [CrossRef] [PubMed]
211. Gurgel Maia, L.H.E.; de Lima Filho, H.L.; Araújo, M.V.A.; de Oliveira Ruellas, A.C.; de Souza Araújo, M.T. Incorporation of Metal and Color Alteration of Enamel in the Presence of Orthodontic Appliances. *Angle Orthod.* **2012**, *82*, 889–893. [CrossRef] [PubMed]
212. Pazzini, C.A.; Pereira, L.J.; Marques, L.S.; Ramos-Jorge, J.; Aparecida da Silva, T.; Paiva, S.M. Nickel-Free vs Conventional Braces for Patients Allergic to Nickel: Gingival and Blood Parameters during and after Treatment. *Am. J. Orthod. Dentofac. Orthop.* **2016**, *150*, 1014–1019. [CrossRef] [PubMed]
213. Martín-Cameán, A.; Jos, A.; Cameán, A.M.; Solano, E.; Iglesias-Linares, A. Genotoxic and Cytotoxic Effects and Gene Expression Changes Induced by Fixed Orthodontic Appliances in Oral Mucosa Cells of Patients: A Systematic Review. *Toxicol. Mech. Methods* **2015**, *25*, 440–447. [CrossRef] [PubMed]
214. Downarowicz, P.; Mikulewicz, M. Trace Metal Ions Release from Fixed Orthodontic Appliances and DNA Damage in Oral Mucosa Cells by in Vivo Studies: A Literature Review. *Adv. Clin. Exp. Med.* **2017**, *26*, 1155–1162. [CrossRef] [PubMed]
215. Burrow, S.J. Friction and Resistance to Sliding in Orthodontics: A Critical Review. *Am. J. Orthod. Dentofac. Orthop.* **2009**, *135*, 442–447. [CrossRef] [PubMed]
216. Kusy, R.P.; Whitley, J.Q. Friction between Different Wire-Bracket Configurations and Materials. *Semin. Orthod.* **1997**, *3*, 166–177. [CrossRef] [PubMed]
217. Kusy, R.P.; Whitley, J.Q. Influence of Archwire and Bracket Dimensions on Sliding Mechanics: Derivations and Determinations of the Critical Contact Angles for Binding. *Eur. J. Orthod.* **1999**, *21*, 199–208. [CrossRef]
218. Articolo, L. Influence of Ceramic and Stainless Steel Brackets on the Notching of Archwires during Clinical Treatment. *Eur. J. Orthod.* **2000**, *22*, 409–425. [CrossRef]
219. Prashant, P.; Nandan, H.; Gopalakrishnan, M. Friction in Orthodontics. *J. Pharm. Bioallied Sci.* **2015**, *7*, 334. [CrossRef]
220. Bishara, S.E.; Barrett, R.D.; Selim, M.I. Biodegradation of Orthodontic Appliances. Part II. Changes in the Blood Level of Nickel. *Am. J. Orthod. Dentofac. Orthop.* **1993**, *103*, 115–119. [CrossRef]

221. Kerosuo, H.; Moe, G.; Hensten-Pettersen, A. Salivary Nickel and Chromium in Subjects with Different Types of Fixed Orthodontic Appliances. *Am. J. Orthod. Dentofac. Orthop.* **1997**, *111*, 595–598. [CrossRef] [PubMed]
222. Eliades, T.; Trapalis, C.; Eliades, G.; Katsavrias, E. Salivary Metal Levels of Orthodontic Patients: A Novel Methodological and Analytical Approach. *Eur. J. Orthod.* **2003**, *25*, 103–106. [CrossRef] [PubMed]
223. Faccioni, F.; Franceschetti, P.; Cerpelloni, M.; Fracasso, M.E. In Vivo Study on Metal Release from Fixed Orthodontic Appliances and DNA Damage in Oral Mucosa Cells. *Am. J. Orthod. Dentofac. Orthop.* **2003**, *124*, 687–693. [CrossRef] [PubMed]
224. Levrini, L.; Lusvardi, G.; Gentile, D. Nickel Ions Release in Patients with Fixed Orthodontic Appliances. *Minerva Stomatol.* **2006**, *55*, 115–121.
225. Menezes, L.M.; Quintão, C.A.; Bolognese, A.M. Urinary Excretion Levels of Nickel in Orthodontic Patients. *Am. J. Orthod. Dentofac. Orthop.* **2007**, *131*, 635–638. [CrossRef]
226. Amini, F.; Borzabadi Farahani, A.; Jafari, A.; Rabbani, M. In Vivo Study of Metal Content of Oral Mucosa Cells in Patients with and without Fixed Orthodontic Appliances. *Orthod. Craniofacial Res.* **2008**, *11*, 51–56. [CrossRef]
227. Fernández-Miñano, E.; Ortiz, C.; Vicente, A.; Calvo, J.L.; Ortiz, A.J. Metallic Ion Content and Damage to the DNA in Oral Mucosa Cells of Children with Fixed Orthodontic Appliances. *BioMetals* **2011**, *24*, 935–941, Correction in *BioMetals* **2018**, *31*, 679–679. [CrossRef]
228. Natarajan, M.; Padmanabhan, S.; Chitharanjan, A.; Narasimhan, M. Evaluation of the Genotoxic Effects of Fixed Appliances on Oral Mucosal Cells and the Relationship to Nickel and Chromium Concentrations: An in-Vivo Study. *Am. J. Orthod. Dentofac. Orthop.* **2011**, *140*, 383–388. [CrossRef]
229. Freitas, M.P.M.; Oshima, H.M.S.; Menezes, L.M. Release of Toxic Ions from Silver Solder Used in Orthodontics: An in-Situ Evaluation. *Am. J. Orthod. Dentofac. Orthop.* **2011**, *140*, 177–181. [CrossRef]
230. Mikulewicz, M.; Chojnacka, K.; Zielińska, A.; Michalak, I. Exposure to Metals from Orthodontic Appliances by Hair Mineral Analysis. *Environ. Toxicol. Pharmacol.* **2011**, *32*, 10–16. [CrossRef]
231. Amini, F.; Jafari, A.; Amini, P.; Sepasi, S. Metal Ion Release from Fixed Orthodontic Appliances—An in Vivo Study. *Eur. J. Orthod.* **2012**, *34*, 126–130. [CrossRef]
232. Ousehal, L.; Lazrak, L. Change in Nickel Levels in the Saliva of Patients with Fixed Orthodontic Appliances. *Int. Orthod.* **2012**, *10*, 190–197. [CrossRef] [PubMed]
233. Amini, F.; Rakhshan, V.; Mesgarzadeh, N. Effects of Long-Term Fixed Orthodontic Treatment on Salivary Nickel and Chromium Levels: A 1-Year Prospective Cohort Study. *Biol. Trace Elem. Res.* **2012**, *150*, 15–20. [CrossRef] [PubMed]
234. Talic, N.F.; Alnahwi, H.H.; Al-Faraj, A.S. Nickel and Chromium Levels in the Saliva of a Saudi Sample Treated with Fixed Orthodontic Appliances. *Saudi Dent. J.* **2013**, *25*, 129–133. [CrossRef] [PubMed]
235. Abtahi, M.; Jahanbin, A.; Yaghoubi, M.; Esmaily, H.; Zare, H. Are More Nickel Ions Accumulated in the Hair of Fixed Orthodontic Patients? *Indian J. Dent. Res.* **2013**, *24*, 298. [CrossRef] [PubMed]
236. Amini, F.; Rahimi, H.; Morad, G.; Mollaei, M. The Effect of Stress on Salivary Metal Ion Content in Orthodontic Patients. *Biol. Trace Elem. Res.* **2013**, *155*, 339–343. [CrossRef] [PubMed]
237. Martín-Cameán, A.; Jos, A.; Calleja, A.; Gil, F.; Iglesias, A.; Solano, E.; Cameán, A.M. Validation of a Method to Quantify Titanium, Vanadium and Zirconium in Oral Mucosa Cells by Inductively Coupled Plasma-Mass Spectrometry (ICP-MS). *Talanta* **2014**, *118*, 238–244. [CrossRef] [PubMed]
238. Martín-Cameán, A.; Jos, A.; Calleja, A.; Gil, F.; Iglesias-Linares, A.; Solano, E.; Cameán, A.M. Development and Validation of an Inductively Coupled Plasma Mass Spectrometry (ICP-MS) Method for the Determination of Cobalt, Chromium, Copper and Nickel in Oral Mucosa Cells. *Microchem. J.* **2014**, *114*, 73–79. [CrossRef]
239. Martín-Cameán, A.; Molina-Villalba, I.; Jos, A.; Iglesias-Linares, A.; Solano, E.; Cameán, A.M.; Gil, F. Biomonitorization of Chromium, Copper, Iron, Manganese and Nickel in Scalp Hair from Orthodontic Patients by Atomic Absorption Spectrometry. *Environ. Toxicol. Pharmacol.* **2014**, *37*, 759–771. [CrossRef]
240. Mikulewicz, M.; Wołowiec, P.; Janeczek, M.; Gedrange, T.; Chojnacka, K. The Release of Metal Ions from Orthodontic Appliances Animal Tests. *Angle Orthod.* **2014**, *84*, 673–679. [CrossRef] [PubMed]
241. Mikulewicz, M.; Wołowiec, P.; Loster, B.; Chojnacka, K. Metal Ions Released from Fixed Orthodontic Appliance Affect Hair Mineral Content. *Biol. Trace Elem. Res.* **2015**, *163*, 11–18. [CrossRef] [PubMed]
242. Amini, F.; Harandi, S.; Mollaei, M.; Rakhshan, V. Effects of Fixed Orthodontic Treatment Using Conventional versus Metal-Injection Molding Brackets on Salivary Nickel and Chromium Levels: A Double-Blind Randomized Clinical Trial. *Eur. J. Orthod.* **2015**, *37*, 522–530. [CrossRef] [PubMed]
243. Amini, F.; Mollaei, M.; Harandi, S.; Rakhshan, V. Effects of Fixed Orthodontic Treatment on Hair Nickel and Chromium Levels: A 6-Month Prospective Preliminary Study. *Biol. Trace Elem. Res.* **2015**, *164*, 12–17. [CrossRef] [PubMed]
244. Saghiri, M.A.; Orangi, J.; Asatourian, A.; Mehriar, P.; Sheibani, N. Effect of Mobile Phone Use on Metal Ion Release from Fixed Orthodontic Appliances. *Am. J. Orthod. Dentofac. Orthop.* **2015**, *147*, 719–724. [CrossRef] [PubMed]
245. Gómez Arcila, V.; Fang Mercado, L.; Herrera Herrera, A.; Díaz Caballero, A. Bioacumulación de Níquel En Encía, Saliva y Hueso Alveolar de Paciente Con Aparatología Ortodóncica Fija: Reporte de Un Caso. *Rev. Clín. Periodoncia Implantol. Rehabil. Oral* **2015**, *8*, 163–166. [CrossRef]

246. Gölz, L.; Knickenberg, A.C.; Keilig, L.; Reimann, S.; Papageorgiou, S.N.; Jäger, A.; Bourauel, C. Nickel Ion Concentrations in the Saliva of Patients Treated with Self-Ligating Fixed Appliances: A Prospective Cohort Study. *J. Orofac. Orthop. Fortschritte Kieferorthopädie* **2016**, *77*, 85–93. [CrossRef]
247. Amini, F.; Shariati, M.; Sobouti, F.; Rakhshan, V. Effects of Fixed Orthodontic Treatment on Nickel and Chromium Levels in Gingival Crevicular Fluid as a Novel Systemic Biomarker of Trace Elements: A Longitudinal Study. *Am. J. Orthod. Dentofac. Orthop.* **2016**, *149*, 666–672. [CrossRef]
248. Kumar, R.V.; Rajvikram, N.; Rajakumar, P.; Saravanan, R.; Deepak, V.A. An Accurate Methodology to Detect Leaching of Nickel and Chromium Ions in the Initial Phase of Orthodontic Treatment: An in Vivo Study. *J. Contemp. Dent. Pract.* **2016**, *17*, 205–210. [CrossRef]
249. Wołowiec, P.; Chojnacka, K.; Loster, B.W.; Mikulewicz, M. Do Dietary Habits Influence Trace Elements Release from Fixed Orthodontic Appliances? *Biol. Trace Elem. Res.* **2017**, *180*, 214–222. [CrossRef]
250. Causado-Vitola, V.; Rumbo-Zubiría, M.; Fang, L.; Diaz-Caballero, A. Nickel Variation in Biofilm, Saliva and Buccal Mucosa During Orthodontic Treatment. *Asian J. Appl. Sci.* **2017**, *10*, 45–49. [CrossRef]
251. Khaneh Masjedi, M.; Haghighat Jahromi, N.; Niknam, O.; Hormozi, E.; Rakhshan, V. Effects of Fixed Orthodontic Treatment Using Conventional (Two-Piece) versus Metal Injection Moulding Brackets on Hair Nickel and Chromium Levels: A Double-Blind Randomized Clinical Trial. *Eur. J. Orthod.* **2017**, *39*, 17–24. [CrossRef] [PubMed]
252. Buczko, P.; Knaś, M.; Grycz, M.; Szarmach, I.; Zalewska, A. Orthodontic Treatment Modifies the Oxidant–Antioxidant Balance in Saliva of Clinically Healthy Subjects. *Adv. Med. Sci.* **2017**, *62*, 129–135. [CrossRef] [PubMed]
253. Lages, R.B.; Bridi, E.C.; Pérez, C.A.; Basting, R.T. Salivary Levels of Nickel, Chromium, Iron, and Copper in Patients Treated with Metal or Esthetic Fixed Orthodontic Appliances: A Retrospective Cohort Study. *J. Trace Elem. Med. Biol.* **2017**, *40*, 67–71, Corrigendum in *J. Trace Elem. Med. Biol.* **2018**, *45*, 189–190. [CrossRef] [PubMed]
254. Nanjannawar, L.G.; Girme, T.S.; Agrawal, J.M.; Agrawal, M.S.; Fulari, S.G.; Shett, S.S.; Kagi, V.A. Effect of Mobile Phone Usage OnNickel Ions Release and PH of Saliva InPatients Undergoing Fixed OrthodonticTreatment. *J. Clin. Diagn. Res.* **2017**, *11*, ZC84–ZC87. [CrossRef] [PubMed]
255. Bhasin, V.; Pustake, S.J.; Joshi, V.; Tiwari, A.; Bhasin, M.; Punia, R.S. Assessment of Changes in Nickel and Chromium Levels in the Gingival Crevicular Fluid during Fixed Orthodontic Treatment. *J. Contemp. Dent. Pract.* **2017**, *18*, 675–678. [CrossRef] [PubMed]
256. Jamshidi, S.; Rahmati Kamel, M.; Mirzaie, M.; Sarrafan, A.; Khafri, S.; Parsian, H. Evaluation of Scalp Hair Nickel and Chromium Level Changes in Patients with Fixed Orthodontic Appliance: A One-Year Follow-up Study. *Acta Odontol. Scand.* **2018**, *76*, 1–5. [CrossRef]
257. Jurela, A.; Verzak, Ž.; Brailo, V.; Škrinjar, I.; Sudarević, K.; Janković, B. Salivary Electrolytes in Patients with Metallic and Ceramic Orthodontic Brackets. *Acta Stomatol. Croat.* **2018**, *52*, 32–36. [CrossRef]
258. Amini, F.; Asadi, E.; Hakimpour, D.; Rakhshan, A. Salivary Nickel and Chromium Levels in Orthodontic Patients with and Without Periodontitis: A Preliminary Historical Cohort Study. *Biol. Trace Elem. Res.* **2019**, *191*, 10–15. [CrossRef]
259. Chitra, P.; Prashantha, G.S.; Rao, A. Long-Term Evaluation of Metal Ion Release in Orthodontic Patients Using Fluoridated Oral Hygiene Agents: An in Vivo Study. *J. World Fed. Orthod.* **2019**, *8*, 107–111. [CrossRef]
260. Quadras, D.D.; Nayak, U.K.; Kumari, N.S.; Priyadarshini, H.R.; Gowda, S.; Fernandes, B. In Vivo Study on the Release of Nickel, Chromium, and Zinc in Saliva and Serum from Patients Treated with Fixed Orthodontic Appliances. *Dent. Res. J.* **2019**, *16*, 209. [CrossRef]
261. Moghadam, M.G.; Hoshyar, R.; Mikulewicz, M.; Chojnacka, K.; Bjørklund, G.; Pen, J.J.; Azadi, N.A.; Pirsaheb, M.; Dashtaki, M.; Mansouri, B. Biomonitorization of Metal Ions in the Serum of Iranian Patients Treated with Fixed Orthodontic Appliances in Comparison with Controls in Eastern Iran. *Environ. Sci. Pollut. Res.* **2019**, *26*, 33373–33386. [CrossRef] [PubMed]
262. Butt, M.; Mengal, N.; Ahmed, M. A Comparison of Nickel Release in Saliva between Orthodontic and Non-Orthodontic Patients. *Pak. Armed Forces Med. J.* **2020**, *70*, 328–332.
263. de Souza Schacher, H.R.; de Menezes, L.M. Metal Ion Quantification in the Saliva of Patients with Lingual Arch Appliances Using Silver Solder, Laser, or TIG Welding. *Clin. Oral Investig.* **2020**, *24*, 2109–2120. [CrossRef] [PubMed]
264. Pritam, A.; Priyadarshini, A.; Hussain, K.; Kumar, A.; Kumar, N.; Malakar, A. Assessment of Nickel and Chromium Level in Gingival Crevicular Fluid in Patients Undergoing Orthodontic Treatment with or without Fluoridated Tooth Paste. *J. Pharm. Bioallied Sci.* **2021**, *13*, 1588. [CrossRef] [PubMed]
265. Campos Zeffa, A.; Dias, B.G.; Silva, D.C.M.S.; Rotta, L.O.; Jussiani, E.I.; Andrello, A.C.; de Paula Ramos, S. Influence of Conventional or Invisalign Orthodontic Treatment on Mineral and Trace Element Salivary Levels: Longitudinal Study with Total Reflection X-Ray Fluorescence. *Biol. Trace Elem. Res.* **2021**, *199*, 2565–2572. [CrossRef] [PubMed]
266. Hamadamin, S.I. In Vivo Kinetic Release of Five Metal Ions (Iron, Titanium, Nickel, Copper, and Chromium) from Fixed Orthodontic Alloys in Erbil City-Kurdistan Region/Iraq. *Environ. Sci. Pollut. Res.* **2022**, *29*, 11730–11735. [CrossRef] [PubMed]
267. Rajendran, R.; Venkatachalapathy, S.; Thiyagarajan, B.; Jeevagan, S.; Chinnasamy, A.; Sivanandham, M. Effect of Mobile Phone with and without Earphones Usage on Nickel Ion Release from Fixed Orthodontic Appliance. *J. Contemp. Dent. Pract.* **2023**, *24*, 303–307. [CrossRef]
268. Vaughan, J.L.; Duncanson, M.G.; Nanda, R.S.; Currier, G.F. Relative Kinetic Frictional Forces between Sintered Stainless Steel Brackets and Orthodontic Wires. *Am. J. Orthod. Dentofac. Orthop.* **1995**, *107*, 20–27. [CrossRef]

269. Dridi, A.; Bensalah, W.; Mezlini, S.; Tobji, S.; Zidi, M. Influence of Bio-Lubricants on the Orthodontic Friction. *J. Mech. Behav. Biomed. Mater.* **2016**, *60*, 1–7. [CrossRef]
270. Fidalgo, T.K.D.S.; Pithon, M.M.; Maciel, J.V.B.; Bolognese, A.M. Friction between Different Wire Bracket Combinations in Artificial Saliva—An in Vitro Evaluation. *J. Appl. Oral Sci.* **2011**, *19*, 57–62. [CrossRef] [PubMed]
271. Ranc, H.; Elkhyat, A.; Servais, C.; Mac-Mary, S.; Launay, B.; Humbert, P. Friction Coefficient and Wettability of Oral Mucosal Tissue: Changes Induced by a Salivary Layer. *Colloids Surf. A Physicochem. Eng. Asp.* **2006**, *276*, 155–161. [CrossRef]
272. Lu, C.; Zheng, Y.; Zhong, Q. Corrosion of Dental Alloys in Artificial Saliva with Streptococcus Mutans. *PLoS ONE* **2017**, *12*, e0174440. [CrossRef] [PubMed]
273. Reichardt, E.; Geraci, J.; Sachse, S.; Rödel, J.; Pfister, W.; Löffler, B.; Wagner, Y.; Eigenthaler, M.; Wolf, M. Qualitative and Quantitative Changes in the Oral Bacterial Flora Occur Shortly after Implementation of Fixed Orthodontic Appliances. *Am. J. Orthod. Dentofac. Orthop.* **2019**, *156*, 735–744. [CrossRef] [PubMed]
274. Mei, L.; Chieng, J.; Wong, C.; Benic, G.; Farella, M. Factors Affecting Dental Biofilm in Patients Wearing Fixed Orthodontic Appliances. *Prog. Orthod.* **2017**, *18*, 4. [CrossRef] [PubMed]
275. Alavi, S.; Yaraghi, N. The Effect of Fluoride Varnish and Chlorhexidine Gel on White Spots and Gingival and Plaque Indices in Fixed Orthodontic Patients: A Placebo-Controlled Study. *Dent. Res. J.* **2018**, *15*, 276–282.
276. Chitra, P.; Prashantha, G.S.; Rao, A. Effect of Fluoride Agents on Surface Characteristics of NiTi Wires. An Ex Vivo Investigation. *J. Oral Biol. Craniofacial Res.* **2020**, *10*, 435–440. [CrossRef] [PubMed]
277. Schiff, N.; Grosgogeat, B.; Lissac, M.; Dalard, F. Influence of Fluoridated Mouthwashes on Corrosion Resistance of Orthodontics Wires. *Biomaterials* **2004**, *25*, 4535–4542. [CrossRef]
278. Nahidh, M.; MH Garma, N.; Jasim, E.S. Assessment of Ions Released from Three Types of Orthodontic Brackets Immersed in Different Mouthwashes: An in Vitro Study. *J. Contemp. Dent. Pract.* **2018**, *19*, 73–80. [CrossRef]
279. Rincic Mlinaric, M.; Karlovic, S.; Ciganj, Z.; Acev, D.P.; Pavlic, A.; Spalj, S. Oral Antiseptics and Nickel–Titanium Alloys: Mechanical and Chemical Effects of Interaction. *Odontology* **2019**, *107*, 150–157. [CrossRef]
280. Condò, R.; Carli, E.; Cioffi, A.; Cataldi, M.E.; Quinzi, V.; Casaglia, A.; Giancotti, A.; Pirelli, P.; Lucarini, I.; Maita, F.; et al. Fluorinated Agents Effects on Orthodontic Alloys: A Descriptive In Vitro Study. *Materials* **2022**, *15*, 4612. [CrossRef] [PubMed]
281. Pastor, F.; Rodríguez, J.C.; Barrera, J.M.; Delgado García-Menocal, J.A.; Brizuela, A.; Puigdollers, A.; Espinar, E.; Gil, J. Effect of Fluoride Content of Mouthwashes on Superelastic Properties of NiTi Orthodontic Archwires. *Materials* **2022**, *15*, 6592. [CrossRef] [PubMed]
282. Sahoo, N.; Bhuyan, L.; Dhull, K.S.; Dash, K.C.; MD, I.; Mishra, P. In Vitro Effect Of Fluoride Prophylactic Agents On Titanium Molybdenum Alloy And Stainless Steel Orthodontic Wires—Scanning Electron Microscope Study. *Bangladesh J. Med. Sci.* **2023**, *22*, 47–51. [CrossRef]
283. Mirhashemi, A.; Jahangiri, S.; Kharrazifard, M. Release of Nickel and Chromium Ions from Orthodontic Wires Following the Use of Teeth Whitening Mouthwashes. *Prog. Orthod.* **2018**, *19*, 4. [CrossRef]
284. Yanisarapan, T.; Thunyakitpisal, P.; Chantarawaratit, P. Corrosion of Metal Orthodontic Brackets and Archwires Caused by Fluoride-Containing Products: Cytotoxicity, Metal Ion Release and Surface Roughness. *Orthod. Waves* **2018**, *77*, 79–89. [CrossRef]
285. Abbassy, M. Fluoride Influences Nickel-Titanium Orthodontic Wires′ Surface Texture and Friction Resistance. *J. Orthod. Sci.* **2016**, *5*, 121. [CrossRef]
286. Rajendran, A.; Sundareswaran, S.; Peediyekkal, L.; Santhakumar, P.; Sathyanadhan, S. Effect of Oral Environment and Prescribed Fluoride Mouthwashes on Different Types of TMA Wires—An in-Vivo Study. *J. Orthod. Sci.* **2019**, *8*, 8. [CrossRef]
287. Li, X.; Wang, J.; Han, E.; Ke, W. Influence of Fluoride and Chloride on Corrosion Behavior of NiTi Orthodontic Wires. *Acta Biomater.* **2007**, *3*, 807–815. [CrossRef]
288. Brandão, G.A.M.; Simas, R.M.; de Almeida, L.M.; da Silva, J.M.; Meneghim, M.d.C.; Pereira, A.C.; de Almeida, H.A.; Brandão, A.M.M. Evaluation of Ionic Degradation and Slot Corrosion of Metallic Brackets by the Action of Different Dentifrices. *Dent. Press J. Orthod.* **2013**, *18*, 86–93. [CrossRef]
289. Chantarawaratit, P.; Yanisarapan, T. Exposure to the Oral Environment Enhances the Corrosion of Metal Orthodontic Appliances Caused by Fluoride-Containing Products: Cytotoxicity, Metal Ion Release, and Surface Roughness. *Am. J. Orthod. Dentofac. Orthop.* **2021**, *160*, 101–112. [CrossRef]
290. Russell, J.S. Current Products and Practice: Aesthetic Orthodontic Brackets. *J. Orthod.* **2005**, *32*, 146–163. [CrossRef] [PubMed]
291. Ali, O.; Makou, M.; Papadopoulos, T.; Eliades, G. Laboratory Evaluation of Modern Plastic Brackets. *Eur. J. Orthod.* **2012**, *34*, 595–602. [CrossRef] [PubMed]
292. Cacciafesta, V.; Sfondrini, M.F.; Scribante, A.; Klersy, C.; Auricchio, F. Evaluation of Friction of Conventional and Metal-Insert Ceramic Brackets in Various Bracket-Archwire Combinations. *Am. J. Orthod. Dentofac. Orthop.* **2003**, *124*, 403–409. [CrossRef] [PubMed]
293. Arici, N.; Akdeniz, B.S.; Arici, S. Comparison of the Frictional Characteristics of Aesthetic Orthodontic Brackets Measured Using a Modified in Vitro Technique. *Korean J. Orthod.* **2015**, *45*, 29. [CrossRef] [PubMed]
294. Varela, J.C.; Velo, M.; Espinar, E.; Llamas, J.M.; Rúperez, E.; Manero, J.M.; Javier Gil, F. Mechanical Properties of a New Thermoplastic Polymer Orthodontic Archwire. *Mater. Sci. Eng. C* **2014**, *42*, 1–6. [CrossRef]
295. Burstone, C.J.; Liebler, S.A.H.; Goldberg, A.J. Polyphenylene Polymers as Esthetic Orthodontic Archwires. *Am. J. Orthod. Dentofac. Orthop.* **2011**, *139*, e391–e398. [CrossRef]

296. Tada, Y.; Hayakawa, T.; Nakamura, Y. Load-Deflection and Friction Properties of PEEK Wires as Alternative Orthodontic Wires. *Materials* **2017**, *10*, 914. [CrossRef]
297. Spendlove, J.; Berzins, D.W.; Pruszynski, J.E.; Ballard, R.W. Investigation of Force Decay in Aesthetic, Fibre-Reinforced Composite Orthodontic Archwires. *Eur. J. Orthod.* **2015**, *37*, 43–48. [CrossRef]
298. Chng, C.K.; Foong, K.; Gandedkar, N.H.; Chan, Y.H.; Chew, C.-L. A New Esthetic Fiber-Reinforced Polymer Composite Resin Archwire: A Comparative Atomic Force Microscope (AFM) and Field-Emission Scanning Electron Microscope (FESEM) Study. *Prog. Orthod.* **2014**, *15*, 39. [CrossRef]
299. Mikulewicz, M.; Gronostajski, Z.; Wielgus, A.; Chojnacka, K. Transparent Orthodontic Archwires: A Systematic Literature Review. *Arch. Civ. Mech. Eng.* **2017**, *17*, 651–657. [CrossRef]
300. Zinelis, S.; Brantley, W. Structure/Property Relationships in Orthodontic Ceramics. In *Orthodontic Applications of Biomaterials*; Eliades, T., Brantley, W.A., Eds.; Elsevier: Amsterdam, The Netherlands, 2017; pp. 61–71.
301. Krauss, J.; Faltermeier, A.; Behr, M.; Proff, P. Evaluation of Alternative Polymer Bracket Materials. *Am. J. Orthod. Dentofac. Orthop.* **2010**, *137*, 362–367. [CrossRef] [PubMed]
302. Faltermeier, A.; Behr, M.; Mussig, D. In Vitro Colour Stability of Aesthetic Brackets. *Eur. J. Orthod.* **2007**, *29*, 354–358. [CrossRef] [PubMed]
303. Faltermeier, A.; Reicheneder, C.; Römer, P.; Castro-Laza, A.; Proff, P. Effect of Ionizing Radiation on Polymer Brackets. *J. Orofac. Orthop.* **2014**, *75*, 334–344. [CrossRef] [PubMed]
304. Retamoso, L.B.; Luz, T.B.; Marinowic, D.R.; Machado, D.C.; De Menezes, L.M.; Freitas, M.P.M.; Oshima, H.M.S. Cytotoxicity of Esthetic, Metallic, and Nickel-Free Orthodontic Brackets: Cellular Behavior and Viability. *Am. J. Orthod. Dentofac. Orthop.* **2012**, *142*, 70–74. [CrossRef] [PubMed]
305. Guignone, B.C.; Silva, L.K.; Soares, R.V.; Akaki, E.; Goiato, M.C.; Pithon, M.M.; Oliveira, D.D. Color Stability of Ceramic Brackets Immersed in Potentially Staining Solutions. *Dent. Press J. Orthod.* **2015**, *20*, 32–38. [CrossRef] [PubMed]
306. Johnson, G.; Walker, M.P.; Kula, K. Fracture Strength of Ceramic Bracket Tie Wings Subjected to Tension. *Angle Orthod.* **2005**, *75*, 95–100. [CrossRef] [PubMed]
307. Sanchez, D.J.; Walker, M.P.; Kula, K.; Williams, K.B.; Eick, J.D. Fluoride Prophylactic Agents Effect on Ceramic Bracket Tie-Wing Fracture Strength. *Angle Orthod.* **2008**, *78*, 524–530. [CrossRef]
308. Guerrero, A.P.; Guariza Filho, O.; Tanaka, O.; Camargo, E.S.; Vieira, S. Evaluation of Frictional Forces between Ceramic Brackets and Archwires of Different Alloys Compared with Metal Brackets. *Braz. Oral Res.* **2010**, *24*, 40–45. [CrossRef]
309. Sahoo, N. Comparison of the Perception of Pain during Fixed Orthodontic Treatment with Metal and Ceramic Brackets. *J. Pharm. Bioallied Sci.* **2019**, *11*, 30. [CrossRef]
310. Matias, M.; Freitas, M.R.D.; Freitas, K.M.S.D.; Janson, G.; Higa, R.H.; Francisconi, M.F. Comparison of Deflection Forces of Esthetic Archwires Combined with Ceramic Brackets. *J. Appl. Oral Sci.* **2018**, *26*, e20170220. [CrossRef]
311. da Silva, D.L.; Mattos, C.T.; de Araújo, M.V.A.; de Oliveira Ruellas, A.C. Color Stability and Fluorescence of Different Orthodontic Esthetic Archwires. *Angle Orthod.* **2013**, *83*, 127–132. [CrossRef] [PubMed]
312. Ortiz, A.J.; Fernández, E.; Vicente, A.; Calvo, J.L.; Ortiz, C. Metallic Ions Released from Stainless Steel, Nickel-Free, and Titanium Orthodontic Alloys: Toxicity and DNA Damage. *Am. J. Orthod. Dentofac. Orthop.* **2011**, *140*, e115–e122. [CrossRef]
313. Pazzini, C.A.; Marques, L.S.; Pereira, L.J.; Corrêa-Faria, P.; Paiva, S.M. Allergic Reactions and Nickel-Free Braces: A Systematic Review. *Braz. Oral Res.* **2011**, *25*, 85–90. [CrossRef] [PubMed]
314. Pantuzo, M.C.G.; Zenóbio, E.G.; Marigo, H.D.A.; Zenóbio, M.A.F. Hypersensitivity to Conventional and to Nickel-Free Orthodontic Brackets. *Braz. Oral Res.* **2007**, *21*, 298–302. [CrossRef] [PubMed]
315. Zheng, Y.F.; Zhang, B.B.; Wang, B.L.; Wang, Y.B.; Li, L.; Yang, Q.B.; Cui, L.S. Introduction of Antibacterial Function into Biomedical TiNi Shape Memory Alloy by the Addition of Element Ag. *Acta Biomater.* **2011**, *7*, 2758–2767. [CrossRef] [PubMed]
316. Baigonakova, G.; Marchenko, E.; Chekalkin, T.; Kang, J.; Weiss, S.; Obrosov, A. Influence of Silver Addition on Structure, Martensite Transformations and Mechanical Properties of TiNi–Ag Alloy Wires for Biomedical Application. *Materials* **2020**, *13*, 4721. [CrossRef]
317. Oh, K.-T.; Joo, U.-H.; Park, G.-H.; Hwang, C.-J.; Kim, K.-N. Effect of Silver Addition on the Properties of Nickel-Titanium Alloys for Dental Application. *J. Biomed. Mater. Res. Part B Appl. Biomater.* **2006**, *76*, 306–314. [CrossRef]
318. He, L.; Cui, Y.; Zhang, C. The Corrosion Resistance, Cytotoxicity, and Antibacterial Properties of Lysozyme Coatings on Orthodontic Composite Arch Wires. *RSC Adv.* **2020**, *10*, 18131–18137. [CrossRef]
319. Iijima, M.; Zinelis, S.; Papageorgiou, S.N.; Brantley, W.; Eliades, T. Orthodontic Brackets. In *Orthodontic Applications of Biomaterials*; Elsevier: Amsterdam, The Netherlands, 2017; pp. 75–96, ISBN 9780081003831.
320. Ntasi, A.; Al Jabbari, Y.; Mueller, W.D.; Eliades, G.; Zinelis, S. Metallurgical and Electrochemical Characterization of Contemporary Silver-Based Soldering Alloys. *Angle Orthod.* **2014**, *84*, 508–515. [CrossRef]
321. Freitas, M.P.M.; Oshima, H.M.S.; Menezes, L.M.; Machado, D.C.; Viezzer, C. Cytotoxicity of Silver Solder Employed in Orthodontics. *Angle Orthod.* **2009**, *79*, 939–944. [CrossRef]
322. Erdogan, A.T.; Nalbantgil, D.; Ulkur, F.; Sahin, F. Metal Ion Release from Silver Soldering and Laser Welding Caused by Different Types of Mouthwash. *Angle Orthod.* **2015**, *85*, 665–672. [CrossRef] [PubMed]

323. da Costa, É.C.S.C.; Neves, J.G.; Borges, L.P.S.; Tsuzuki, F.M.; Correr, A.B.; Correr-Sobrinho, L.; Costa, A.R. Comparison of the Physico-Chemical Impact of Chlorhexidine and Silver Nanoparticles on Orthodontic Appliances Made with Laser and Silver Solder: An in Vitro Study. *Int. Orthod.* **2022**, *20*, 100631. [CrossRef] [PubMed]
324. Rodrigues, P.F.; Fernandes, F.M.B.; Magalhães, R.; Camacho, E.; Lopes, A.; Paula, A.S.; Basu, R.; Schell, N. Thermo-Mechanical Characterization of NiTi Orthodontic Archwires with Graded Actuating Forces. *J. Mech. Behav. Biomed. Mater.* **2020**, *107*, 103747. [CrossRef] [PubMed]
325. Javaid, M.; Haleem, A. Current Status and Applications of Additive Manufacturing in Dentistry: A Literature-Based Review. *J. Oral Biol. Craniofacial Res.* **2019**, *9*, 179–185. [CrossRef] [PubMed]
326. Panayi, N.C. 3D Printing of in Office Custom-Made Brackets: Concept, Design, Production and Evidence. *Semin. Orthod.* **2023**, *29*, 11–16. [CrossRef]
327. Graf, S.; Thakkar, D.; Hansa, I.; Muthuswamy Pandian, S.; Adel, S.M. 3D Metal Printing in Orthodontics: Current Trends, Biomaterials, Workflows and Clinical Implications. *Semin. Orthod.* **2023**, *29*, 34–42. [CrossRef]
328. Abalos, C.; Paúl, A.; Mendoza, A.; Solano, E.; Gil, F.J. Influence of Topographical Features on the Fluoride Corrosion of Ni–Ti Orthodontic Archwires. *J. Mater. Sci. Mater. Med.* **2011**, *22*, 2813–2821. [CrossRef]
329. Wichelhaus, A.; Geserick, M.; Hibst, R.; Sander, F.G. The Effect of Surface Treatment and Clinical Use on Friction in NiTi Orthodontic Wires. *Dent. Mater.* **2005**, *21*, 938–945. [CrossRef]
330. Neumann, P.; Bourauel, C.; Jäger, A. Corrosion and Permanent Fracture Resistance of Coated and Conventional Orthodontic Wires. *J. Mater. Sci. Mater. Med.* **2002**, *13*, 141–147. [CrossRef]
331. Chu, P.; Chen, J.Y.; Wang, L.P.; Huang, N. Plasma-Surface Modification of Biomaterials. *Mater. Sci. Eng. R Rep.* **2002**, *36*, 143–206. [CrossRef]
332. Katic, V.; Curkovic, L.; Bosnjak, M.; Peros, K.; Mandic, D.; Spals, S. Effect of PH, Fluoride and Hydrofluoric Acid Concentration on Ion Release from NiTi Wires with Various Coatings. *Dent. Mater. J.* **2017**, *36*, 149–156. [CrossRef] [PubMed]
333. de Albuquerque, C.G.; Correr, A.B.; Venezian, G.C.; Santamaria Jr, M.; Tubel, C.A.; Vedovello, S.A.S. Deflection and Flexural Strength Effects on the Roughness of Aesthetic-Coated Orthodontic Wires. *Braz. Dent. J.* **2017**, *28*, 40–45. [CrossRef] [PubMed]
334. Iijima, M.; Muguruma, T.; Brantley, W.; Choe, H.-C.; Nakagaki, S.; Alapati, S.B.; Mizoguchi, I. Effect of Coating on Properties of Esthetic Orthodontic Nickel-Titanium Wires. *Angle Orthod.* **2012**, *82*, 319–325. [CrossRef] [PubMed]
335. Alsanea, J.; Al Shehri, H. Evaluation of Nanomechanical Properties, Surface Roughness, and Color Stability of Esthetic Nickel-Titanium Orthodontic Archwires. *J. Int. Soc. Prev. Community Dent.* **2019**, *9*, 33. [CrossRef]
336. Pinzan-Vercelino, C.R.M.; de Araújo Gurgel, J.; Bramante, F.S.; Pessoa, T.F.; Albertin, S.A.; Fialho, M.P.N.; Pinzan, A. Coated Nickel Titanium Archwires and Their Uncoated Counterparts: In-Vitro Contemporary Comparison of Unloading Forces. *Int. Orthod.* **2020**, *18*, 374–379. [CrossRef]
337. Batista, C.H.; Neves, J.G.; Terossi de Godoi, A.P.; Veroni Degan, V.; Custódio, W.; Furletti, V.; Vedovello Filho, M. Comparison of Force Delivery of Thermally Activated Aesthetic and Non-Aesthetic Ni-Ti Wires: An in-Vitro Study. *Int. Orthod.* **2020**, *18*, 359–365. [CrossRef]
338. Jusufi Osmani, Z.; Poljšak, B.; Zelenika, S.; Kamenar, E.; Marković, K.; Perčić, M.; Katić, V. Ion Release and Surface Changes of Nickel–Titanium Archwires Induced by Changes in the PH Value of the Saliva—Significance for Human Health Risk Assessment. *Materials* **2022**, *15*, 1994. [CrossRef]
339. Katić, V.; Ćurković, L.; Ujević Bošnjak, M.; Špalj, S. Determination of Corrosion Rate of Orthodontic Wires Based on Nickel-Titanium Alloy in Artificial Saliva. *Materwiss. Werksttech.* **2014**, *45*, 99–105. [CrossRef]
340. Katić, V.; Ćurković, H.O.; Semenski, D.; Baršić, G.; Marušić, K.; Špalj, S. Influence of Surface Layer on Mechanical and Corrosion Properties of Nickel-Titanium Orthodontic Wires. *Angle Orthod.* **2014**, *84*, 1041–1048. [CrossRef]
341. Katić, V.; Buljan, Z.I.; Špalj, S.; Ćurković, H.O. Corrosion Behavior of Coated and Uncoated Nickel-Titanium Orthodontic Wires in Artificial Saliva with Short-Term Prophylactic Fluoride Treatment. *Int. J. Electrochem. Sci.* **2018**, *13*, 4160–4170. [CrossRef]
342. de Amorim, M.C.; da Rocha Gomes, S.; da Silva, B.P.; Aoki, I.V.; Basting, R.T. Surface Micromorphology, Ion Release and Resistance to Corrosion of Orthodontic Wires Aesthetic Coating Subject to Degradation. *J. Bio-Tribo-Corros.* **2022**, *8*, 22. [CrossRef]
343. Nsaif, Y.A.; Mahmood, A.B. Effect of Fluoride Agent on the Load Deflection of Rhodium-Coated Arch Wires; An In-Vitro Study. *Indian J. Public Health Res. Dev.* **2019**, *10*, 823. [CrossRef]
344. Musa Trolić, I.; Turco, G.; Contardo, L.; Serdarević, N.L.; Ćurković, H.O.; Špalj, S. Corrosion of Nickel-Titanium Orthodontic Archwires in Saliva and Oral Probiotic Supplements. *Acta Stomatol. Croat.* **2017**, *51*, 316–325. [CrossRef] [PubMed]
345. Musa Trolic, I.; Serdarevic, N.L.; Todoric, Z.; Budimir, A.; Spalj, S.; Otmacic Curkovic, H. Corrosion of Orthodontic Archwires in Artificial Saliva in the Presence of Lactobacillus Reuteri. *Surf. Coat. Technol.* **2019**, *370*, 44–52. [CrossRef]
346. Musa Trolic, I.; Todoric, Z.; Pop Acev, D.; Makreski, P.; Pejova, B.; Spalj, S. Effects of the Presence of Probiotic Bacteria in the Aging Medium on the Surface Roughness and Chemical Composition of Two Dental Alloys. *Microsc. Res. Tech.* **2019**, *82*, 1384–1391. [CrossRef]
347. Costa Lima, K.C.; Benini Paschoal, M.A.; de Araújo Gurgel, J.; Salvatore Freitas, K.M.; Maio Pinzan-Vercelino, C.R. Comparative Analysis of Microorganism Adhesion on Coated, Partially Coated, and Uncoated Orthodontic Archwires: A Prospective Clinical Study. *Am. J. Orthod. Dentofac. Orthop.* **2019**, *156*, 611–616. [CrossRef]
348. Usui, T.; Iwata, T.; Miyake, S.; Otsuka, T.; Koizumi, S.; Shirakawa, N.; Kawata, T. Mechanical and Frictional Properties of Aesthetic Orthodontic Wires Obtained by Hard Chrome Carbide Plating. *J. Dent. Sci.* **2018**, *13*, 151–159. [CrossRef]

349. Albawardi, A.; Warunek, S.; Makowka, S.; Al-Jewair, T. Friction Forces Generated by Aesthetic Gummetal® (Ti-Nb) Orthodontic Archwires: A Comparative in Vitro Study. *Int. Orthod.* **2022**, *20*, 100683. [CrossRef]
350. Ramasamy, M.; Prabhakar, R.; Thirunavukarasu, R.; Saravanan, R.; Rajvikram, N.; Sowndarya, V. Evalution of Color Stability of Coated Aesthetic Arch Wires. *Eur. J. Mol. Clin. Med.* **2020**, *07*, 5194–5200.
351. Cowley, A.; Woodward, B. A Healthy Future: Platinum in Medical Applications. *Platin. Met. Rev.* **2011**, *55*, 98–107. [CrossRef]
352. Johnson, A.; Shiraishi, T. Biocompatibility of Precious Metals for Medical Applications. In *Precious Metals for Biomedical Applications*; Elsevier: Amsterdam, The Netherlands, 2014; pp. 37–55, ISBN 9780857094346.
353. Khonsari, S.K.; Towhidi, N.; Cheraghi, M.S.; Allahkaram, S.R.; Rabizadeh, T. Pt Nanoparticles Coating on Orthodontic Ni-Ti Wires Using Pulse Current. In Proceedings of the 2011 IEEE Nanotechnology Materials and Devices Conference, Jeju, Republic of Korea, 18–21 October 2011; pp. 502–505.
354. Sim, W.; Barnard, R.; Blaskovich, M.A.T.; Ziora, Z. Antimicrobial Silver in Medicinal and Consumer Applications: A Patent Review of the Past Decade (2007–2017). *Antibiotics* **2018**, *7*, 93. [CrossRef] [PubMed]
355. Talapko, J.; Matijević, T.; Juzbašić, M.; Antolović-Požgain, A.; Škrlec, I. Antibacterial Activity of Silver and Its Application in Dentistry, Cardiology and Dermatology. *Microorganisms* **2020**, *8*, 1400. [CrossRef] [PubMed]
356. Metin-Gürsoy, G.; Taner, L.; Barış, E. Biocompatibility of Nanosilver-Coated Orthodontic Brackets: An in Vivo Study. *Prog. Orthod.* **2016**, *17*, 39. [CrossRef] [PubMed]
357. Mhaske, A.R.; Shetty, P.C.; Bhat, N.S.; Ramachandra, C.S.; Laxmikanth, S.M.; Nagarahalli, K.; Tekale, P.D. Antiadherent and Antibacterial Properties of Stainless Steel and NiTi Orthodontic Wires Coated with Silver against Lactobacillus Acidophilus—An in Vitro Study. *Prog. Orthod.* **2015**, *16*, 40. [CrossRef]
358. Ghasemi, T.; Arash, V.; Rabiee, S.M.; Rajabnia, R.; Pourzare, A.; Rakhshan, V. Antimicrobial Effect, Frictional Resistance, and Surface Roughness of Stainless Steel Orthodontic Brackets Coated with Nanofilms of Silver and Titanium Oxide: A Preliminary Study. *Microsc. Res. Tech.* **2017**, *80*, 599–607. [CrossRef]
359. Meyer-Kobbe, V.; Doll, K.; Stiesch, M.; Schwestka-Polly, R.; Demling, A. Comparison of Intraoral Biofilm Reduction on Silver-Coated and Silver Ion-Implanted Stainless Steel Bracket Material. *J. Orofac. Orthop. Fortschritte Kieferorthopädie* **2019**, *80*, 32–43. [CrossRef]
360. Metin-Gürsoy, G.; Taner, L.; Akca, G. Nanosilver Coated Orthodontic Brackets: In Vivo Antibacterial Properties and Ion Release. *Eur. J. Orthod.* **2017**, *39*, 9–16. [CrossRef]
361. Espinosa-Cristóbal, L.F.; López-Ruiz, N.; Cabada-Tarín, D.; Reyes-López, S.Y.; Zaragoza-Contreras, A.; Constandse-Cortéz, D.; Donohué-Cornejo, A.; Tovar-Carrillo, K.; Cuevas-González, J.C.; Kobayashi, T. Antiadherence and Antimicrobial Properties of Silver Nanoparticles against Streptococcus Mutans on Brackets and Wires Used for Orthodontic Treatments. *J. Nanomater.* **2018**, *2018*, 9248527. [CrossRef]
362. Bindu, S.H.; Kala Vani, S.V.; Nirisha, G.; Madhuri, N.; Sai Deepa, B.; Hemadri, S. Evaluation of Antibacterial Effect of Silver Nanoparticle Coated Stainless Steel Band Material—An In Vitro Study. *Orthod. J. Nepal* **2019**, *9*, 13–19. [CrossRef]
363. Prabha, R.D.; Kandasamy, R.; Sivaraman, U.S.; Nandkumar, M.A.; Nair, P.D. Antibacterial Nanosilver Coated Orthodontic Bands with Potential Implications in Dentistry. *Indian J. Med. Res.* **2016**, *144*, 580–586. [CrossRef]
364. Bahrami, R.; Pourhajibagher, M.; Badiei, A.; Masaeli, R.; Tanbakuchi, B. Evaluation of the Cell Viability and Antimicrobial Effects of Orthodontic Bands Coated with Silver or Zinc Oxide Nanoparticles: An in Vitro Study. *Korean J. Orthod.* **2023**, *53*, 16–25. [CrossRef] [PubMed]
365. Gonçalves, I.S.; Viale, A.B.; Sormani, N.N.; Pizzol, K.E.D.C.; Araujo-Nobre, A.R.D.; Oliveira, P.C.S.D.; Barud, H.G.D.O.; Antonio, S.G.; Barud, H.D.S. Antimicrobial Orthodontic Wires Coated with Silver Nanoparticles. *Braz. Arch. Biol. Technol.* **2020**, *63*, e20190339. [CrossRef]
366. Gil, F.J.; Espinar-Escalona, E.; Clusellas, N.; Fernandez-Bozal, J.; Artes-Ribas, M.; Puigdollers, A. New Bactericide Orthodonthic Archwire: NiTi with Silver Nanoparticles. *Metals* **2020**, *10*, 702. [CrossRef]
367. Zeidan, N.K.; Enany, N.M.; Mohamed, G.G.; Marzouk, E.S. The Antibacterial Effect of Silver, Zinc-Oxide and Combination of Silver/ Zinc Oxide Nanoparticles Coating of Orthodontic Brackets (an in Vitro Study). *BMC Oral Health* **2022**, *22*, 230. [CrossRef]
368. Anand, B.G.; Shejale, K.P.; Rajesh Kumar, R.; Thangam, R.; Prajapati, K.P.; Kar, K.; Mala, R. Bioactivation of an Orthodontic Wire Using Multifunctional Nanomaterials to Prevent Plaque Accumulation. *Biomater. Adv.* **2023**, *148*, 213346. [CrossRef]
369. Ryu, H.S.; Bae, I.H.; Lee, K.G.; Hwang, H.S.; Lee, K.H.; Koh, J.T.; Cho, J.H. Antibacterial Effect of Silver-Platinum Coating for Orthodontic Appliances. *Angle Orthod.* **2012**, *82*, 151–157. [CrossRef] [PubMed]
370. Shah, P.; Sharma, P.; Goje, S. Comparative Evaluation of Frictional Resistance of Silver-Coated Stainless Steel Wires with Uncoated Stainless Steel Wires: An In Vitro Study. *Contemp. Clin. Dent.* **2018**, *9*, 331. [CrossRef]
371. Shah, P.; Sharma, P.; Naik, H.; Patel, K.; Panchal, C. A Comparative Evaluation of Frictional Resistance and Surface Roughness of Silver Coated and Uncoated Stainless-Steel Bracket Wire Assembly- An in-Vitro Study. *J. Clin. Exp. Dent.* **2023**, *15*, e411–e419. [CrossRef]
372. Arash, V.; Anoush, K.; Rabiee, S.M.; Rahmatei, M.; Tavanafar, S. The Effects of Silver Coating on Friction Coefficient and Shear Bond Strength of Steel Orthodontic Brackets. *Scanning* **2015**, *37*, 294–299. [CrossRef]
373. Shirakawa, N.; Iwata, T.; Miyake, S.; Otsuka, T.; Koizumi, S.; Usui, T.; Kawata, T. Development of the Aesthetic Orthodontic Appliances Using a Silver Plating Process: The Report on Peel Resistance. *Biomed. Res.* **2017**, *28*, 3217–3221.

374. Usui, T.; Miyake, S.; Iwata, T.; Otsuka, T.; Koizumi, S.; Shirakawa, N.; Kawata, T. Aesthetic Characteristics of the Orthodontic Wire with Silver Plating. *Biomed. Res.* **2017**, *28*, 4937–4941.
375. Ozeki, K.; Yuhta, T.; Aoki, H.; Asaoka, T.; Daisaku, T.; Fukui, Y. Deterioration in the Superelasticity of Ti Sputter Coated on NiTi Orthodontic Wire. *Biomed. Mater. Eng.* **2003**, *13*, 355–362. [PubMed]
376. Anuradha, P.; Varma, N.K.S.; Balakrishnan, A. Reliability Performance of Titanium Sputter Coated Ni–Ti Arch Wires: Mechanical Performance and Nickel Release Evaluation. *Biomed. Mater. Eng.* **2015**, *26*, 67–77. [CrossRef] [PubMed]
377. Ozeki, K.; Yuhta, T.; Aoki, H.; Fukui, Y. Inhibition of Ni Release from NiTi Alloy by Hydroxyapatite, Alumina, and Titanium Sputtered Coatings. *Biomed. Mater. Eng.* **2003**, *13*, 271–279.
378. Prashanth, L.; Kattapagari, K.; Chitturi, R.; Baddam, V.R.; Prasad, L. A Review on Role of Essential Trace Elements in Health and Disease. *J. Dr. NTR Univ. Health Sci.* **2015**, *4*, 75. [CrossRef]
379. Tapiero, H.; Tew, K.D. Trace Elements in Human Physiology and Pathology: Zinc and Metallothioneins. *Biomed. Pharmacother.* **2003**, *57*, 399–411. [CrossRef] [PubMed]
380. Kogan, S.; Sood, A.; Garnick, M.S. Zinc and Wound Healing: A Review of Zinc Physiology and Clinical Applications. *Wounds* **2017**, *29*, 102–106.
381. Karandish, M.; Pakshir, M.; Moghimi, M.; Jafarpour, D. Evaluating the Mechanical Properties of Zinc-Coated Stainless Steel Orthodontic Wires Using Physical Vapor Deposition. *Int. J. Dent.* **2021**, *2021*, 6651289. [CrossRef]
382. Mostaed, E.; Sikora-Jasinska, M.; Drelich, J.W.; Vedani, M. Zinc-Based Alloys for Degradable Vascular Stent Applications. *Acta Biomater.* **2018**, *71*, 1–23. [CrossRef]
383. Abendrot, M.; Kalinowska-Lis, U. Zinc-Containing Compounds for Personal Care Applications. *Int. J. Cosmet. Sci.* **2018**, *40*, 319–327. [CrossRef]
384. Hernández-Escobar, D.; Champagne, S.; Yilmazer, H.; Dikici, B.; Boehlert, C.J.; Hermawan, H. Current Status and Perspectives of Zinc-Based Absorbable Alloys for Biomedical Applications. *Acta Biomater.* **2019**, *97*, 1–22. [CrossRef] [PubMed]
385. Sousa, V.F.C.; Silva, F.J.G. Recent Advances on Coated Milling Tool Technology-a Comprehensive Review. *Coatings* **2020**, *10*, 235. [CrossRef]
386. Simka, W.; Kaczmarek, M.; Baron-Wiecheć, A.; Nawrat, G.; Marciniak, J.; Żak, J. Electropolishing and Passivation of NiTi Shape Memory Alloy. *Electrochim. Acta* **2010**, *55*, 2437–2441. [CrossRef]
387. Espinar, E.; Llamas, J.M.; Michiardi, A.; Ginebra, M.P.; Gil, F.J. Reduction of Ni Release and Improvement of the Friction Behaviour of NiTi Orthodontic Archwires by Oxidation Treatments. *J. Mater. Sci. Mater. Med.* **2011**, *22*, 1119–1125. [CrossRef] [PubMed]
388. HORIUCHI, Y.; HORIUCHI, M.; HANAWA, T.; SOMA, K. Effect of Surface Modification on the Photocatalysis of Ti-Ni Alloy in Orthodontics. *Dent. Mater. J.* **2007**, *26*, 924–929. [CrossRef] [PubMed]
389. Chun, M.J.; Shim, E.; Kho, E.H.; Park, K.J.; Jung, J.; Kim, J.M.; Kim, B.; Lee, K.H.; Cho, D.L.; Bai, D.H.; et al. Surface Modification of Orthodontic Wires with Photocatalytic Titanium Oxide for Its Antiadherent and Antibacterial Properties. *Angle Orthod.* **2007**, *77*, 483–488. [CrossRef] [PubMed]
390. Chhattani, S.; Shetty, P.C.; Laxmikant, S.; Ramachandra, C. In Vitro Assessment of Photocatalytic Titanium Oxide Surface-Modified Stainless Steel and Nickel Titanium Orthodontic Wires for Its Antiadherent and Antibacterial Properties against Streptococcus Mutans. *J. Indian Orthod. Soc.* **2014**, *48*, 82–87. [CrossRef]
391. Awata, M.; Okada, M.; Nambu, T.; Matsumoto, N. Improvement of Photocatalytic Activity of TiO_2 Coating by the Modified Sol-Gel Method. *Nano Biomed.* **2015**, *7*, 51–62. [CrossRef]
392. Özyildz, F.; Uzel, A.; Hazar, A.S.; Güden, M.; Ölmez, S.; Aras, I.; Karaboz, İ. Photocatalytic Antimicrobial Effect of TiO2 Anatase Thin-Film–Coated Orthodontic Arch Wires on 3 Oral Pathogens. *Turk. J. Biol.* **2014**, *38*, 289–295. [CrossRef]
393. Salehi, P.; Babanouri, N.; Roein-Peikar, M.; Zare, F. Long-Term Antimicrobial Assessment of Orthodontic Brackets Coated with Nitrogen-Doped Titanium Dioxide against *Streptococcus mutans*. *Prog. Orthod.* **2018**, *19*, 35. [CrossRef]
394. Cao, B.; Wang, Y.; Li, N.; Liu, B.; Zhang, Y. Preparation of an Orthodontic Bracket Coated with an Nitrogen-Doped TiO2-XNy Thin Film and Examination of Its Antimicrobial Performance. *Dent. Mater. J.* **2013**, *32*, 311–316. [CrossRef] [PubMed]
395. Shah, A.G.; Shetty, P.C.; Ramachandra, C.S.; Bhat, N.S.; Laxmikanth, S.M. In Vitro Assessment of Photocatalytic Titanium Oxide Surface Modified Stainless Steel Orthodontic Brackets for Antiadherent and Antibacterial Properties against Lactobacillus Acidophilus. *Angle Orthod.* **2011**, *81*, 1028–1035. [CrossRef] [PubMed]
396. Kaliaraj, G.S.; Ramadoss, A.; Sundaram, M.; Balasubramanian, S.; Muthirulandi, J. Studies of Calcium-Precipitating Oral Bacterial Adhesion on TiN, TiO2 Single Layer, and TiN/TiO_2 Multilayer-Coated 316L SS. *J. Mater. Sci.* **2014**, *49*, 7172–7180. [CrossRef]
397. Baby, R.D.; Subramaniam, S.; Arumugam, I.; Padmanabhan, S. Assessment of Antibacterial and Cytotoxic Effects of Orthodontic Stainless Steel Brackets Coated with Different Phases of Titanium Oxide: An in-Vitro Study. *Am. J. Orthod. Dentofac. Orthop.* **2017**, *151*, 678–684. [CrossRef] [PubMed]
398. Fatani, E.J.; Almutairi, H.H.; Alharbi, A.O.; Alnakhli, Y.O.; Divakar, D.D.; Muzaheed; Alkheraif, A.A.; Khan, A.A. In Vitro Assessment of Stainless Steel Orthodontic Brackets Coated with Titanium Oxide Mixed Ag for Anti-Adherent and Antibacterial Properties against *Streptococcus mutans* and *Porphyromonas gingivalis*. *Microb. Pathog.* **2017**, *112*, 190–194. [CrossRef]
399. Zhang, R.; Zhang, W.; Bai, X.; Song, X.; Wang, C.; Gao, X.; Tian, X.; Liu, F. Discussion on the Development of Nano Ag/TiO_2 Coating Bracket and Its Antibacterial Property and Biocompatibility in Orthodontic Treatment. *Pak. J. Pharm. Sci.* **2015**, *28*, 807–810.

400. Kielan-Grabowska, Z.; Bącela, J.; Zięty, A.; Seremak, W.; Gawlik-Maj, M.; Kawala, B.; Borak, B.; Detyna, J.; Sarul, M. Improvement of Properties of Stainless Steel Orthodontic Archwire Using TiO_2:Ag Coating. *Symmetry* **2021**, *13*, 1734. [CrossRef]
401. Bącela, J.J.; Kielan-Grabowska, Z.; Borak, B.; Sobieszczańska, B.; Walczuk, U.; Kawala, B.; Zięty, A.; Detyna, J.; Sarul, M. Antiadherent and Antibacterial Properties of TiO_2-Coated and TiO_2:Ag-Coated Stainless Steel Orthodontic Wires against *S. mutans* Bacteria. *Acta Bioeng. Biomech.* **2022**, *24*, 107–118. [CrossRef]
402. Li, N.; Zhou, H.J.; Han, B.; Zhang, Y.J.; Cao, B.C. Friction Properties of Orthodontic Brackets Coated with TiO_2-XNy. *Chin. J. Tissue Eng. Res.* **2014**, *18*, 7621–7626. [CrossRef]
403. Mollabashi, V.; Farmany, A.; Alikhani, M.Y.; Sattari, M.; Soltanian, A.R.; Kahvand, P.; Banisafar, Z. Effects of TiO_2-Coated Stainless Steel Orthodontic Wires on Streptococcus Mutans Bacteria: A Clinical Study. *Int. J. Nanomed.* **2020**, *15*, 8759–8766. [CrossRef]
404. Jung, O.; Becker, J.-P.; Smeets, R.; Gosau, M.; Becker, G.; Kahl-Nieke, B.; Jung, A.-K.; Heiland, M.; Kopp, A.; Barbeck, M.; et al. Surface Characteristics of Esthetic Nickel–Titanium and Beta-Titanium Orthodontic Archwires Produced by Plasma Electrolytic Oxidation (PEO)—Primary Results. *Materials* **2019**, *12*, 1403. [CrossRef] [PubMed]
405. Liu, F.; Xu, J.L.; Yu, D.Z.; Wang, F.P.; Zhao, L.C. Wear Resistance of Micro-Arc Oxidation Coatings on Biomedical NiTi Alloy. *J. Alloys Compd.* **2009**, *487*, 391–394. [CrossRef]
406. Campeol, D.A.; Fontoura, C.P.; Rodrigues, M.M.; Aguzzoli, C. Assessment of Mechanical and Corrosion Properties of Plasma Oxidized Medical Grade NiTi Wire. *Vacuum* **2020**, *171*, 109013. [CrossRef]
407. Supriadi, S.; Suharno, B.; Nugraha, N.K.; Yasinta, A.O.; Annur, D. Adhesiveness of TiO_2 PVD Coating on Electropolished Stainless Steel 17–4 PH Orthodontic Bracket. *Mater. Res. Express* **2019**, *6*, 094003. [CrossRef]
408. Fu, T.; Wen, C.S.; Lu, J.; Zhou, Y.M.; Ma, S.G.; Dong, B.H.; Liu, B.G. Sol-Gel Derived TiO2 Coating on Plasma Nitrided 316L Stainless Steel. *Vacuum* **2012**, *86*, 1402–1407. [CrossRef]
409. Liu, J.; Lou, Y.; Zhang, C.; Yin, S.; Li, H.; Sun, D.; Sun, X. Improved Corrosion Resistance and Antibacterial Properties of Composite Arch-Wires by N-Doped TiO_2 Coating. *RSC Adv.* **2017**, *7*, 43938–43949. [CrossRef]
410. Solanki, L.A.; Dinesh, S.P.S.; Jain, R.K.; Balasubramaniam, A. Effects of Titanium Oxide Coating on the Antimicrobial Properties, Surface Characteristics, and Cytotoxicity of Orthodontic Brackets—A Systematic Review and Meta Analysis of in-Vitro Studies. *J. Oral Biol. Craniofacial Res.* **2023**, *13*, 553–562. [CrossRef]
411. Asiry, M.A.; AlShahrani, I.; Almoammar, S.; Durgesh, B.H.; Al Kheraif, A.A.; Hashem, M.I. Influence of Epoxy, Polytetrafluoroethylene (PTFE) and Rhodium Surface Coatings on Surface Roughness, Nano-Mechanical Properties and Biofilm Adhesion of Nickel Titanium (Ni-Ti) Archwires. *Mater. Res. Express* **2018**, *5*, 026511. [CrossRef]
412. Madasamy, R.; Prabhakar, R.; Ramanadhan, T.; Ramachandran, S.; Natrajan, R.; Vasudevan, S. Assessment of Roughness of Fiber-Reinforced Polymer Composite Wires and Other Coated Esthetic Archwires. *World J. Dent.* **2021**, *12*, 156–159. [CrossRef]
413. Tawakal, M.S.; Abdelghany Metwally, A.M.; El-Wassefy, N.A.; Tawfik, M.A.; Shamaa, M.S. Static Friction, Surface Roughness, and Antibacterial Activity of Orthodontic Brackets Coated with Silver and Silver Chitosan Nanoparticles. *J. World Fed. Orthod.* **2023**. [CrossRef]
414. Setiyorini, Y.; Pintowantoro, S. Biocompatibility Improvement of NiTi Orthodontic Wire from Various Coatings. *Adv. Mater. Res.* **2013**, *789*, 225–231. [CrossRef]
415. Pintowantoro, S.; Setiyorini, Y. Reduction of Nickel Ion Release on a TiO_2 Coated onto an Orthodontic Wire. *Adv. Mater. Res.* **2013**, *789*, 204–209. [CrossRef]
416. Supriadi, S.; Ovilia, A.; Ilmaniar, N.; Suharno, B. Formation of Tio2 Thin Film on Antibacterial Metal Injection Molding Stainless Steel Orthodontic Bracket 17-4 Ph Using Physical Vapor Deposition Method. *Key Eng. Mater.* **2020**, *846*, 169–174. [CrossRef]
417. Kurtoğlu, S.F.; Yağcı, M.B.; Uzun, A.; Ünal, U.; Canadinc, D. Enhancing Biocompatibility of NiTi Shape Memory Alloys by Simple NH3 Treatments. *Appl. Surf. Sci.* **2020**, *525*, 146547. [CrossRef]
418. Math, M.; Shah, A.G.; Gangurde, P.; Karandikar, A.G.; Gheware, A.; Jadhav, B.S. In-Vitro Comparative Assessment of Antibacterial and Anti-Adherent Effect of Two Types of Surface Modificants on Stainless Steel Orthodontic Brackets Against Streptococcus Mutans. *J. Indian Orthod. Soc.* **2021**, *56*, 282–289. [CrossRef]
419. Kim, H.; Johnson, J.W. Corrosion of Stainless-Steel, Nickel-Titanium, Coated Nickel-Titanium, and Titanium Orthodontic Wires. *Angle Orthod.* **1999**, *69*, 39–44. [CrossRef]
420. Kao, C.T.; Ding, S.J.; Chen, Y.C.; Huang, T.H. The Anticorrosion Ability of Titanium Nitride (TiN) Plating on an Orthodontic Metal Bracket and Its Biocompatibility. *J. Biomed. Mater. Res.* **2002**, *63*, 786–792. [CrossRef]
421. Gil, F.J.; Solano, E.; Mendoza, A.; Pena, J. Inhibition of Ni Release from NiTi and NiTiCu Orthodontic Archwires by Nitrogen Diffusion Treatment. *J. Appl. Biomater. Biomech.* **2004**, *2*, 151–155. [CrossRef]
422. Iijima, M.; Yuasa, T.; Endo, K.; Muguruma, T.; Ohno, H.; Mizoguchi, I. Corrosion Behavior of Ion Implanted Nickel-Titanium Orthodontic Wire in Fluoride Mouth Rinse Solutions. *Dent. Mater. J.* **2010**, *29*, 53–58. [CrossRef]
423. Huang, T.-H.; Guo, J.-U.; Kao, C.T. A Comparison of the Friction Associated with Diamond-like Carbon (DLC) or Titanium Nitride (TiN) Plating Metal Brackets. *Surf. Coat. Technol.* **2010**, *205*, 1917–1921. [CrossRef]
424. Kao, C.-T.; Guo, J.-U.; Huang, T.-H. Comparison of Friction Force between Corroded and Noncorroded Titanium Nitride Plating of Metal Brackets. *Am. J. Orthod. Dentofac. Orthop.* **2011**, *139*, 594–600. [CrossRef]
425. Rongo, R.; Ametrano, G.; Gloria, A.; Spagnuolo, G.; Galeotti, A.; Paduano, S.; Valletta, R.; D'Antò, V. Effects of Intraoral Aging on Surface Properties of Coated Nickel-Titanium Archwires. *Angle Orthod.* **2014**, *84*, 665–672. [CrossRef] [PubMed]

426. Zuo, J.; Xie, Y.; Zhang, J.; Wei, Q.; Zhou, B.; Luo, J.; Wang, Y.; Yu, Z.M.; Tang, Z.G. TiN Coated Stainless Steel Bracket: Tribological, Corrosion Resistance, Biocompatibility and Mechanical Performance. *Surf. Coat. Technol.* **2015**, *277*, 227–233. [CrossRef]
427. Rongo, R.; Valletta, R.; Bucci, R.; Rivieccio, V.; Galeotti, A.; Michelotti, A.; D'Antò, V. In Vitro Biocompatibility of Nickel-Titanium Esthetic Orthodontic Archwires. *Angle Orthod.* **2016**, *86*, 789–795. [CrossRef] [PubMed]
428. Sugisawa, H.; Kitaura, H.; Ueda, K.; Kimura, K.; Ishida, M.; Ochi, Y.; Kishikawa, A.; Ogawa, S.; Takano-Yamamoto, T. Corrosion Resistance and Mechanical Properties of Titanium Nitride Plating on Orthodontic Wires. *Dent. Mater. J.* **2018**, *37*, 286–292. [CrossRef]
429. Arici, N.; Akdeniz, B.S.; Oz, A.A.; Gencer, Y.; Tarakci, M.; Arici, S. Effectiveness of Medical Coating Materials in Decreasing Friction between Orthodontic Brackets and Archwires. *Korean J. Orthod.* **2021**, *51*, 270–281. [CrossRef] [PubMed]
430. Teixeira, L.P.; Gontijo, L.C.; Franco Júnior, A.R.; Pereira, M.F.; Schuenck, R.P.; Malacarne-Zanon, J. Evaluation of Antimicrobial Potential and Surface Morphology in Thin Films of Titanium Nitride and Calcium Phosphate on Orthodontic Brackets. *Am. J. Orthod. Dentofac. Orthop.* **2021**, *160*, 209–214. [CrossRef]
431. Liu, J.K.; Liu, I.H.; Liu, C.; Chang, C.J.; Kung, K.C.; Liu, Y.T.; Lee, T.M.; Jou, J.L. Effect of Titanium Nitride/Titanium Coatings on the Stress Corrosion of Nickel-Titanium Orthodontic Archwires in Artificial Saliva. *Appl. Surf. Sci.* **2014**, *317*, 974–981. [CrossRef]
432. Krishnan, V.; Krishnan, A.; Remya, R.; Ravikumar, K.K.; Nair, S.A.; Shibli, S.M.A.; Varma, H.K.; Sukumaran, K.; Kumar, K.J. Development and Evaluation of Two PVD-Coated β-Titanium Orthodontic Archwires for Fluoride-Induced Corrosion Protection. *Acta Biomater.* **2011**, *7*, 1913–1927. [CrossRef]
433. Krishnan, V.; Ravikumar, K.K.; Sukumaran, K.; Jyothindra Kumar, K. In Vitro Evaluation of Physical Vapor Deposition Coated Beta Titanium Orthodontic Archwires. *Angle Orthod.* **2012**, *82*, 22–29. [CrossRef]
434. Golshah, A.; Feyli, S. Effect of Zirconium Oxide Nano-Coating on Frictional Resistance of Orthodontic Wires. *J. Orthod. Sci.* **2022**, *11*, 35. [CrossRef]
435. Wu, H.; Yang, J.; Yan, Y.; Zheng, B.; Algahefi, A.L.; Ma, S.; Liu, Y. Study of Al–SiO_2 Aesthetic Composite Coating on Orthodontic Metal Archwire. *Coatings* **2022**, *12*, 746. [CrossRef]
436. Krishnan, M.; Seema, S.; Sukumaran, K.; Pawar, V. Phase Transitions in Coated Nickel Titanium Arch Wires: A Differential Scanning Calorimetric and X-Ray Diffraction Analysis. *Bull. Mater. Sci.* **2012**, *35*, 905–911. [CrossRef]
437. Ramazanzadeh, B.; Jahanbin, A.; Yaghoubi, M.; Shahtahmassbi, N.; Ghazvini, K.; Shakeri, M.; Shafaee, H. Comparison of Antibacterial Effects of ZnO and CuO Nanoparticles Coated Brackets against Streptococcus Mutans. *J. Dent.* **2015**, *16*, 200–205.
438. Kachoei, M.; Divband, B.; Eskandarinejad, F.; Khatamian, M. Deposition of ZnO Nano Particles on Stainless Steel Orthodontic Wires by Chemical Solution Method for Friction Reduction Propose. *Res. J. Pharm. Biol. Chem. Sci.* **2015**, *6*, 104–112.
439. Behroozian, A.; Kachoei, M.; Khatamian, M.; Divband, B. The Effect of ZnO Nanoparticle Coating on the Frictionalresistance between Orthodontic Wires and Ceramic Brackets. *J. Dent. Res. Dent. Clin. Dent. Prospect.* **2016**, *10*, 106–111. [CrossRef] [PubMed]
440. Kachoei, M.; Nourian, A.; Divband, B.; Kachoei, Z.; Shirazi, S. Zinc-Oxide Nanocoating for Improvement of the Antibacterial and Frictional Behavior of Nickel-Titanium Alloy. *Nanomedicine* **2016**, *11*, 2511–2527. [CrossRef]
441. Hammad, S.M.; El-Wassefy, N.A.; Shamaa, M.S.; Fathy, A. Evaluation of Zinc-Oxide Nanocoating on the Characteristics and Antibacterial Behavior of Nickel-Titanium Alloy. *Dent. Press J. Orthod.* **2020**, *25*, 51–58. [CrossRef]
442. Gholami, M.; Esmaeilzadeh, M.; Kachoei, Z.; Kachoei, M.; Divband, B. Influence of Physical Dimension and Morphological-Dependent Antibacterial Characteristics of ZnO Nanoparticles Coated on Orthodontic NiTi Wires. *Biomed Res. Int.* **2021**, *2021*, 6397698. [CrossRef]
443. Elhelbawy, N.; Ellaithy, M. Comparative Evaluation of Stainless-Steel Wires and Brackets Coated with Nanoparticles of Chitosan or Zinc Oxide upon Friction: An in Vitro Study. *Int. Orthod.* **2021**, *19*, 274–280. [CrossRef]
444. Palanivel, J.; Srinivasan, D.; Chakravathy, N.C.S. Comparison of the Frictional Resistance and Optical Properties of Aluminum Oxide and Zinc Oxide Coated Nickel Titanium Archwires—An in Vitro Study. *APOS Trends Orthod.* **2022**, *12*, 168. [CrossRef]
445. Tanbakuchi, B.; Kharrazi, S.; Nikfarjam, M.; Akhoundi, M.S.A.; Shahroudi, A.S. Comparative Assessment of the Orthodontic Wire's Friction Coated with Zinc Oxide Nanoparticles by Two Methods of Chemical Precipitation and Hydrothermal Process. *Folia Med.* **2022**, *64*, 945–952. [CrossRef] [PubMed]
446. Ameli, N.; Ghorbani, R.; Asadi, S.; Zarrinzade, Z. Investigation of the Effects of Orthodontic Brackets Coated by Silver Hydroxyapatite, Copper Oxide, and Titanium Oxide Nanoparticles on Wire-Bracket Friction. *APOS Trends Orthod.* **2022**, *12*, 27. [CrossRef]
447. Ameli, N.; Asadi, S.; Ghorbani, R.; Mohebi, S.; Hans, M. Comparative Antibacterial Efficacy of Orthodontic Brackets Coated with Titanium Dioxide, Copper Oxide, and Hydroxyapatite-Silver Nanoparticles against *Streptococcus mutans*. *Middle East J. Rehabil. Health Stud.* **2022**, *9*, e119536. [CrossRef]
448. Venkatesan, K.; Kailasam, V.; Padmanabhan, S. Evaluation of Titanium Dioxide Coating on Surface Roughness of Nickel-Titanium Archwires and Its Influence on Streptococcus Mutans Adhesion and Enamel Mineralization: A Prospective Clinical Study. *Am. J. Orthod. Dentofac. Orthop.* **2020**, *158*, 199–208. [CrossRef]
449. Silveira, R.E.; Elias, C.N.; Amaral, F.L.B. Assessment of Frictional Resistance and Surface Roughness in Orthodontic Wires Coated with Two Different Nanoparticles. *Microsc. Res. Tech.* **2022**, *85*, 1884–1890. [CrossRef]
450. Chaturvedi, T.; Indumathi, P.; Sharma, V.; Agrawal, A.; Singh, D.; Upadhyay, C. Evaluation of Surface-Modified Orthodontic Wires by Different Concentration and Dipping Duration of Titanium Oxide (TiO_2) Nanoparticles. *J. Orthod. Sci.* **2023**, *12*, 3. [CrossRef]

451. Redlich, M.; Gorodnev, A.; Feldman, Y.; Kaplan-Ashiri, I.; Tenne, R.; Fleischer, N.; Genut, M.; Feuerstein, N. Friction Reduction and Wear Resistance of Electro-Co-Deposited Inorganic Fullerene-like WS 2 Coating for Improved Stainless Steel Orthodontic Wires. *J. Mater. Res.* **2008**, *23*, 2909–2915. [CrossRef]
452. Amorodnitzky-Naveh, G.R.S.; Redlich, M.; Katz, A.; Adini, A.R.; Gorodnev, A.; Rapoport, L.; Moshkovich, A.; Cohen, S.R.; Rosentsveig, R.; Moshonov, J.; et al. Towards Medical Applications of Self-Lubricating Coatings with Fullerene-like (IF) WS2 Nanoparticles. *Int. J. Nano Biomater.* **2010**, *3*, 140. [CrossRef]
453. Katz, A.; Redlich, M.; Rapoport, L.; Wagner, H.D.; Tenne, R. Self-Lubricating Coatings Containing Fullerene-like WS2 Nanoparticles for Orthodontic Wires and Other Possible Medical Applications. *Tribol. Lett.* **2006**, *21*, 135–139. [CrossRef]
454. Redlich, M.; KATZ, A.; RAPOPORT, L.; WAGNER, H.; FELDMAN, Y.; TENNE, R. Improved Orthodontic Stainless Steel Wires Coated with Inorganic Fullerene-like Nanoparticles of WS2 Impregnated in Electroless Nickel–Phosphorous Film. *Dent. Mater.* **2008**, *24*, 1640–1646. [CrossRef] [PubMed]
455. Gracco, A.; Dandrea, M.; Deflorian, F.; Zanella, C.; De Stefani, A.; Bruno, G.; Stellini, E. Application of a Molybdenum and Tungsten Disulfide Coating to Improve Tribological Properties of Orthodontic Archwires. *Nanomaterials* **2019**, *9*, 753. [CrossRef] [PubMed]
456. Samorodnitzky-Naveh, G.R.; Redlich, M.; Rapoport, L.; Feldman, Y.; Tenne, R. Inorganic Fullerene-like Tungsten Disulfide Nanocoating for Friction Reduction of Nickel–Titanium Alloys. *Nanomedicine* **2009**, *4*, 943–950. [CrossRef] [PubMed]
457. Fu, Y.; Wu, X.; Wang, Y.; Li, B.; Yang, S. Study of Corrosion Resistance Property and Microstructure of TiNi Shape Memory Alloy Modified by Pulsed High-Energy Density Plasma. *Appl. Surf. Sci.* **2000**, *157*, 167–177. [CrossRef]
458. Shevchenko, N.; Pham, M.-T.; Maitz, M.F. Studies of Surface Modified NiTi Alloy. *Appl. Surf. Sci.* **2004**, *235*, 126–131. [CrossRef]
459. Zhang, M.; Liu, X.; Shang, H.; Lin, J. Comparison of TiN and CNx Coatings on Orthodontic Stainless Steel: Tribological and Biological Evaluation. *Surf. Coat. Technol.* **2019**, *362*, 381–387. [CrossRef]
460. Jin, S.; Zhang, Y.; Wang, Q.; Zhang, D.; Zhang, S. Influence of TiN Coating on the Biocompatibility of Medical NiTi Alloy. *Colloids Surf. B Biointerfaces* **2013**, *101*, 343–349. [CrossRef] [PubMed]
461. Zhang, J.; Lou, J.; He, H.; Xie, Y. Comparative Investigation on the Tribological Performances of TiN, TiCN, and Ti-DLC Film-Coated Stainless Steel. *JOM* **2019**, *71*, 4872–4879. [CrossRef]
462. Wang, X.; Bai, S.; Li, F.; Li, D.; Zhang, J.; Tian, M.; Zhang, Q.; Tong, Y.; Zhang, Z.; Wang, G.; et al. Effect of Plasma Nitriding and Titanium Nitride Coating on the Corrosion Resistance of Titanium. *J. Prosthet. Dent.* **2016**, *116*, 450–456. [CrossRef]
463. Shukla, K.; Rane, R.; Alphonsa, J.; Maity, P.; Mukherjee, S. Structural, Mechanical and Corrosion Resistance Properties of Ti/TiN Bilayers Deposited by Magnetron Sputtering on AISI 316L. *Surf. Coat. Technol.* **2017**, *324*, 167–174. [CrossRef]
464. Bellini, H.; Moyano, J.; Gil, J.; Puigdollers, A. Comparison of the Superelasticity of Different Nickel–Titanium Orthodontic Archwires and the Loss of Their Properties by Heat Treatment. *J. Mater. Sci. Mater. Med.* **2016**, *27*, 158. [CrossRef] [PubMed]
465. Rahmati, M.; Mozafari, M. Biocompatibility of Alumina-based Biomaterials–A Review. *J. Cell. Physiol.* **2019**, *234*, 3321–3335. [CrossRef] [PubMed]
466. Gautam, C.; Joyner, J.; Gautam, A.; Rao, J.; Vajtai, R. Zirconia Based Dental Ceramics: Structure, Mechanical Properties, Biocompatibility and Applications. *Dalt. Trans.* **2016**, *45*, 19194–19215. [CrossRef]
467. Kaliaraj, G.S.; Vishwakarma, V.; Kirubaharan, K.; Dharini, T.; Ramachandran, D.; Muthaiah, B. Corrosion and Biocompatibility Behaviour of Zirconia Coating by EBPVD for Biomedical Applications. *Surf. Coat. Technol.* **2018**, *334*, 336–343. [CrossRef]
468. Louro, C.S.; Fróis, A.; Cerqueira, M. Are W Based Coatings a Surface Alternative to the Adverse Ni Release in Orthodontic Applications? (Poster). In Proceedings of the Iberian Vacuum Conference—RIVA-X, Bilbao, Spain, 4–6 October 2017.
469. Sánchez-López, E.; Gomes, D.; Esteruelas, G.; Bonilla, L.; Lopez-Machado, A.L.; Galindo, R.; Cano, A.; Espina, M.; Ettcheto, M.; Camins, A.; et al. Metal-Based Nanoparticles as Antimicrobial Agents: An Overview. *Nanomaterials* **2020**, *10*, 292. [CrossRef] [PubMed]
470. Moradpoor, H.; Safaei, M.; Mozaffari, H.R.; Sharifi, R.; Imani, M.M.; Golshah, A.; Bashardoust, N. An Overview of Recent Progress in Dental Applications of Zinc Oxide Nanoparticles. *RSC Adv.* **2021**, *11*, 21189–21206. [CrossRef]
471. Puts, G.J.; Crouse, P.; Ameduri, B.M. Polytetrafluoroethylene: Synthesis and Characterization of the Original Extreme Polymer. *Chem. Rev.* **2019**, *119*, 1763–1805. [CrossRef]
472. Roina, Y.; Auber, F.; Hocquet, D.; Herlem, G. EPTFE Functionalization for Medical Applications. *Mater. Today Chem.* **2021**, *20*, 100412. [CrossRef]
473. Kameda, T.; Sato, H.; Oka, S.; Miyazaki, A.; Ohkuma, K.; Terada, K. Low Temperature Polytetrafluoroethylene (PTFE) Coating Improves the Appearance of Orthodontic Wires without Changing Their Mechanical Properties. *Dent. Mater. J.* **2020**, *39*, 721–734. [CrossRef]
474. Abdulkader, Y.C.; Kamaruddin, A.F.; Mydin, R.B.S.M.N. Effects of Salivary PH on Coating Durability of Two Different Aesthetic Archwire Coatings under a Simulated Intraoral Environment. *Saudi Dent. J.* **2020**, *32*, 306–313. [CrossRef]
475. Mareci, D.; Earar, K.; Zetu, I.; Bolat, G.; Crimu, C.; Istrate, B.; Munteanu, C.; Matei, M. Comparative Electrochemical Behaviour of Uncoated and Coated NiTi for Dental Orthodontic Wires. *Mater. Plast.* **2015**, *2*, 150–153.
476. Earar, K.; Matei, M.N.; Mareci, D.; Trinca, L.C.; Ariton, M.A.; Arotaritei, D.; Cerghizan, D.; Bica, C.; Ciupilan, C. The Electrochemical Behaviour of NiTi Orthodontic Wires with Polytetrafluoroethylene Coating in Red Ruffled Pimiento Peppers Paste. *Rev. Chim.* **2016**, *67*, 1850–1853.

477. Matei, M.N.; Earar, K.; Trinca, L.C.; Mareci, D.; Fotea, L.; Peptu, C.A.; Bica, C. Degradation Characteristics of Poly-Tetrafluoroethylene Coatings on Stainless Steel Orthodontic Wires Immersed in Tuna Fish Derived Products. *Rev. Chim.* **2016**, *67*, 800–807.
478. Farronato, G.; Maijer, R.; Carìa, M.P.; Esposito, L.; Alberzoni, D.; Cacciatore, G. The Effect of Teflon Coating on the Resistance to Sliding of Orthodontic Archwires. *Eur. J. Orthod.* **2012**, *34*, 410–417. [CrossRef] [PubMed]
479. Husmann, P.; Bourauel, C.; Wessinger, M.; Jäger, A. The Frictional Behavior of Coated Guiding Archwires. *J. Orofac. Orthop. Fortschritte Kieferorthopädie* **2002**, *63*, 199–211. [CrossRef] [PubMed]
480. Demling, A.; Elter, C.; Heidenblut, T.; Bach, F.-W.; Hahn, A.; Schwestka-Polly, R.; Stiesch, M.; Heuer, W. Reduction of Biofilm on Orthodontic Brackets with the Use of a Polytetrafluoroethylene Coating. *Eur. J. Orthod.* **2010**, *32*, 414–418. [CrossRef] [PubMed]
481. Ryu, S.H.; Lim, B.S.; Kwak, E.J.; Lee, G.J.; Choi, S.; Park, K.H. Surface Ultrastructure and Mechanical Properties of Three Different White-Coated NiTi Archwires. *Scanning* **2015**, *37*, 414–421. [CrossRef]
482. Elsaka, S.; Hassan, A.; Elnaghy, A. Effect of Gastric Acids on Surface Topography and Bending Properties of Esthetic Coated Nickel-Titanium Orthodontic Archwires. *Clin. Oral Investig.* **2021**, *25*, 1319–1326. [CrossRef]
483. Dokku, A.; Peddu, R.; Prakash, A.S.; Padhmanabhan, J.; Kalyani, M.; Devikanth, L. Surface and Mechanical Properties of Different Coated Orthodontic Archwires. *J. Indian Orthod. Soc.* **2018**, *52*, 238–242. [CrossRef]
484. Choi, S.; Park, D.-J.; Kim, K.-A.; Park, K.-H.; Park, H.-K.; Park, Y.-G. In Vitro Sliding-Driven Morphological Changes in Representative Esthetic NiTi Archwire Surfaces. *Microsc. Res. Tech.* **2015**, *78*, 926–934. [CrossRef]
485. Rego, M.V.N.N.D.; Lau, G.W.T.; Araújo, Y.C.; Silva, R.M. e Color Stability of Esthetic Coatings Applied to Nickel-Titanium Archwires. *Rev. Odontol. UNESP* **2017**, *46*, 307–311. [CrossRef]
486. Jejurikar, H.; Contractor, T.; Nene, S.; Kalia, A.; Patil, W.; Khan, N. A Comparison of Surface Characteristics, Coating Stability and Friction Coefficients of Esthetic Archwires: A Comparative Study. *J. Indian Orthod. Soc.* **2021**, *55*, 56–63. [CrossRef]
487. Shahabi, M.; Salari, S.; Poosti, M.; Abtahi, M. Static and Kinetic Frictional Forces of Silica-Insert Ceramic Brackets with Coated Archwires in Artificial Saliva. *Dent. Res. J.* **2017**, *14*, 412. [CrossRef]
488. Argalji, N.; Da Silva, E.M.; Cury-Saramago, A.; Mattos, C.T. Characterization and Coating Stability Evaluation of Nickel-Titanium Orthodontic Esthetic Wires: An in Vivo Study. *Braz. Oral Res.* **2017**, *31*, e68. [CrossRef] [PubMed]
489. Abdulkhabeer, H.; Alyassiri, A.; Farhan Kamaruddin, A.; Ismail, K.; Shafiai, N.A.A.; Rahman, N.A.; Amir, W.M.; Ahmad, W. Preliminary Result of Randomised Controlled Trial of Three Different Coated Archwires Part 1: Tooth Alignment and Coating Loss. *Malays. J. Med. Health Sci.* **2020**, *16*, 2636–9346.
490. Zegan, G.; Sodor, A.; Munteanu, C. Surface Characteristics of Retrieved Coated and Nickel-Titanium Orthodontic Archwires. *Rom. J. Morphol. Embryol.* **2012**, *53*, 935–939. [PubMed]
491. da Silva, D.L.; Santos, E.; de Souza Camargo, S.; de Oliveira Ruellas, A.C. Infrared Spectroscopy, Nano-Mechanical Properties, and Scratch Resistance of Esthetic Orthodontic Coated Archwires. *Angle Orthod.* **2015**, *85*, 777–783. [CrossRef]
492. Lin, C.; Kim, J.S.; Lin, E.T.; Lin, E.T. Prolonged Water Immersion Alters Resistance to Sliding of Aesthetic Orthodontic Coated Wires. *Orthod. Craniofac. Res.* **2021**, *24*, 111–120. [CrossRef]
493. Rudawska, A.; Sarna-Boś, K.; Rudawska, A.; Olewnik-Kruszkowska, E.; Frigione, M. Biological Effects and Toxicity of Compounds Based on Cured Epoxy Resins. *Polymers* **2022**, *14*, 4915. [CrossRef]
494. Raji, S.H.; Shojaei, H.; Ghorani, P.S.; Rafiei, E. Bacterial Colonization on Coated and Uncoated Orthodontic Wires: A Prospective Clinical Trial. *Dent. Res. J.* **2014**, *11*, 680–683.
495. Shamohammadi, M.; Hormozi, E.; Moradinezhad, M.; Moradi, M.; Skini, M.; Rakhshan, V. Surface Topography of Plain Nickel-Titanium (NiTi), as-Received Aesthetic (Coated) NiTi, and Aesthetic NiTi Archwires Sterilized by Autoclaving or Glutaraldehyde Immersion: A Profilometry/SEM/AFM Study. *Int. Orthod.* **2019**, *17*, 60–72. [CrossRef] [PubMed]
496. Hosseini, N.; Alavi, S. Load-Deflection and Surface Properties of Coated and Conventional Superelastic Orthodontic Archwires in Conventional and Metal-Insert Ceramic Brackets. *Dent. Res. J.* **2012**, *9*, 133. [CrossRef]
497. Pop, S.I.; Dudescu, M.; Bratu, D.C.; Merie, V.V.; Pacurar, M. Effect of Esthetic Coating on the Load Deflection and Surface Characteristics of the NiTi Orthodontic Archwires. *Rev. Chim.* **2015**, *66*, 364–367.
498. Elayyan, F.; Silikas, N.; Bearn, D. Mechanical Properties of Coated Superelastic Archwires in Conventional and Self-Ligating Orthodontic Brackets. *Am. J. Orthod. Dentofac. Orthop.* **2010**, *137*, 213–217. [CrossRef] [PubMed]
499. Dragomirescu, A.O.; Teodorescu, E.; Tarmure, V.; Baluta, A.; Pacurar, M.; Nenovici, D.; Chibelean, M.; Ionescu, E. Variation of Static Frictional Forces in the Fixed Orthodontic System. *Rev. Chim.* **2019**, *70*, 3954–3956. [CrossRef]
500. Aboalnaga, A.A.; Shahawi, A.M. El Comparison of Surface Roughness and Hardness of Three Different Brands of Esthetic Coated NiTi Archwires: Invitro Study. *BMC Oral Health* **2023**, *23*, 816. [CrossRef]
501. Elayyan, F.; Silikas, N.; Bearn, D. Ex Vivo Surface and Mechanical Properties of Coated Orthodontic Archwires. *Eur. J. Orthod.* **2008**, *30*, 661–667. [CrossRef]
502. Ping, S.; Xue, F.; Jie He, S.; Wei, C.; Tao, W.; Wei, M. Color Schemes and Biocompatibility of Epoxy Resin/Polytetrafluorethylene Coat on the Surface of Tini Arth Wires. *Int. J. Mod. Phys. B* **2009**, *23*, 1578–1583.
503. Sheiko, N.; Kékicheff, P.; Marie, P.; Schmutz, M.; Jacomine, L.; Perrin-Schmitt, F. PEEK (Polyether-Ether-Ketone)-Coated Nitinol Wire: Film Stability for Biocompatibility Applications. *Appl. Surf. Sci.* **2016**, *389*, 651–665. [CrossRef]
504. Shirakawa, N.; Iwata, T.; Miyake, S.; Otuka, T.; Koizumi, S.; Kawata, T. Mechanical Properties of Orthodontic Wires Covered with a Polyether Ether Ketone Tube. *Angle Orthod.* **2018**, *88*, 442–449. [CrossRef]

505. do Rego, M.V.N.N.; de Araújo, G.M.M.; Martinez, E.F.; de Sousa Lima, K.R.; Fortes, P.T.F.; Leal, L.M.P. Influence of Aesthetic Coating on the Load-Deflection Ratio of Nickel–Titanium Archwires. *Braz. J. Oral Sci.* **2017**, *15*, 293. [CrossRef]
506. Zhou, A.; Makowka, S.; Warunek, S.; Chen, M.-Y.; Al-Jewair, T. Effects of Various Coating Methods on the Mechanical, Physical, and Aesthetic Properties of GUMMETAL® Archwires: In Vitro Study. *Int. Orthod.* **2023**, *21*, 100753. [CrossRef] [PubMed]
507. Bravo, L.A.; de Cabañes, A.G.; Manero, J.M.; Rúperez, E.; Gil, F.J. NiTi Superelastic Orthodontic Archwires with Polyamide Coating. *J. Mater. Sci. Mater. Med.* **2014**, *25*, 555–560. [CrossRef] [PubMed]
508. Tupinambá, R.A.; Claro, C.A.D.A.; Pereira, C.A.; Nobrega, C.J.P.; Claro, A.P.R.A. Bacterial Adhesion on Conventional and Self-Ligating Metallic Brackets after Surface Treatment with Plasma-Polymerized Hexamethyldisiloxane. *Dent. Press J. Orthod.* **2017**, *22*, 77–85. [CrossRef] [PubMed]
509. Oliveira, A.S.; Kaizer, M.R.; Azevedo, M.S.; Ogliari, F.A.; Cenci, M.S.; Moraes, R.R. (Super)Hydrophobic Coating of Orthodontic Dental Devices and Reduction of Early Oral Biofilm Retention. *Biomed. Mater.* **2015**, *10*, 065004. [CrossRef]
510. Kunimatsu, R.; Tsuka, Y.; Nakajima, K.; Sumi, K.; Yoshimi, Y.; Kado, I.; Inada, A.; Kiritoshi, Y.; Tanimoto, K. The Influence of 2-Methacryloyloxyethyl Phosphorylcholine Polymer Materials on Orthodontic Friction and Attachment of Oral Bacteria. *Materials* **2022**, *15*, 5770. [CrossRef] [PubMed]
511. Wang, Y.; Xu, J.; Yu, C.; Zhou, X.; Chang, L.; Liu, J.; Peng, Q. Prevention of Bacterial Biofilm Formation on Orthodontic Brackets by Non-Crosslinked Chitosan Coating. *Int. J. Biol. Macromol.* **2023**, *251*, 126283. [CrossRef]
512. Kumarasinghe, L.S.; Ninan, N.; Dabare, P.R.L.; Cavallaro, A.; Doğramacı, E.J.; Rossi-Fedele, G.; Dreyer, C.; Vasilev, K.; Zilm, P. Bioactive Plasma Coatings on Orthodontic Brackets: In Vitro Metal Ion Release and Cytotoxicity. *Coatings* **2021**, *11*, 857. [CrossRef]
513. Liu, X.; Peng, L.; Meng, J.; Zhu, Z.; Han, B.; Wang, S. Protein-Mediated Anti-Adhesion Surface against Oral Bacteria. *Nanoscale* **2018**, *10*, 2711–2714. [CrossRef]
514. Li, M.; Tang, C.; Yu, X.; Shi, X.; Yu, H.; Yin, H.; You, M.; Chen, Q.; Ding, X. Tough Adhesion and Antifouling Poly(Vinyl Alcohol) Hydrogel Coating on the Arch Wire for Antibacterial Adhesion. *Sci. China Technol. Sci.* **2023**, *66*, 2786–2796. [CrossRef]
515. Wang, Y.; Ding, C.; Ge, Z.; Li, Z.; Chen, L.; Guo, X.; Dong, G.; Zhou, P. A Novel Antibacterial and Fluorescent Coating Composed of Polydopamine and Carbon Dots on the Surface of Orthodontic Brackets. *J. Mater. Sci. Mater. Med.* **2023**, *34*, 10. [CrossRef]
516. Ahmed, R.A.; Farghali, R.A.; Alshahrani, W.A. Influence of Fluoride and/or Bovine Albumin on Electrochemical Properties of Bare and Ionic Liquid- Coated Ni47Ti49Co4 Orthodontic Archwires in Artificial Saliva Solution. *Int. J. Electrochem. Sci.* **2021**, *16*, 211230. [CrossRef]
517. Xu, J.L.; Lai, T.; Luo, J.M. Preparation and Characterization of the Aesthetic Coating on Nickel-Titanium Orthodontic Archwire by Electrophoretic Deposition. *Prog. Org. Coat.* **2019**, *137*, 105271. [CrossRef]
518. Lee, B.-S.; Lin, Y.-C.; Hsu, W.-C.; Hou, C.-H.; Shyue, J.-J.; Hsiao, S.-Y.; Wu, P.-J.; Lee, Y.-T.; Luo, S.-C. Engineering Antifouling and Antibacterial Stainless Steel for Orthodontic Appliances through Layer-by-Layer Deposition of Nanocomposite Coatings. *ACS Appl. Bio Mater.* **2020**, *3*, 486–494. [CrossRef] [PubMed]
519. Szcześ, A.; Hołysz, L.; Chibowski, E. Synthesis of Hydroxyapatite for Biomedical Applications. *Adv. Colloid Interface Sci.* **2017**, *249*, 321–330. [CrossRef]
520. Chen, L.; Al-Bayatee, S.; Khurshid, Z.; Shavandi, A.; Brunton, P.; Ratnayake, J. Hydroxyapatite in Oral Care Products—A Review. *Materials* **2021**, *14*, 4865. [CrossRef] [PubMed]
521. Jiang, H.C.; Rong, L.J. Effect of Hydroxyapatite Coating on Nickel Release of the Porous NiTi Shape Memory Alloy Fabricated by SHS Method. *Surf. Coat. Technol.* **2006**, *201*, 1017–1021. [CrossRef]
522. Kocijan, A. Electrodeposition of a Hydroxyapatite Coating on a Biocompatible NiTi Alloy. *Mater. Tehnol.* **2018**, *52*, 67–70. [CrossRef]
523. Surmenev, R.A.; Ryabtseva, M.A.; Shesterikov, E.V.; Pichugin, V.F.; Peitsch, T.; Epple, M. The Release of Nickel from Nickel–Titanium (NiTi) Is Strongly Reduced by a Sub-Micrometer Thin Layer of Calcium Phosphate Deposited by Rf-Magnetron Sputtering. *J. Mater. Sci. Mater. Med.* **2010**, *21*, 1233–1239. [CrossRef]
524. Dimasruhin, F.; Ichwan, M.; Nur, A.; Yuwana, M. Electrodeposition of Hydroxyapatite-SiO2 Composite Particles for Biomedical Applications. *AIP Conf. Proc.* **2014**, *1586*, 119–123.
525. Kawaguchi, K.; Iijima, M.; Endo, K.; Mizoguchi, I. Electrophoretic Deposition as a New Bioactive Glass Coating Process for Orthodontic Stainless Steel. *Coatings* **2017**, *7*, 199. [CrossRef]
526. Kawaguchi, K.; Iijima, M.; Muguruma, T.; Endo, K.; Mizoguchi, I. Effects of Bioactive Glass Coating by Electrophoretic Deposition on Esthetical, Bending, and Frictional Performance of Orthodontic Stainless Steel Wire. *Dent. Mater. J.* **2020**, *39*, 593–600. [CrossRef] [PubMed]
527. Rapiejko, C.; Fouvry, S.; Grosgogeat, B.; Wendler, B. A Representative Ex-Situ Fretting Wear Investigation of Orthodontic Arch-Wire/Bracket Contacts. *Wear* **2009**, *266*, 850–858. [CrossRef]
528. *VDI 2840*; Carbon Films—Basic Knowledge, Film Types and Properties. VDI-Rischtlinien 2012; VDI (Verein Deutscher Ingenieure): Düsseldorf, Germany, 2012.
529. Aisenberg, S.; Chabot, R. Ion-Beam Deposition of Thin Films of Diamondlike Carbon. *J. Appl. Phys.* **1971**, *42*, 2953–2958. [CrossRef]
530. Grill, A. Diamond-like Carbon Coatings as Biocompatible Materials—An Overview. *Diam. Relat. Mater.* **2003**, *12*, 166–170. [CrossRef]

531. Hauert, R.; Thorwarth, K.; Thorwarth, G. An Overview on Diamond-like Carbon Coatings in Medical Applications. *Surf. Coat. Technol.* **2013**, *233*, 119–130. [CrossRef]
532. Fedel, M. Blood Compatibility of Diamond-like Carbon (DLC) Coatings. In *Diamond-Based Materials for Biomedical Applications*; Narayan, R., Ed.; Woodhead Publishing Limited: Sawston, UK, 2013; pp. 71–102, ISBN 9780857093400.
533. Love, C.A.; Cook, R.B.; Harvey, T.J.; Dearnley, P.A.; Wood, R.J.K. Diamond like Carbon Coatings for Potential Application in Biological Implants—A Review. *Tribol. Int.* **2013**, *63*, 141–150. [CrossRef]
534. Robertson, J. Diamond-like Amorphous Carbon. *Mater. Sci. Eng. R Rep.* **2002**, *37*, 129–281. [CrossRef]
535. Kobayashi, S.; Ohgoe, Y.; Ozeki, K.; Sato, K.; Sumiya, T.; Hirakuri, K.K.; Aoki, H. Diamond-like Carbon Coatings on Orthodontic Archwires. *Diam. Relat. Mater.* **2005**, *14*, 1094–1097. [CrossRef]
536. Kobayashi, S.; Ozeki, K.; Ohgoe, Y.; Gei, L.; Hirakuri, K.K.; Aoki, H. Biocompatibility of Diamond-Like Carbon Coated NiTi Orthodontic Wire and Acrylic Resin Teeth. *Key Eng. Mater.* **2005**, *284–286*, 783–786. [CrossRef]
537. Ohgoe, Y.; Hirakuri, K.K.; Ozeki, K.; Fukui, Y. Investigation of Diamond-like Carbon Coating for Orthodontic Archwire. *New Diam. Front. Carbon Technol.* **2007**, *17*, 281–288.
538. Ohgoe, Y.; Kobayashi, S.; Ozeki, K.; Aoki, H.; Nakamori, H.; Hirakuri, K.K.; Miyashita, O. Reduction Effect of Nickel Ion Release on a Diamond-like Carbon Film Coated onto an Orthodontic Archwire. *Thin Solid Film.* **2006**, *497*, 218–222. [CrossRef]
539. Huang, S.Y.; Huang, J.J.; Kang, T.; Diao, D.F.; Duan, Y.Z. Coating NiTi Archwires with Diamond-like Carbon Films: Reducing Fluoride-Induced Corrosion and Improving Frictional Properties. *J. Mater. Sci. Mater. Med.* **2013**, *24*, 2287–2292. [CrossRef] [PubMed]
540. Muguruma, T.; Iijima, M.; Brantley, W.A.; Nakagaki, S.; Endo, K.; Mizoguchi, I. Frictional and Mechanical Properties of Diamond-like Carbon-Coated Orthodontic Brackets. *Eur. J. Orthod.* **2013**, *35*, 216–222. [CrossRef] [PubMed]
541. Akaike, S.; Kobayashi, D.; Aono, Y.; Hiratsuka, M.; Hirata, A.; Hayakawa, T.; Nakamura, Y. Relationship between Static Friction and Surface Wettability of Orthodontic Brackets Coated with Diamond-like Carbon (DLC), Fluorine- or Silicone-Doped DLC Coatings. *Diam. Relat. Mater.* **2016**, *61*, 109–114. [CrossRef]
542. Muguruma, T.; Iijima, M.; Brantley, W.A.; Mizoguchi, I. Effects of a Diamond-like Carbon Coating on the Frictional Properties of Orthodontic Wires. *Angle Orthod.* **2011**, *81*, 141–148. [CrossRef]
543. Akaike, S.; Hayakawa, T.; Kobayashi, D.; Aono, Y.; Hirata, A.; Hiratsuka, M.; Nakamura, Y. Reduction in Static Friction by Deposition of a Homogeneous Diamond-like Carbon (DLC) Coating on Orthodontic Brackets. *Dent. Mater. J.* **2015**, *34*, 888–895. [CrossRef]
544. Muguruma, T.; Iijima, M.; Kawaguchi, M.; Mizoguchi, I. Effects of Sp2/Sp3 Ratio and Hydrogen Content on In Vitro Bending and Frictional Performance of DLC-Coated Orthodontic Stainless Steels. *Coatings* **2018**, *8*, 199. [CrossRef]
545. Kobayashi, S.; Ohgoe, Y.; Ozeki, K.; Hirakuri, K.; Aoki, H. Dissolution Effect and Cytotoxicity of Diamond-like Carbon Coatings on Orthodontic Archwires. *J. Mater. Sci. Mater. Med.* **2007**, *18*, 2263–2268. [CrossRef]
546. Zhang, H.; Guo, S.; Wang, D.; Zhou, T.; Wang, L.; Ma, J. Effects of Nanostructured, Diamondlike, Carbon Coating and Nitrocarburizing on the Frictional Properties and Biocompatibility of Orthodontic Stainless Steel Wires. *Angle Orthod.* **2016**, *86*, 782–788. [CrossRef]
547. Tantiwinyupong, N.; Chintavalakorn, R.; Santiwong, P.; Khantachawana, A. Frictional and Mechanical Properties of Surface Modified Nickel-Titanium Orthodontic Archwires. *Key Eng. Mater.* **2019**, *801*, 39–43. [CrossRef]
548. Kang, T.; Huang, S.-Y.; Huang, J.-J.; Li, Q.-H.; Diao, D.-F.; Duan, Y.-Z. The Effects of Diamond-Like Carbon Films on Fretting Wear Behavior of Orthodontic Archwire-Bracket Contacts. *J. Nanosci. Nanotechnol.* **2015**, *15*, 4641–4647. [CrossRef] [PubMed]
549. Danisman, H.; Celebi, F.; Danisman, S.; Bicakci, A.A. Effects of Diamond-like Carbon Coating on Frictional and Mechanical Properties of Orthodontic Brackets: An in Vitro Study. *APOS Trends Orthod.* **2021**, *12*, 13. [CrossRef]
550. Almaguer-Flores, A.; Olivares-Navarrete, R.; Lechuga-Bernal, A.; Ximénez-Fyvie, L.A.; Rodil, S.E. Oral Bacterial Adhesion on Amorphous Carbon Films. *Diam. Relat. Mater.* **2009**, *18*, 1179–1185. [CrossRef]
551. Carvalho, I.; Rodrigues, L.; Lima, M.J.; Carvalho, S.; Cruz, S.M.A. Overview on the Antimicrobial Activity and Biocompatibility of Sputtered Carbon-Based Coatings. *Processes* **2021**, *9*, 1428. [CrossRef]
552. Almaguer-Flores, A.; Olivares-Navarrete, R.; Ximénez-Fyvie, L.A.; García-Zarco, O.; Rodil, S.E. Biocompatibility and Anti-Microbial Properties of Silver Modified Amorphous Carbon Films. *MRS Proc.* **2009**, *1244*, 2. [CrossRef]
553. Wei, S.; Shao, T.; Ding, P. Study of CNx Films on 316L Stainless Steel for Orthodontic Application. *Diam. Relat. Mater.* **2010**, *19*, 648–653. [CrossRef]
554. Wei, S.; Shao, T.; Ding, P. Improvement of Orthodontic Friction by Coating Archwire with Carbon Nitride Film. *Appl. Surf. Sci.* **2011**, *257*, 10333–10337. [CrossRef]
555. Pan, Z.; Zhou, Q.; Wang, P.; Diao, D. Robust Low Friction Performance of Graphene Sheets Embedded Carbon Films Coated Orthodontic Stainless Steel Archwires. *Friction* **2022**, *10*, 142–158. [CrossRef]
556. Wang, P.; Luo, X.; Qin, J.; Pan, Z.; Zhou, K. Effect of Graphene Sheets Embedded Carbon Films on the Fretting Wear Behaviors of Orthodontic Archwire–Bracket Contacts. *Nanomaterials* **2022**, *12*, 3430. [CrossRef]

MDPI AG
Grosspeteranlage 5
4052 Basel
Switzerland
Tel.: +41 61 683 77 34

Metals Editorial Office
E-mail: metals@mdpi.com
www.mdpi.com/journal/metals

www.ingramcontent.com/pod-product-compliance
Lightning Source LLC
Chambersburg PA
CBHW040746250926
51850CB00007BB/1421
* 9 7 8 3 7 2 5 8 5 8 8 7 3 *